SENSORY EVOLUTION *on the* THRESHOLD

SENSORY EVOLUTION
on the THRESHOLD

Adaptations in Secondarily Aquatic Vertebrates

Edited by

J. G. M. Thewissen
and Sirpa Nummela

中

UNIVERSITY OF CALIFORNIA PRESS

Berkeley   Los Angeles   London

University of California Press, one of the most distinguished university presses in the United States, enriches lives around the world by advancing scholarship in the humanities, social sciences, and natural sciences. Its activities are supported by the UC Press Foundation and by philanthropic contributions from individuals and institutions. For more information, visit www.ucpress.edu.

University of California Press
Berkeley and Los Angeles, California

University of California Press, Ltd.
London, England

Library of Congress Cataloging-in-Publication Data

Sensory evolution on the threshold : adaptations in secondarily aquatic vertebrates / edited by J. G. M. Thewissen and Sirpa Nummela.
    p.   cm.
  Includes bibliographical references and index.
  ISBN 978-0-520-25278-3 (case : alk. paper)
    1. Aquatic animals—Sense organs.   2. Aquatic animals—Adaptation.
3. Sense organs—Evolution.   I. Thewissen, J. G. M.   II. Nummela, Sirpa.
  QL120.S47 2008
  591.4—dc22

                                                           2007039518

Manufactured in the United States of America
10   09   08
10 9 8 7 6 5 4 3 2 1

The paper used in this publication meets the minimum requirements of ANSI/NISO Z39.48-1992 (R 1997) (*Permanence of Paper*).

Cover photograph: Ian Murphy / Getty Images

CONTENTS

CONTRIBUTORS

GUIDO DEHNHARDT, University of Rostock, Institute for Biosciences, Sensory & Cognitive Ecology, Rostock, Germany

HEATHER L. EISTHEN, Department of Zoology, Michigan State University, East Lansing

JUSTIN A. GEORGI, Doctoral Program of Anatomical Sciences, Stony Brook University, New York

SIMO HEMILÄ, Department of Biological and Environmental Sciences, University of Helsinki, Finland

THOMAS HETHERINGTON, Department of Evolution, Ecology, and Organismal Biology, The Ohio State University, Columbus

TOBIN L. HIERONYMUS, Department of Biological Sciences, Ohio University, Athens

MICHAEL H. HOFMANN, Department of Biology and Center for Neurodynamics, University of Missouri-St. Louis

GADI KATZIR, Department of Biology, University of Haifa at Oranim, Israel

RONALD H.H. KRÖGER, Department of Cell and Organism Biology, Lund University, Sweden

BJÖRN MAUCK, Institute of Biology, University of Southern Denmark, Odense, Denmark

SIRPA NUMMELA, Department of Biological and Environmental Sciences, University of Helsinki, Finland

LEO PEICHL, Department of Neuroanatomy, Max Planck Institute for Brain Research, Frankfurt am Main, Germany

HENRY PIHLSTRÖM, Department of Biological and Environmental Sciences, University of Helsinki, Finland

JOHN O. REISS, Department of Biological Sciences, Humboldt State University, Arcata, California

TOM REUTER, Department of Biological and Environmental Sciences, University of Helsinki, Finland

KURT SCHWENK, Department of Ecology and Evolutionary Biology, University of Connecticut, Storrs

JUSTIN S. SIPLA, Department of Rehabilitation Sciences, University of Texas at El Paso

FRED SPOOR, Department of Anatomy and Developmental Biology, University College London, London, UK

J.G.M. THEWISSEN, Department of Anatomy, Northeastern Ohio Universities College of Medicine, Rootstown

LON A. WILKENS, Department of Biology and Center for Neurodynamics, University of Missouri-St. Louis

1

Introduction

ON BECOMING AQUATIC

J. G. M. Thewissen and Sirpa Nummela

Vertebrate life became terrestrial about 370 million years ago when a lobe-finned fish evolved into the giant-salamander-like shape of a labyrinthodont amphibian. The transition is well documented in the fossil record, and important discoveries continue to fill out its details. Over the eons subsequent to the water-to-land transition, vertebrates became more and more independent from water. The new land vertebrates are called tetrapods, a group that includes modern amphibians, reptiles, birds, and mammals. The term refers to their extremities: the replacement of four paired fins by four paired legs. Amphibians still return to the water to avoid dehydration of their eggs and larvae and have skin that is permeable to water, restricting them

to live in humid environments. With the origin of amniotes (reptiles, birds, and mammals) around 320 million years ago, the transition to land was complete: the embryo is protected from dehydration by being bathed in a fluid bubble surrounded by membranes. Moreover, amniote skin is covered by keratin, a waterproof protein that minimizes water loss through evaporation, allowing amniotes to live away from humidity. Dry land offered a host of opportunities and challenges to the newly terrestrial tetrapods.

In the period since the origin of tetrapod land life, many vertebrates have returned to the water. Some, such as crocodiles, became amphibious and never left these transitional habitats. Others, such as whales, returned to the oceans completely and are unable to live on land. In spite of their deep watery roots, these secondarily aquatic vertebrates started their evolutionary journey with bodies that were adapted to live on land and in air. Occasionally they evolved adaptations similar to those of their fish ancestors, such as the multirayed, multisegmented forelimbs of ichthyosaurs. In most cases, though, their amniote body plan was specialized and did not revert to its ancestral state. Most amniotes evolved solutions for the challenges of aquatic life that were different from those of their fish-like ancestors.

NATURAL EXPERIMENTS

The entire range of secondarily aquatic tetrapods adapted to handle similar problems inflicted by their new environment. They commenced their journey to the water with different phylogenetic backgrounds and body plans. This makes becoming aquatic one of the greatest natural experiments: evolutionary hypotheses about specific adaptations can be tested in other aquatic groups that are not closely related phylogenetically. This book focuses on the aquatic adaptations in some of the most complex organ systems of a vertebrate's body, the sense organs. Though often not the first organ to change in the journey from land to water, most sense organs had to change pervasively to adapt to the new environment. In many cases, a clade that relied on one sense organ for gathering information about the outside world changed to another organ after entering the water. The origin of echolocation in toothed whales is an example of such a change. At other times, it was the dominant sense that was retained and improved in order to meet the challenges of living in water. Vision in penguins is an example of this phenomenon. Maybe most interestingly, vertebrates from varying phylogenetic backgrounds that are ecologically similar may have vastly different sensory specializations. For instance, toothed whales, sharks, and ichthyosaurs are all predators and are a classic example of adaptive convergence in body form and locomotor behavior: all have sleek and streamlined bodies and are fast pursuit predators (Fig. 1.1). Sensorywise, they could hardly be more different. Dolphins and porpoises use hearing as their main sense organ in locating prey, ichthyosaurs used vision, and in sharks mechanoreception, electrore-

FIGURE 1.1. A cetacean, an ichthyosaur, and a shark. These animals are textbook examples of convergent evolution in locomotor and foraging behavior. However, their main sensory organs differ greatly: toothed whales (odontocete cetaceans) hunt mostly with their ears, ichthyosaurs with their eyes, and sharks with a combination of chemical, electric, and tactile senses.

ception, and chemoreception are all important. The sense organs are excellent examples of natural experiments: they are complex and functionally well understood, most fossilize well, and there are well established procedures for experimenting with the senses in modern animals.

AQUATIC TETRAPODS

Vertebrates have become secondarily aquatic many times in their evolution. The timing of origin of the back-to-the-water events varied greatly among these forms, and Figure 1.2 summarizes the approximate time of the return to the water for some of the major clades, as well as the amount of time it took to become obligately aquatic. Figures 1.3 to 1.7 and Tables 1.1 to 1.4 summarize the phylogeny and diversity of the major groups of secondarily aquatic tetrapods, focusing on those animals that are discussed in the chapters of this book.

Amphibians (Fig. 1.3; Table 1.1) assume a special position in this book, because most of them never completely left the water, and they retained an aquatic larval stage. On the other hand, evolutionarily, modern amphibians are probably derived from a terrestrial ancestor, more terrestrial than most modern forms. In this volume, they are included for those sense organs that form a useful comparison to those of amniotes (such as chemical senses and hearing), but not when their sense organs resemble those of fishes and may represent a

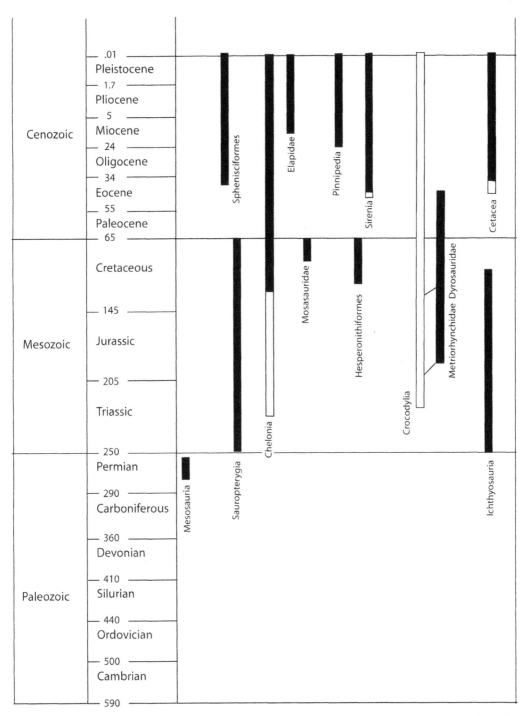

FIGURE 1.2. Temporal ranges of some important groups of aquatic tetrapods. *White bars* indicate that most members of a clade were transitional or amphibious, *black bars* indicate that the clade was mostly fully aquatic. Terms such as amphibious, semiaquatic, and transitional cannot be unambiguously defined, making figures such as this suggestive of a pattern rather than explicit statements.

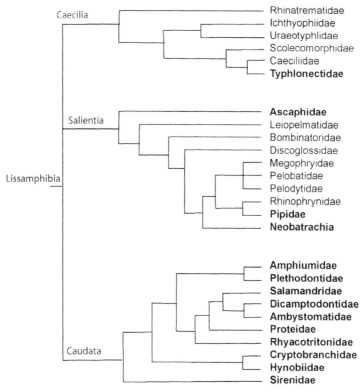

FIGURE 1.3. Generalized phylogeny of extant amphibians emphasizing taxa discussed in this volume. *Bold* indicates amphibians with aquatic adult stages. See Table 1.1 for details. By T. Hetherington.

primitive retention from a larval form (such as mechanoreception and balance).

SENSES IN EVOLUTION

Sensory biology is a large field positioned on the crossroads of neurobiology, physiology, and cell biology. Research in sensory biology is rich and multidisciplinary, and many questions about the basic function of the sense organs are asked and answered. However, evolutionary studies of the sense organs are less common. The reason for this is may be that most sensory scientists are not trained in evolution and most evolutionary scientists lack background in sensory systems. We believe that the senses offer a remarkable opportunity to study evolution in action, and we hope that this book will help in the cross-fertilization of evolutionary and sensory biology.

GOAL AND SCOPE

The purpose of this book is then to bring together basic information about the comparative evolutionary aspects of all the sense organs in secondarily aquatic tetrapods. For the sense organs that have been studied extensively from a comparative perspective, the focus has been on summarizing past research, whereas for sense organs that have received less attention, primary research is reported here. However, as editors, we felt strongly that this book should form a coherent work that covers all the basics, allowing those newly interested in sensory evolution to use it as a primer for boosting their background knowledge and as a sounding board for determining where new research would be most fruitful. We have encouraged authors to be explicit about where new research will answer important questions, as well as to boldly generalize evolutionary patterns, thus generating new hypotheses that

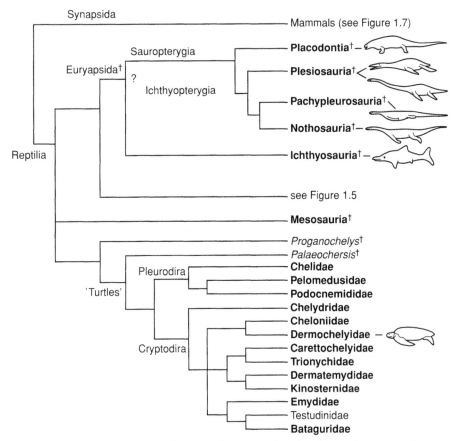

FIGURE 1.4. Generalized phylogeny of some reptiles (euryapsids and turtles) emphasizing taxa discussed in this volume. *Crosses* indicate extinct groups, *bold* indicates groups with aquatic representatives. See Table 1.2 for details. By Kurt Schwenk and J. G. M. Thewissen.

can be tested by future generations. We have also encouraged them to not get lost in details and jargon; this is a book for those interested in biology in the broadest sense of the word.

The sense organs can be understood only if their stimulus is understood, and the difference between stimuli in air and water is of great importance. For this reason, for most of the senses in this book, there is a chapter that discusses the difference between how they are stimulated in air and water.

Finally, it is our belief that all evolutionary studies need to be done against a systematic background, and we therefore have included simplified phylogenetic information in this chapter and have encouraged authors to propose daring hypotheses of sensory evolution in their chapters. These hypotheses are explicit and can therefore be tested relatively straight-

forwardly. The phylogenetic information in this chapter is condensed so that finding information about unfamiliar taxa does not distract from the central theme of the book: the sense organs as they move on either side of the threshold between land and water.

We will consider this volume a success if it is used by those sensory biologists who are interested in evolution, and by those evolutionary researchers that are interested in the senses. In addition, we hope that it will form the basis for much research in the future, facilitating the entry of young people into this field.

AMPHIBIANS

Thomas Hetherington and J. G. M. Thewissen

There are three modern groups of amphibians, often referred to as Lissamphibia (Fig. 1.3):

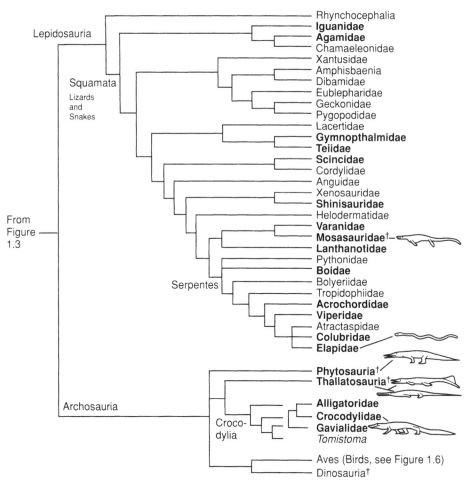

FIGURE 1.5. Generalized phylogeny of some reptiles (lepidosaurs and archosaurs excluding birds) emphasizing taxa discussed in this volume. *Crosses* indicate extinct groups, *bold* indicates groups with aquatic representatives. See Table 1.2 for details. By Kurt Schwenk and J. G. M. Thewissen.

Anura (toads and frogs), Caudata (or Salientia, salamanders and newts), and Gymnophiona (caecilians). All three of these have representatives whose adults are aquatic. During the Paleozoic Era tetrapods radiated into several lineages, many of which are called "amphibians" but are not closely related to the surviving modern amphibians (Lissamphibia). Lissamphibians usually are considered to form a monophyletic group (Trueb and Cloutier, 1991; Hedges and Maxson, 1993; Feller and Hedges, 1998). They are not closely related to the other surviving lineage of tetrapods, the amniotes. Lissamphibians and amniotes last shared a common ancestor more than 300 million years

ago and possibly even arose independently from different fishlike ancestors.

Living amphibians and amniotes, therefore, do not share a long history of common terrestrial ancestors. Although the oldest fossil lissamphibians are late Triassic in age, lissamphibians likely share a common ancestor that lived more than 270 million years ago (Clack, 2002). The earliest lissamphibians lived in relatively dry environments and likely had terrestrial adult stages (Bolt, 1991), so it is appropriate to consider the more aquatic modern amphibians as secondarily aquatic. The first known largely aquatic representatives appeared approximately 200 million years ago.

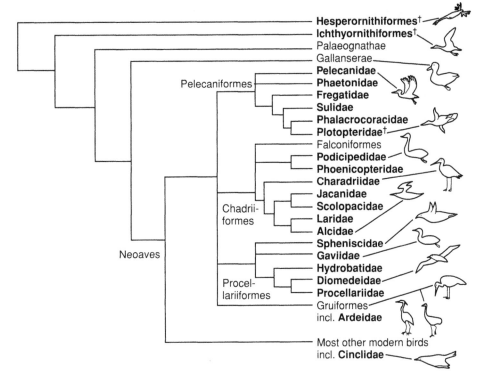

FIGURE 1.6. Generalized phylogeny of aquatic birds *(bold)* emphasizing taxa discussed in this volume. *Crosses* indicate extinct groups. See Table 1.3 for further details. By J. G. M. Thewissen and Tobin Hieronymus.

Most lissamphibians return to the water to lay eggs that hatch into aquatic larvae. Some amphibians, however, have no aquatic larval stage but rather lay terrestrial eggs that hatch into fully terrestrial juveniles. In many amphibians that have aquatic larval stages, the larvae usually metamorphose into more terrestrial forms, but in several taxa the adults remain completely or largely aquatic (Table 1.1). Except for a few species of frogs that live in brackish water (Duellman and Trueb, 1994), amphibians are restricted to freshwater habitats.

LITERATURE CITED

Bolt, J. R. 1991. Lissamphibian origins; pp. 194–222 in Origins of the Major Groups of Tetrapods: Controversies and Concensus, H. P. Schultze and L. Trueb (eds.). Cornell University Press, Ithaca, NY.

Clack, J. A. 2002. Gaining Ground: The Origin and Evolution of Tetrapods. Indiana University Press, Bloomington.

Duellman, W. E., and L. Trueb. 1994. Biology of Amphibians. Johns Hopkins University Press, Baltimore.

Feller, A. E., and S. B. Hedges. 1998. Molecular evidence for the early history of living amphibians. Molecular Phylogenetics and Evolution 9:509–516.

Hedges, S. B., and L. R. Maxson. 1993. A molecular perspective on lissamphibian phylogeny. Herpetological Monographs 7:27–42.

Trueb, L., and R. Cloutier. 1991. Toward an understanding of the amphibians: two centuries of systematic history; pp. 175–193 in Origins of the Major Groups of Tetrapods: Controversies and Concensus, H. P. Schultze and L. Trueb (eds.). Cornell University Press, Ithaca, NY.

AQUATIC AND SEMIAQUATIC REPTILES
Kurt Schwenk and J. G. M. Thewissen

The reptiles comprise a diverse and phylogenetically ancient group of amniote vertebrates (Figs. 1.4 and 1.5; Table 1.2). Reptiles are a paraphyletic group if birds are excluded. Phylogenetically,

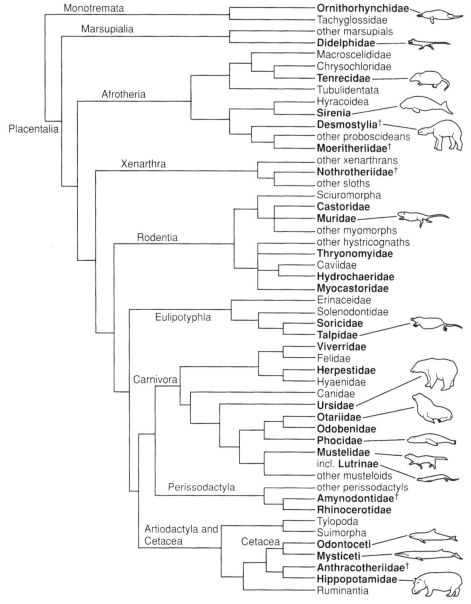

FIGURE 1.7. Generalized phylogeny of mammals *(aquatic in bold)* emphasizing taxa discussed in this volume. *Crosses* indicate extinct groups. See Table 1.4 for details. By Henry Pihlström.

birds are included with the dinosaurs. There are three principal lineages of reptiles with extant representatives: Chelonia (turtles), Lepidosauria (tuatara, lizards, and snakes), and Archosauria (crocodylians and birds). Each of these major extant clades of modern reptiles contains aquatic taxa, including some of the most fully water adapted species among tetrapod vertebrates. Nonetheless, aquatic forms constitute only about 8% of living species. Indeed, the major aquatic reptile lineages, turtles and crocodylians, are also the least speciose. Most reptile diversity (in terms of species numbers) is represented by the Squamata (lizards and snakes), and within this group, specialized aquatic habits have arisen infrequently and are phylogenetically dispersed (with several notable exceptions; see below). In addition to

TABLE 1.1

Overview of Amphibians in Which the Adults Have Aquatic Habits, and Other Taxa Discussed in This Volume

CLASSIFICATION	EXAMPLES OF AQUATIC GENERA	HABITAT	DEGREE	NOTES
Caudata (salamanders and newts)				
Sirenidae	Siren	F	1	
Hymobiidae	Ranodon, Onychodactylus	F	1, 2, 3	
Cryptobranchidae	Andrias, Cryptobranchus	F	1	Includes giant form
Ambystomatidae	Ambystoma	F	1, 3	Terrestrial adults except for completely aquatic neotenic forms such as the axolotl
Dicamptodontidae	Dicamptodon	F	1, 3	Aquatic neotenic adults common
Rhyacotritonidae	Rhyacotriton	F	2	
Salamandridae	Cynops, Notophthalmus, Pleurodeles, Triturus	F	1, 2, 3, 4	
Amphiumidae	Amphiuma	F	1	
Proteidae	Proteus, Necturus	F	1	
Plethodontidae	Desmognathus, Eurycea, Gyrinophilus, Pseudotriton, Stereochilus	F	1, 2, 3, 4	Fully aquatic species are neotenic. Listed are aquatic genera (1, 2), most plethodontid genera are fully terrestrial (4).
Gymnophiona (caecilians)				
Typhlonectidae	Typhlonectes	F	1	Live-bearing
Anura (frogs and toads)				
Ascaphidae	Ascaphus	F	2	
Pipidae	Hymenochirus, Pipa, Xenopus	F	1	Capable of moving on land but rarely do
Discoglossidae	Bombina	F	2	
Leptodactylidae	Telmatobius	F	1, 2, 3, 4	Most leptodactylids are 4 and lay directly developing eggs on land, Telmatobius is 1 or 2.
Ranidae	Rana	C, F	2, 3	

NOTE: Habitat; F, freshwater; C, coastal. Degree to which aquatic: 1, never leave water; 2, forage both on land and in water; 3, forage mostly on land; 4, completely terrestrial; lay eggs on land; that hatch into terrestrial juveniles or give birth to terrestrial juveniles. Table by Thomas Hetherington and J. G. M. Thewissen.

	EXAMPLE AQUATIC TAXA	HABITAT	DEGREE	NOTES
Mesosauria	*Mesosaurus*	MC?	2?	Extinct, Permian of South America and Africa. Limb structure indicates that mesosaurs could walk on land.
Placodontia	*Placodus*	MC	2?	Extinct, Triassic of Europe and Africa.
Pachypleurosauria	*Pachypleurosaurus*	MC	2?	Extinct, Triassic of Europe. Sometimes included in Nothosauria.
Nothosauria	*Nothosaurus* *Ceresiosaurus*	M	2?	Extinct, mostly Triassic of Europe and Asia.
Plesiosauria	*Cryptocleidus* *Liopleurodon*	M	2?	Extinct, Mesozoic. Well known skeletally. Morphological intermediates between nothosaurs and plesiosaurs are known.
Ichthyosauria	*Mixosaurus* *Cymbospondylus* *Ophthalmosaurus* *Temnodontosaurus*	M	1	Extinct, Mesozoic. Well known skeletally, but all fully aquatic.
Chelonia				
Chelidae	*Chelodina* *Chelus* *Elusor* *Platemys* *Rheodytes*	FP, FS	2, 3	*Elusor* and *Rheodytes* are highly aquatic, rarely leaving the water; no terrestrial representatives.
Pelomedusidae	*Pelomedusa* *Pelusios*	FP, FS	3	Often forage by bottom-walking; no terrestrial representatives.
Podocnemididae	*Erymnochelys* *Peltocephalus* *Podocnemis*	FS	3	Sometimes considered a subfamily of Pelomedusidae; primarily stream and river turtles; no terrestrial representatives.
Chelydridae	*Chelydra* *Macrochelys* *Platysternon*	FP, FS	2, 3	*Chelydra* rarely leaves water except to lay eggs; *Macrochelys* uses a wormlike lingual lure to capture prey underwater; no terrestrial representatives.
Cheloniidae	*Caretta* *Chelonia* *Eretmochelys* *Lepidochelys*	M	2	The sea turtles; the most aquatic of all turtles; pelagic, often swim at depth; no terrestrial representatives.

TABLE 1.2 (*continued*)

	EXAMPLE AQUATIC TAXA	HABITAT	DEGREE	NOTES
Dermochelyidae	*Dermochelys*	M	2	The leatherback sea turtle; pelagic, often swim at depth; no terrestrial representatives.
Carettochelyidae	*Carettochelys*	FS	2	The pig-nosed turtle has a snorkellike nose and forelimbs modified as flippers, like sea turtles; no terrestrial representatives.
Trionychidae	*Lissemys* *Amyda* *Apalone* *Aspideretes* *Trionyx*	FP, FS	2	Many with snorkellike snouts; spend most of their time on the bottom; no terrestrial representatives.
Dermatemydidae	*Dermatemys*	FP, FS	2	Monotypic; highly aquatic, restricted to slow-moving water.
Kinosternidae	*Kinosternon* *Sternotherus*	FP, FS	2, 3	Musk glands may be used for communication, as well as for defense; no terrestrial representatives.
Emydidae	*Chrysemys* *Clemmys* *Emys* *Graptemys* *Malaclemys* *Pseudemys*	BE, FP, FS	3	Highly variable in aquatic habits—some species highly aquatic, some completely terrestrial; most are semiaquatic pond and stream turtles; only *Malaclemys* is estuarine.
Bataguridae	*Batagur* *Cuora* *Chinemys* *Heosemys* *Kachuga* *Mauremys* *Rhinoclemmys*	BE, FP, FS	3	Sometimes included within Emydidae; similarly variable in aquatic habits.
Crocodylia				
Crocodylidae	*Crocodylus* *Osteolaemus* *Tomistoma*	MC, BE, FP, FS	3	The saltwater crocodile, *C. porosus*, often estuarine, sometimes travels long distances in the ocean; no terrestrial representatives.
Alligatoridae	*Alligator* *Caiman* *Melanosuchus* *Paleosuchus*	FP, FS	3	All amphibious; no terrestrial representatives.
Gavialidae	*Gavialis*	FS	2	The most aquatic of the crocodylians, gharials rarely leave the water; no terrestrial representatives.
		M	2	Extinct marine crocodylians that left the water only to reproduce.

(*continued*)

TABLE 1.2 (*continued*)

	EXAMPLE AQUATIC TAXA	HABITAT	DEGREE	NOTES
Metriorhynchidae Squamata, snakes				
Boidae	*Eunectes*	FS	2, 3	Green anacondas, *E. murinus*, are highly aquatic, spending most time submerged; with dorsal eyes and nostrils; most boids are terrestrial.
Acrochordidae	*Acrochordus*	MC, BE, FP, FS	1	Rarely observed outside of water, file snakes are some of the most aquatic-adapted snakes; exceptionally long tines of the forked tongue, dorsal nostrils; tongue-flick underwater; coastal species swim at depth; no terrestrial representatives.
Colubridae	*Bitia* *Erpeton* *Helicops* *Homalopsis* *Hydrodynastes* *Liophis* *Lycodonomorphus* *Natrix* *Nerodia* *Regina* *Thamnophis*	BE, FP, FS	2, 3	Natricines, homolopsines, and some xenodontines are among the most aquatic of the colubrids; some homolopsines, in particular, are highly aquatic and inhababit coastal areas—*B. hydroides* resembles a true seasnake (Elapidae) in body form and coloration; *Erpeton* has two mechanoreceptive tentacles it probably uses to locate fish prey; most colubrids are terrestrial.
Elapidae	*Emydocephalus* *Enhydrina* *Enhydris* *Hydrelaps* *Hydrophis* *Pelamis* *Laticauda*	BE, MC, M	1, 2	Laticaudinae (seakraits) and Hydrophiinae (seasnakes) are sometimes elevated to family status (Hydrophiidae); some species are the most aquatic of any reptile, never leaving the water; frequently tongue-flick underwater; many species swim at depth; most elapids are terrestrial.
Viperidae	*Agkistrodon*	BE, FP, FS	3	The cottonmouth, *A. piscivorus*, is highly aquatic and is tolerant of brackish water; all other viperids are terrestrial.

TABLE 1.2 (*continued*)

	EXAMPLE AQUATIC TAXA	HABITAT	DEGREE	NOTES
Squamata, Lizards				
Iguanidae	*Amblyrhynchus Basiliscus*	MC, FP, FS	3	*A. cristatus* is the only marine lizard; it often grazes on submerged algae at depth; *Basiliscus* use water as an avenue of escape; occasionally catch prey in water; usually run or swim across surface of water; occasional water escape in other iguanid species, but most are terrestrial.
Agamidae	*Hydrosaurus Physignathus*	FP, FS	3	Water as escape; good swimmers; *Hydrosaurus* with laterally compressed tail and tail fin; most agamids are terrestrial.
Scincidae	*Eulamprus Tropidophorus*	FP, FS	3	*Eulamprus* known as the "water skinks"; most scincids are terrestrial.
Teiidae	*Crocodilurus Dracaena*	FP, FS	3	Swim or walk on bottom, foraging; *Dracaena* observed tongue-flicking underwater; most teiids are terrestrial.
	Alopoglossus	FS	3	Swim through water; most gymnopthalmids are terrestrial.
Gymnophthalmidae	*Neusticurus*	FS	3	Sometimes placed in its own family; water as escape; the other genus in the family, *Xenosaurus* is terrestrial.
Xenosauridae	*Shinisaurus*			
Lanthanotidae	*Lanthanotus*	FP, FS	3	Monotypic; highly aquatic, but very poorly known; sometimes collected in fish seins.
Varanidae	*Varanus*	FP, FS	3	Many species enter the water readily, where they swim and forage or spend time submerged; some species specialize on aquatic prey; most varanids are terrestrial.
Mosasauridae	*Clidastes Tylosaurus Plotosaurus*	MC	2	Extinct, Cretaceous, well known skeletally. A possible intermediate between the fully aquatic mosasaurs and terrestrial lepidosaurs are the aigialosaurs, which combine terrestrial and aquatic features.

(*continued*)

TABLE 1.2 (*continued*)

	EXAMPLE AQUATIC TAXA	HABITAT	DEGREE	NOTES
Phytosauria	*Phytosaurus* *Rutiodon*	MC	2?	Extinct, mostly Triassic of Northern Hemisphere.
Thalattosauria	*Askeptosaurus*	MC	2?	Extinct, Triassic of Europe and North America.

SOURCES: Neill, 1971; Carroll, 1988; Ernst and Barbour, 1989; Greer, 1989; Glasby et al., 1993; Greene, 1997; Cann, 1998; Zug et al., 2001; Pianka and Vitt, 2003; and see text.
NOTE: Habitat: M, marine; MC, marine. coastal; BE, brackish/estuarine; FP, freshwater, ponds, lakes, marshes; FS, freshwater, streams, rivers. Degree to which aquatic: 1, fully aquatic, never returns to land; 2, mostly aquatic, returns to land only to reproduce; 3, semiaquatic, spends time on land for purposes other than reproduction, e.g., foraging, basking. Table by Kurt Schwenk and J. G. M. Thewissen.

the extant clades, there are several diverse groups of extinct reptiles, some with predominantly or exclusively aquatic members.

Members of the crown group Reptilia probably arose during the Middle to Late Pennsylvanian (approximately 290 million years ago). However, the split between the stem groups Sauropsida and Synapsida (Fig. 1.4) occurred earlier, during the Early Pennsylvanian (approximately 300 to 320 million years ago) (Carroll, 1988). Interestingly, many basal reptiles are aquatic, such as mesosaurs, chelonians, and euryapsids. Seymour (1982) suggested that low metabolic rate and body temperature, and a tolerance of low-oxygen conditions are useful features in aquatic animals and may have facilitated life in water among early reptiles.

MESOSAURS The affinities of mesosaurs are unclear; they are usually included with the sauropsids but may have instead branched off before the sauropsid-synapsid split (Rossman, 2000). They were up to 1 meter long with a long snout and long teeth, possibly for filter-feeding, and a bilaterally flat tail (Huene, 1941).

TURTLES Living turtles comprise approximately 300 extant species that are divided into two lineages: the Pleurodira, or side-necked turtles, and the Cryptodira, or hidden-neck turtles (Shaffer et al., 1997; Zug et al., 2001). The origin of turtles is contentiously debated. The traditional view is that turtles are basal sauropsids that exhibit the ancestral anapsid amniote skull condition (i.e., lacking a temporal fenestra) (Carroll, 1988). Reisz and Laurin (1991) suggested that turtles might not be reptiles in the cladistic sense. However, recent studies are divided on whether turtles are primitively anapsid reptiles that arose within a basal group known as parareptiles (Lee, 1995, 1996), or whether they are primitively diapsid (having two temporal fenestrae), allied with either lepidosaurs or archosaurs (Rieppel, 1994/1995; Rieppel and deBraga, 1996; deBraga and Rieppel, 1997; Rieppel and Reisz, 1999; Zardoya and Meyer, 2001).

Although the earliest turtle ancestors must have been terrestrial, it is now well established that the common ancestor of all living turtles was aquatic (Gaffney et al., 1987; Rieppel and Reisz, 1999; Joyce and Gauthier, 2003). Nearly all extant turtles are semi- or fully aquatic. The sea turtles (Cheloniidae and Dermatochelyidae) are highly adapted for marine life and emerge from the water only to lay eggs. Several freshwater lineages are almost as specialized, including the Carettochelyidae and Trionychidae. The only fully terrestrial family is the Testudinidae (the tortoises). Systematic study of chelonians argues that terrestriality in testudinids must have evolved secondarily from an aquatic ancestry. A recent mitochondrial DNA study also suggested that the terrestrial box turtles (Terrepene), of the diverse family Emydidae, arose from within a wholly aquatic or semiaquatic

TABLE 1.3

Overview of Aquatic Birds (English Names are Generalized), Focusing on Taxa Discussed in This Volume

	SELECTED GENERA	HABITAT	DEGREE	NOTES
Galloanserae				Mostly terrestrial forms, such as fowl.
Anatidae (ducks, geese)	*Somateria, Histrionicus, Anas*	C, F	2, 3	Some species are partly pelagic (*Somateria*).
Pelecaniformes Pelecanidae (pelicans)	*Pelecanus*	C, F	2	Most species gather food while surface swimming and bill dipping; one species plunge-dives (*P. occidentalis*).
Phaethontidae (tropic birds)	*Phaethon*	M	4	*Phaeton* gathers food in flight over water; can barely walk and is poor diver.
Fregatidae (frigate birds)	*Fregata*	M	4	*Fregata* gathers food flying over water and rarely dives.
Sulidae (boobies, gannets)	*Morus, Papasula, Sula*	C	2	Plunge-divers.
Phalacrocoracidae (cormorants)	*Phalacrocorax*	C, F	2	Surface divers.
Anhingidae	Anhinga	C, F	2	Surface divers.
Plotopteridae		C, M	2?	Extinct birds strongly resembling penguins, but unrelated. Known only from the North Pacific.
Podicipediformes Podicipedidae (grebes)	Around 10 species	C, F	2	Mostly pursuit divers from position on surface freshwater.
Phoenicopteriformes Phoenicopteridae (flamingos)	*Phoenicopterus Phoeniconaias Phoenicoparrus*	C, F	2	Use their bill to filter feed while wading.
Charadriiformes				Mostly nonaquatic birds.
Charadriidae (plovers)		C, F	3	Some species probe wet sand with their bills feeding in the intertidal zone, most not aquatic.
Jacanidae (Jacanas)	*Jacana*	F	3	
Scolopacidae (sandpipers)	Many genera	C, F	3	Some species wade while probing bottom for buried food.
Laridae (gulls and terns)	*Larus* *Sterna* (tern)	M, C, F C	2–3	Not all gull species are aquatic. Gulls are dietary opportunists. *Sterna* plunge-dives for food.
Rhynchopidae	*Rhynchops*	F	4	Skim water, dipping lower jaw.
Alcidae (auks)	*Alca, Uria, Fratercula*	C	2	Excellent pursuit divers, stay under water for minutes.
Falconiformes	*Haliaeetus, Pandion*	F	4	Hunt swimming prey.

(continued)

TABLE 1.3. (*continued*)

	SELECTED GENERA	HABITAT	DEGREE	NOTES
Sphenisciformes				
Spheniscidae (penguins)	Six genera	M	2	The most aquatic birds, some species spend months without leaving the water. No species fly, all are fast pursuit divers, and land locomotion is poor.
Gaviiformes				
Gaviidae (loons)	*Gavia*	C, F	2	Pursuit divers from water surface.
Procellariiformes				All live on the open ocean and feed mostly while flying closely over the water surface.
Hydrobatidae (storm petrels)	*Oceanites, Fregetta, Hydrobates*	M	4	Occasional divers.
Diomedeidae (albatroses)	*Diomedea, Phoebetria*	M	4	
Procellariidae (shearwaters, petrels)	*Pterodroma, Puffinus*	M	2, 4	*Pelecanoides* is a wing-propelled diver.
Gruiformes				Mostly nonaquatic birds.
Ardeidae (Herons)		F	3	Striker hunters, breaking the water surface with their bill after spotting prey.
Passeriformes				Mostly nonaquatic birds.
Cinclidae (dippers)		F	3	Forage while walking on the bottom of streams, fully submerged.
Coraciiformes				Mostly nonaquatic birds.
Alcedinidae (Kingfishers)	*Ceyx, Alcedo*	F	3	Plunge divers.

NOTE: Habitat: M, marine (coastal for reproduction); C, coastal; F, freshwater. Degree to which aquatic: 2, highly aquatic, always feed in water, return to land only for reproduction; 3, moderately aquatic, occasionally forage on land; flying; 4, usually feed over water, but remain airborne. Table by J. G. M. Thewissen and T. Hieronymus.

clade (Feldman and Parham, 2002), once again supporting the secondary derivation of terrestrial habits in living turtles.

SAUROPTERYGIA The sauropterygians are a diverse group of extinct marine reptiles known only from the Mesozoic (250 to 65 million years ago), with the well-known genus *Claudiosaurus* near its base (Carroll, 1981). *Claudiosaurus* was aquatic, as indicated by its paddlelike, weakly muscled forelimbs and its poorly ossified skeleton, and had large hind limbs with which it swam. However, it displays fewer aquatic adaptations than most sauropterygians. Later sauropterygians include nothosaurs (Kuhn-Snyder, 1963) and pachypleurosaurs (Carroll and Gaskill, 1985), which are tail-propelled forms with weak limbs. Nothosaurs had partly pachyostotic skeletons, suggesting a partly submerged life. Most plesiosaurs had a small head on a long, mobile neck with very large hands and feet, but the plesiosaur family Pliosauridae had short necks with large heads specialized for different aquatic modes of life (Brown, 1981). Placodonts were heavy, slow moving, putative mollusk-eaters (Drevermann, 1933; Peyer and Kuhn-Snyder, 1955).

CLASSIFICATION	GENERA	HABITAT	DEGREE	NOTES
Monotremata				
Ornithorhynchidae	*Ornithorhynchus*	F	2	*O. anatinus* has an electric organ for locating prey underwater.
(platypus)				
Didelphimorphia				
Didelphidae	*Chironectes*	F	3	*C. minimus* is the only extant semiaquatic marsupial.
(American				
opossums)				
Afrosoricida				
Tenrecidae (tenrecs)				
Oryzorictinae	*Limnogale*	F	3	
(rice tenrecs)				
Potamogalinae				
(otter shrews)	*Potamogale*	F	3	
	Micropotamogale	F	3	
Sirenia				
Dugongidae	*Dugong*	C	1	
(dugongs)	*Hydrodamalis*	C	1	The recently extinct *H. gigas* was the only nontropical sirenian, living near the Arctic Circle.
Trichechidae	*Trichechus*	C, F	1	
(manatees)				
Rodentia				
Castoridae (beavers)	*Castor*	F	3	Semiaquatic, but feed mainly on terrestrial plants.
Muridae (mice)				Largest extant mammal family, several semiaquatic lineages.
Sigmodontinae	*Oryzomys*	F	3	This genus has 36 terrestrial and 1 semiaquatic species, *O. palustris*.
(New World rats	*Nectomys*	F	3	
and mice)	*Amphinectomys*	F	3	
	Holochilus	F	3	
	Scapteromys	F	3	
	Neusticomys	F	3	
	Chibchanomys	F	3	
	Anotomys	F	3	
	Ichthyomys	F	3	
	Rheomys	F	3	
Arvicolinae (voles)	*Arvicola*	F	3	
	Microtus	F	3	This genus has 64 terrestrial and 1 semiaquatic species: *M. richardsoni*.
	Neofiber	F	3	
	Ondatra	F	3	

(continued)

TABLE 1.4 (*continued*)

CLASSIFICATION	GENERA	HABITAT	DEGREE	NOTES
Murinae (Old World rats and mice)	*Colomys*	F	3	
	Nilopegamys	F	3	
Hydromyinae (water rats)	*Hydromys*	F	3	This subfamily is known as water rats, although most of its genera are terrestrial.
	Crossomys	F	3	
Thryonomyidae (cane rats)	*Thryonomys*	F	3	This genus has one terrestrial and one semiaquatic species: *T. swinderianus.*
Hydrochaeridae (capybara)	*Hydrochaeris*	F	3	
Myocastoridae (nutria)	*Myocastor*	F, C	3	
Eulipotyphla				
Soricidae (shrews)	*Sorex*	F	3	This genus has 66 terrestrial and 2 semiaquatic species: *S. palustris* and *S. bendirii.*
	Neomys	F	3	
	Nectogale	F	3	
	Chimarrogale	F	3	
Talpidae (moles)	*Desmana*	F	3	
	Galemys	F	3	
	Condylura	F	3	
Carnivora				
Ursidae (bears)	*Ursus*	C	3	This genus has five terrestrial and one semiaquatic species: *U. maritimus,* the polar bear.
Mustelidae (weasels, mink, otters, etc.)	*Mustela*	F, C	3	This genus has 14 terrestrial and 3 semiaquatic species: *M. lutreola,* *M. vison,* and the recently extinct *M. macrodon.*
	Lutra	F, C	3	
	Lutrogale	F, C	3	
	Lontra	F, C	3	
	Pteronura	F	3	
	Aonyx	F, C	3	
	Enhydra	M, C	2	*E. lutris* is the most aquatic otter, some individuals apparently never come ashore.
Otariidae (fur seals, sea lions)	*Callorhinus*	M, C	2	All species in this family are semiaquatic.
	Arctocephalus	M, C	2	
	Zalophus	M, C	2	
	Phocarctos	M, C	2	
	Neophoca	M, C	2	
	Otaria	M, C	2	
	Eumetopias	M, C	2	
Odobenidae (walrus)	*Odobenus*	M, C	2	All species in this family are semiaquatic.

TABLE 1.4 (*continued*)

CLASSIFICATION	GENERA	HABITAT	DEGREE	NOTES
Phocidae (earless seals)	*Monachus*	M, C	2	All species in this family are semiaquatic.
	Lobodon	M, C	2	
	Hydrurga	M, C	2	
	Leptonychotes	M, C	2	
	Ommatophoca	M, C	2	
	Mirounga	M, C	2	
	Erignathus	M, C	2	
	Cystophora	M, C	2	
	Phoca	M, C, F	2	Some species and subspecies in this genus live permanently in freshwater.
Viverridae (civets, genets)	*Osbornictis*	F	3	
	Cynogale	F	3	
Herpestidae (mongoose)	*Atilax*	F	3	
Cetacea/Odontoceti				
Platanistidae (Indian river dolphin)	*Platanista*	F	1	Lives in murky waters; eyes small, and vision presumably poor.
Lipotidae (baji)	*Lipotes*	F	1	
Pontoporiidae (franciscana)	*Pontoporia*	C	1	
Iniidae (Amazon River dolphin)	*Inia*	F	1	
Monodontidae (narwhal, beluga)	*Delphinapterus*	M, C	1	
	Monodon	M	1	
Phocoenidae (porpoises)	*Phocoena*	M, C	1	
	Neophocaena	C, F	1	
	Phocoenoides	M, C	1	
Delphinidae (marine dolphins)	*Steno*	M	1	
	Sousa	C, F	1	
	Sotalia	C, F	1	
	Lagenorhynchus	M	1	
	Grampus	M	1	
	Tursiops	M	1	
	Stenella	M	1	
	Delphinus	M	1	
	Lagenodelphis	M	1	
	Lissodelphis	M	1	
	Orcaella	C, F	1	
	Cephalorhynchus	M, C	1	
	Peponocephala	M	1	
	Feresa	M	1	
	Pseudorca	M	1	
	Orcinus	M	1	
	Globicephala	M	1	

(*continued*)

TABLE 1.4 (*continued*)

CLASSIFICATION	GENERA	HABITAT	DEGREE	NOTES
Ziphiidae (beaked whales)	*Berardius*	M	1	This family is highly specialized for deep-sea diving.
	Ziphius	M	1	
	Tasmacetus	M	1	
	Hyperoodon	M	1	
	Indopacetus	M	1	
	Mesoplodon	M	1	
Physeteridae (sperm whales)	*Kogia*	M	1	This family is highly specialized for deep-sea diving.
	Physeter	M	1	
Cetacea/Mysticeti				This suborder includes highly specialized filter feeders.
Eschrichtiidae (gray whale)	*Eschrichtius*	M	1	
Neobalaenidae (pygmy right whale)	*Caperea*	M	1	
Balaenidae (right whales)	*Eubalaena*	M	1	
	Balaena	M	1	
Balaenopteridae (rorquals)	*Balaenoptera*	M	1	
	Megaptera	M	1	
Artiodactyla				
Hippopotamidae (hippopotami)	*Hippopotamus*	F	3	This genus has one living (*H. amphibius*) and three recently extinct species. Hippopotamids are semiaquatic but feed on terrestrial plants.
	Choeropsis	F	3	

NOTE: Habitat: M, marine; C, coastal; F, freshwater. Degree to which aquatic: 1, obligately aquatic; 2, mostly aquatic, return to land for reproduction and resting; 3, moderately aquatic, fully capable of terrestrial locomotion, may forage on land. Table by H. Pihlström.

ICHTHYOPTERYGIA Ichthyosaurs were obligate marine swimmers with short limbs, a long snout, and a tail fin (Kuhn-Snyder, 1963; Mazin, 1981). Consistent with life in water, they were live-bearing, and the external nares of many forms are located posteriorly on the snout. Even the earliest ichthyosaur genera are fully adapted for aquatic life (Shikama et al., 1978).

ARCHOSAURIANS Archosaurs include crocodylians (discussed below) and dinosaurs (including birds, which are discussed below). The radiation of extinct, Mesozoic dinosaurs is enormous, and it is surprising that only a single dinosaur has been identified with aquatic adaptations (Li et al., 2004).

CROCODYLIANS The Crocodylia (Fig. 1.5) comprises only 23 extant species (Zug et al., 2001) of aquatic reptiles arrayed within three lineages/families (Alligatoridae, Crocodylidae, Gavialidae). There is some disagreement about the relationships among these families, centering around the placement of two monotypic genera, *Tomistoma* and *Gavialis*. Although the distant archosaurian ancestors of crocodylians were terrestrial, it is clear that aquatic habits are primitive for the living species (Benton and Clark, 1988).

Among modern crocodylians, gharials (*Gavialis*) are the most highly aquatic, rarely leaving the water. Most crocodylians are freshwater species, but many can tolerate saltwater for extended periods of time. The saltwater crocodile (*Crocodylus porosus*) is especially tolerant of hyperosmotic environments and is often found in coastal areas, sometimes venturing far out to sea (Grigg and Gans, 1993; Richardson et al., 2002). Indeed, biogeographic, phylogenetic, and physiological evidence supports transoceanic migrations in crocodylian history (Brochu, 2001). One family of fossil Mesozoic crocodilians, the metriorhynchids, was fully marine and had a tailfin.

SQUAMATES There are approximately 7200 species of squamate reptiles divided almost equally between snakes and lizards (Zug et al., 2001). Snakes (Serpentes) form a monophyletic radiation that, in all likelihood, is derived from within lizards, so that the latter group is paraphyletic and not recognized in formal taxonomies (e.g., Fig. 1.5). The only extant, non-squamate lepidosaurs are two species of terrestrial tuatara (*Sphenodon*), relics of a once diverse group of reptiles now restricted to a few small islands off the coast of New Zealand.

Fossils that can be assigned to Recent families date from the Cretaceous (approximately 100 million years ago), but as a group, squamates arose in the Triassic, or even Late Permian (approximately 250 million years ago) (Estes, 1983; Carroll, 1988). Thus, modern squamate diversity reflects a very ancient radiation. It is thus surprising that, relatively, so few species have become specialized for aquatic environments. Even more striking is the failure of squamates and reptiles, in general, to invade marine habitats in significant numbers. Only about 75 species (approximately 1%) of all living reptiles spend time in saltwater, and more than half of these belong to a group of marine snakes in the family Elapidae (Greene, 1997). Only a few other snake groups include estuarine or marine species, and there is only one species of marine lizard. The remaining aquatic squamates are all freshwater species, and few of these evince obvious adaptations for life in water (Schwenk, chapter 5 in this volume).

At least one lineage of extinct squamates became marine predators: the mosasaurs (Russell, 1967). Mosasaurs are known only from the Upper Cretaceous and reached large sizes; some are 10 meters long. Mosasaurs are varanoid lizards (Lee, 2005), with limbs and sometimes tail modified for swimming. Although well known from many fossils, no mosasaurs are known with fetuses in their bodies, suggesting that the females hauled out of the water to lay eggs.

Relationships within the monophyletic Serpentes (snakes) are poorly resolved, although much of this controversy revolves around the relationships of the basal snake groups. Most of these taxa are fossorial and none are aquatic and therefore do not concern us here. All aquatic species, including the marine elapids, are members of the advanced snakes known as the macrostomatans, in reference to their ability to increase gape by separating the mandibular rami and swinging the quadrate bones laterally, permitting them to swallow prey that are very large relative to head size. As noted, the most highly aquatic snakes in the world, and probably the most adaptively committed aquatic reptiles of any kind, are the marine elapids belonging to the subfamilies Laticaudinae and Hydrophiinae. Two subfamilies of the diverse and speciose Colubridae are also highly aquatic, the Natricinae and Homolopsinae. Aquatic species are infrequently distributed among other taxa.

OTHER DIAPSIDS Phytosaurs (Colbert, 1947) were probably similar to crocodiles in habitat and ecology and are, in fact, related to archaic crocodiles. Thallatosauria were a small radiation of long-tailed diapsids that swam with their elongate tails. Their limbs are not specialized for aquatic locomotion (Fraas, 1902).

LITERATURE CITED

Benton, M. J., and J. Clark. 1988. Archosaur phylogeny and the relationships of the Crocodylia; pp. 295–338 in M. J. Benton (ed.), The Phylogeny and Classification of the Tetrapods, Vol. 1, Amphibians, Reptiles, Birds. Clarendon Press, Oxford.

Brochu, C. A. 2001. Congruence between physiology, phylogenetics and the fossil record on crocodylian historical biogeography; pp. 9–28 in G. C. Grigg, F. Seebacher, and C. E. Franklin (eds.), Crocodilian Biology and Evolution. Surrey Beatty & Sons, Chipping Norton, NSW, Australia.

Brown, D. S. 1981. The English Upper Jurassic Plesiosauroidea (Reptilia), and a review of the phylogeny and classification of the Plesiosauria. Bulletin of the British Museum (Natural History) Geology 35:253–347.

Cann, J. 1998. Australian Freshwater Turtles. Beaumont Publ., Singapore.

Carroll, R. L. 1981. Plesiosaur ancestors from the Upper Permian of Madagascar. Palaeontographica 170:139–200.

Carroll, R. L. 1988. Vertebrate Paleontology and Evolution. W. H. Freeman, New York.

Carroll, R. L., and P. Gaskill. 1985. The nothosaur *Pachypleurosaurus* and the origin of plesiosaurs. Philosophical Transactions of the Royal Society B 309:343–393.

Colbert, E. H. 1947. Studies of the phytosaurs *Machaeroprosopus* and *Rutiodon*. American Museum of Natural History Bulletin 88:53–96.

deBraga, M., and O. Rieppel. 1997. Reptile phylogeny and the interrelationships of turtles. Zoological Journal of the Linnean Society 120:281–354.

Drevermann, F. 1933. Die Placodontier. 3. Das Skelett von *Placodus gigas* Agassiz im Senckenberg Museum. Abhandlungen Senckenbergische Naturforschende Gesellschaft 38:319–364.

Ernst, C. H., and R. W. Barbour. 1989. Turtles of the World. Smithsonian Inst. Press, Washington, D. C.

Estes, R. 1983. The fossil record and the early distribution of lizards; pp. 365–398 in A. G. J. Rhodin and K. Miyata (eds.), Advances in Herpetology. Essays in Honor of Ernest E. Williams. Museum of Comparative Zoology, Harvard University, Cambridge, MA.

Feldman, C. R., and J. F. Parham. 2002. Molecular phylogenetics of emydine turtles: taxonomic revision and the evolution of shell kinesis. Molecular Phylogenetics and Evolution 22:388–398.

Fraas, E. 1902. Die Meer-Krokodilier (Thallatosuchia) des oberen Jura unter spezieller Beriicksichtigung von *Dracosaurus* und *Geosaurus*. Palaeontographica 49:1–72.

Gaffney, E. S., J. H. Hutchinson, F. A. Jenkins, Jr., and L. J. Meeker. 1987. Modern turtle origins: the oldest known cryptodire. Science 237:289–291.

Glasby, C. J., G. J. B. Ross, and P. L. Beesley (eds.). 1993. Fauna of Australia, Vol. 2A, Amphibia & Reptilia. Australian Government Publ. Service, Canberra.

Greene, H. W. 1997. Snakes. The Evolution of Mystery in Nature. University of California Press, Berkeley.

Greer, A. E. 1989. The Biology and Evolution of Australian Lizards. Surrey Beatty and Sons, Chipping Norton, NSW, Australia.

Grigg, G., and C. Gans. 1993. Morphology and physiology of the Crocodylia; pp. 326–336 in C. J. Glasby, G. J. B. Ross, and P. L. Beesley (eds.), Fauna of Australia, Vol. 2A, Amphibia and Reptilia. Australian Government Publishing Service, Canberra.

Huene, F. v. 1941. Osteologie und systematische Stellung von *Mesosaurus*. Palaeontographica A 92:45–58.

Joyce, W. G., and J. A. Gauthier. 2003. Palaeoecology of Triassic stem turtles sheds new light on turtle origins. Proceedings of the Royal Society of London B 271:1–5.

Kuhn-Snyder, E. 1963. I Sauri del Monte San Giorgi. Communicazioni dell'Instituto Paleontologia dell'Universita di Zurigo 20:811–854.

Lee, M. S. Y. 1995. Historical burden in systematics and the interrelationships of "parareptiles." Biological Reviews 70:459–547.

Lee, M. S. Y. 1996. Correlated progression and the origin of turtles. Nature 379:812–815.

Lee, M. S. Y. 2005. Squamate phylogeny, taxon sampling, and data congruence. Organisms Diversity and Evolution 5:25–45.

Li, C., O. Rieppel, and M. C. LaBarbera. 2004. A Triassic aquatic protorosaur with an extremely long neck. Science 305:1931–1932.

Mazin, J.-M. 1981. *Grippia longirostris* Wiman, 1929, un ichthyopterygia primitif du Trias inferieur de Spitsbergen. Bulletin du Muséum National d'Histoire Naturelle Paris C 2:243–263.

Neill, W. T. 1971. The Last of the Ruling Reptiles: Alligators, Crocodiles, and Their Kin. Columbia University Press, New York.

Peyer, B., and E. Kuhn-Snyder. 1955. Placodontia; pp. 458–486 in J. Piveteau (ed.), Traite de Paléontologie. Masson, Paris.

Pianka, E. R., and L. J. Vitt. 2003. Lizards: Windows to the Evolution of Diversity. University of California Press, Berkeley.

Reisz, R. R., and M. Laurin. 1991. *Owenetta* and the origin of turtles. Nature 349:324–326.

Richardson, K. C., G. J. W. Webb, and S. C. Manolis. 2002. Crocodiles: Inside Out. A Guide to the Crocodilians and Their Functional Morphology. Surrey Beatty and Sons, Chipping Norton, NSW, Australia.

Rieppel, O. 1994/1995. Studies on skeleton formation in reptiles: implications for turtle relationships. Zoology 98:298–308.

Rieppel, O., and M. deBraga. 1996. Turtles ad diapsid reptiles. Nature 384:453–455.

Rieppel, O., and R. R. Reisz. 1999. The origin and early evolution of turtles. Annual Review of Ecology and Systematics 30:1–22.

Rossman, T. 2000. Studies on mesosaurs (Amniota inc. sed.: Mesosauridae). 2. New information on their anatomy, with a special remark on the taxonomic status of *Mesosaurus pleurogaster* (Seely). Senckenbergiana-Lethae Lethaea 80:13–28.

Russell, D. A. 1967. Systematics and morphology of American mosasaurs (Reptilia, Sauria). Peabody Museum of Natural History, Bulletin 23:1–237.

Seymour, R. S. 1982. Physiological adaptations to aquatic life; pp. 1–51 in C. Gans and F. H. Plough, Biology of the Reptilia, Physiological Ecology, Vol. 13. Academic Press, San Diego.

Shaffer, H. B., P. Meylan, and M. L. McKnight. 1997. Tests of turtle phylogeny: molecular, morphological, and paleontological approaches. Systematic Biology 46:235–268.

Shikama, T., T, Kamei, and M. Murata. 1978. Early Triassic Ichthyosaurus, *Utatsusaurus hataii* gen. et sp. nov., from the Kitakami Massif, Northeast Japan. Science Reports of the Tohoku University Series 2 48:77–97.

Zardoya, R., and A. Meyer. 2001. The evolutionary position of turtles revisited. Naturwissenschaften 88:193–200.

Zug, G. R., L. J. Vitt and J. P. Caldwell. 2001. Herpetology. Academic Press, San Diego.

AQUATIC AND SEMIAQUATIC BIRDS

J. G. M. Thewissen and Tobin Hieronymus

With the plethora of discoveries of Mesozoic birds in the last decade (reviewed in Chiappe and Dyke, 2002; Hou et al., 2003), it has become clear that early birds were nearly as diverse as modern birds and included aquatic taxa. The recent discovery that an Early Cretaceous (110 million year old) aquatic bird is closely related to modern birds (You et al., 2006) has led to speculations that modern birds (Neornithes, Fig. 1.6) had aquatic ancestors. It is also clear that many Cretaceous birds were strongly specialized for life in water and were unable to fly (e.g., the hesperornithiform *Hesperornis* [Marsh, 1880]). Even among the modern Neornithes, a large number of orders display a variety of aquatic specializations (Table 1.3).

The phylogeny of Chiappe and Dyke (2002) is used here to describe the relationships of the fossil groups to living birds (Neornithes). While the relationship of the basal modern bird groups Palaeognathae and Galloanserae (including the aquatic Anseriformes) to Neoaves are well understood, relationships within Neoaves have proved difficult to resolve (Cracraft et al., 2004). Because of this, many of the higher-order comparisons presented in Figure 1.6 are tentative and have only heuristic value. This figure uses the phylogeny of Cracraft et al. (2004) to resolve higher-order relationships within Neoaves. Relationships within groups of aquatic birds are shown after those proposed by Bertelli and Giannini (2005) for Sphenisciformes; Kennedy and Page (2002) for Procellariiformes; Kennedy et al. (2000) and Kennedy and Spencer (2004) for Pelecaniformes; Thomas et al. (2004a, 2004b) and van Tuinen et al. (2004) for Charadriiformes; and van Tuinen et al. (2001) for Podicipediformes.

LITERATURE CITED

Bertelli, S., and N. P. Giannini. 2005. A phylogeny of extant penguins (Aves: Sphenisciformes) combining morphology and mitochondrial sequences. Cladistics 21:209–239.

Chiappe, L. M., and G. J. Dyke. 2002. The Mesozoic radiation of birds. Annual Review of Ecology and Systematics 33:91–124.

Cracraft, J., F. K. Barker, M. Braun, J. Harshman, G. J. Dyke, J. Feinstein, S. Stanley, A. Cibois, P. Schikler, P. Beresford, J. García-Moreno, M. D. Sorenson, T. Yuri, and D. P. Mindell. 2004. Phylogenetic relationships among modern birds (Neornithes): toward an avian tree of life; pp. 468–489 in J. Cracraft and M. J. Donoghue (eds.), Assembling the Tree of Life. Oxford University Press, Oxford, U.K.

Hou, L., C. Chuong, C., Y. Anderson, X. Zeng, and J. Hou. 2003. Fossil Birds of China. Yunnan Science and Technology Press, Kumming, China.

Kennedy, M., and R. D. M. Page. 2002. Seabird supertrees: combining partial estimates of procellariiform phylogeny. Auk 119:88–108.

Kennedy, M., and H. G. Spencer. 2004. Phylogenies of the frigatebirds (Fregatidae) and tropicbirds (Phaethonidae), two divergent groups of the traditional order Pelecaniformes, inferred from mitochondrial DNA sequences. Molecular Phylogenetics and Evolution 31:31–38.

Kennedy, M., R. D. Gray, and H. G. Spencer. 2000. The phylogenetic relationships of the shags and cormorants: can sequence data resolve a disagreement between behavior and morphology? Molecular Phylogenetics and Evolution 17: 345–359.

Marsh, O. C. 1880. Odontornithes: a monograph on the extinct toothed birds of North America. U.S. Geological Explorations, 40th Parallel, Washington, DC.

Thomas, G. H., M. A. Wills, and T. Székely. 2004a. Phylogeny of shorebirds, gulls, and alcids (Aves: Charadrii) from the cytochrome-b gene: parsimony, Bayesian inference, minimum evolution, and quartet puzzling. Molecular Phylogenetics and Evolution 30:516–526.

Thomas, G. H., M. A. Wills, and T. Székely. 2004b. A supertree approach to shorebird phylogeny. BMC Evolutionary Biology 4:28.

Tuinen, M. v., D. B. Butvill, J. A. W. Kirsch, and S. B. Hedges. 2001. Convergence and divergence in the evolution of aquatic birds. Proceedings of the Royal Society of London B 268:1–6.

Tuinen, M. v., D. Waterhouse, and G. J. Dyke. 2004. Avian molecular systematics on the rebound: a fresh look at modern shorebird phylogenetic relationships. Journal of Avian Biology 35:191–194.

You, H.-I., M. C. Lamanna, J. D. Harris, L. M. Chiappe, J. O'Connor, S.-A. Ji, J.-C. Lü, C.-X. Yuan, D.-Q. Li, X. Zhang, K. J. Lacovara, P. Dodson, and Q. Ji. 2006. A nearly modern amphibious bird from the Early Cretaceous of northwestern China. Science 312:1640–1643.

AQUATIC AND SEMIAQUATIC MAMMALS
Henry Pihlström

Table 1.4 and Figure 1.7 present an overview of orders, families, and genera of extant aquatic mammals. Table 1.4 is a consensus classification and indicates habitat and degree of aquatic specialization of these mammals, and Figure 1.7 is a cladogram. The higher-level phylogeny of Table 1.4 and Figure 1.7 are identical except that the traditional orders Cetacea and Artiodactyla have been retained in the former, which is invalid from a purist phylogenetic standpoint (Price et al., 2005). Habitat information,

as well as nomenclature for genera and species, follows Nowak (1999), with these exceptions: the gray seal genus *Halichoerus* is included in the genus *Phoca* (following Arnason et al. [1995] and Davis et al. [2004]), and the genus name *Choeropsis* is used for the pygmy hippopotamus (following Boisserie [2005]).

Figure 1.7 is compiled from Murphy et al. (2001) for overall phylogeny; Cardillo et al. (2004) for marsupials; Stanhope et al. (1998), Douady et al. (2002), Clementz et al. (2003), Grenyer and Purvis (2003), and Shoshani and Tassy (2005) for afrotherians; Gaudin (2004) for xenarthrans; Adkins et al. (2001) and Huchon et al. (2002) for rodents; Grenyer and Purvis (2003) for Eulipotyphla; Flynn et al. (2005) for Carnivora; Prothero et al. (1986) for amynodontid and rhinocerotid perissodactyls; Price et al. (2005) and Boisserie et al. (2005) for whales and artiodactyls.

LITERATURE CITED

Adkins, R. M., E. L. Gelke, D. Rowe, and R. L. Honeycutt. 2001. Molecular phylogeny and divergence time estimates for major rodent groups: evidence from multiple genes. Molecular Biology and Evolution 18:777–791.

Arnason, U., K. Bodin, A. Gullberg, C. Ledje, and S. Mouchaty. 1995. A molecular view of pinniped relationships with particular emphasis on the true seals. Journal of Molecular Evolution 40:78–85.

Boisserie, J.-R. 2005. The phylogeny and taxonomy of Hippopotamidae (Mammalia: Artiodactyla): a review based on morphology and cladistic analysis. Zoological Journal of the Linnean Society 143:1–26.

Boisserie, J.-R., F. Lihoreau, and M. Brunet. 2005. The position of Hippopotamidae within Cetartiodactyla. Proceedings of the National Academy of Sciences USA 102:1537–1541.

Cardillo, M., O. R. P. Bininda-Emonds, E. Boakes, and A. Purvis. 2004. A species-level phylogenetic supertree of marsupials. Journal of Zoology (London) 264:11–31.

Clementz, M. T., K. A. Hoppe, and P. L. Koch. 2003. A paleoecological paradox: the habitat and dietary preferences of the extinct tethythere *Desmostylus*, inferred from stable isotope analysis. Paleobiology 29:506–519.

Davis, C. S., I. Delisle, I. Stirling, D. B. Siniff, and C. Strobeck. 2004. A phylogeny of the extant Phocidae inferred from complete mitochondrial DNA coding regions. Molecular Phylogenetics and Evolution 33:363–377.

Douady, C. J., P. I. Chatelier, O. Madsen, W. W. de Jong, F. Catzeflis, M. S. Springer, and M. J. Stanhope. 2002. Molecular phylogenetic evidence confirming the Eulipotyphla concept and in support of hedgehogs as the sister group to shrews. Molecular Phylogenetics and Evolution 25:200–209.

Flynn, J. J., J. A. Finarelli, S. Zehr, J. Hsu, and M. A. Nedbal. 2005. Molecular phylogeny of the Carnivora (Mammalia): assessing the impact of increased sampling on resolving enigmatic relationships. Systematic Biology 54:317–337.

Gaudin, T. J. 2004. Phylogenetic relationships among sloths (Mammalia, Xenarthra, Tardigrada): the craniodental evidence. Zoological Journal of the Linnean Society 140:255–305.

Grenyer, R., and A. Purvis. 2003. A composite species-level phylogeny of the "Insectivora" (Mammalia: Order Lipotyphla Haeckel, 1866). Journal of Zoology (London) 260:245–257.

Huchon, D., O. Madsen, M. J. J. B. Sibbald, K. Ament, M. J. Stanhope, F. Catzeflis, W. W. de Jong, and E. J. P. Douzery. 2002. Rodent phylogeny and a timescale for the evolution of Glires: evidence from an extensive taxon sampling using three nuclear genes. Molecular Biology and Evolution 19:1053–1065.

Murphy, W. J., E. Eizirik, S. J. O'Brien, O. Madsen, M. Scally, C. J. Douady, E. Teeling, O. A. Ryder, M. J. Stanhope, W. W. de Jong, and M. S. Springer. 2001. Resolution of the early placental mammal radiation using Bayesian phylogenetics. Science 294:2348–2351.

Nowak, R. M. 1999. Walker's Mammals of the World, 6th edition, Volumes I and II. Johns Hopkins University Press, Baltimore.

Price, S. A., O. R. P. Bininda-Emonds, and J. L. Gittleman. 2005. A complete phylogeny of the whales, dolphins and even-toed hoofed mammals (Cetartiodactyla). Biological Reviews 80:445–473.

Prothero, D. R., E. Manning, and C. B. Hanson. 1986. The phylogeny of the Rhinocerotoidea (Mammalia, Perissodactyla). Zoological Journal of the Linnean Society 87:341–366.

Shoshani, J., and P. Tassy. 2005. Advances in proboscidean taxonomy and classification, anatomy and physiology, and ecology and behavior. Quaternary International 126–128:5–20.

Stanhope, M. J., V. G. Waddell, O. Madsen, W. W. de Jong, S. B. Hedges, G. C. Cleven, D. Kao, and M. S. Springer. 1998. Molecular evidence for multiple origins of Insectivora and for a new order of endemic African insectivore mammals. Proceedings of the National Academy of Sciences USA 95:9967–9972.

Chemical Senses

2

The Physics and Biology
of Olfaction and Taste

Simo Hemilä and Tom Reuter

Diffusion

Odorants and Their Perception

Stimulus Strength and Detection

Traditionally, the chemical senses include olfaction and gustation: olfaction (smell) reports on gaseous substances carried by air over long distances, and gustation (taste) provides information about food material already in the mouth. However, in recent times behavioral, physiological, histological, and molecular studies have increased our understanding of vertebrate chemoreception, and four chemical senses are recognized by some authors, reflecting the diversity in function and physiology (Dulac and Axel, 1995; Eisthen, 1997; Zufall and Munger, 2001; Mombaerts, 2004; Breer et al., 2005). For the purpose of this volume, three chemical senses are important: olfaction, vomeronasal sense, and gustation (Fig. 7.1 in this volume).

It is difficult to distinguish sharp borders between the various chemosensory systems, especially in aquatic vertebrates. For example, the catfish *Ictalurus natalis* carries on its barbels densely packed taste buds. These taste buds

appear to report on dissolved substances in the surrounding water (Caprio, 1988). Histologically, the barbels' sensory buds serve taste, but functionally their role seems to overlap with olfaction: catfish successfully orient toward distant "olfactory stimuli" even after bilateral destruction of the olfactory tract (Bardach et al., 1967). A similar quasi-olfactory role has also been described for the tongue of the dolphin *Tursiops truncatus* (Kuznetzov, 1990), and Schmidt and Wöhrmann-Repenning (2004) have reported histological observations indicating a close interaction between taste buds and vomeronasal organ.

Molecular transduction mechanisms of the sensory cells of the olfactory, vomeronasal, and gustatory system are also similar, as are the microstructures of the distal parts of the receptor cells (Eisthen, 1992, 1997; Reiss and Eisthen, chapter 4 in this volume). For mammals a classification based on the molecular genetics of the receptor proteins may be within reach (Mombaerts, 2004; Grus et al., 2005). However, this volume adheres to an anatomical basis for distinguishing between chemical sense organs, as it embraces the inclusion of amphibian, reptile, and bird chemical senses, for which the molecular genetics of olfactory receptors are less well known. As a result, this volume uses a classification based on the innervation by the cranial nerves of the sensory organs.

The sensory cells of the olfactory and vomeronasal systems send out axons passing through the perforated part of the ethmoid bone (the cribriform plate) and terminate in specific glomeruli of the olfactory bulb (Vassar et al., 1993; Belluscio et al., 2002; Pihlström, chapter 7 in this volume). The axons from the main olfactory epithelium project to the main olfactory bulb, while the axons from the vomeronasal epithelia project to the accessory olfactory bulb (Zufall and Munger, 2001; Breer et al., 2005). Afferent axons leaving the main and accessory olfactory bulbs unite into bundles forming the olfactory tract (cranial nerve I). More centrally the olfactory tract divides into many projections contacting different brain nuclei (Shepherd, 1994). The final brain projections from the main olfactory system and the vomeronasal system differ (Smith, 2000).

Receptor cells of the taste buds on the tongue do not form centrally projecting axons; instead they are innervated by thin dendritic sensory endings projecting from specific brain neurons. The taste receptors of the anterior part of the human tongue are innervated by a branch of the facial nerve (cranial nerve VII), while the rear of the tongue is served by the glossopharyngeal nerve (cranial nerve IX). The taste cells of the roof of the oral cavity and the upper esophagus are innervated by branches of the vagus nerve (cranial nerve X). Although diffuse distally, taste is an integrated sensory system as indicated by the termination of nerve fibers in all these taste pathways in the solitary nuclear complex in the medulla (Smith, 2000).

In this chapter, we briefly review the basics of diffusion, the process that underlies the perception of all chemical stimuli. Then, we discuss odorants, and finally the information

carried by chemical stimuli. This discussion includes the dimensionless Reynolds number, useful when considering odorant-carrying fluid flows.

The term *odorant* is often used for biologically relevant molecules such as pheromones and alarm substances, and for metabolites unintentionally released by predators or prey animals. Here, we extend the use of the term to include all substances spread in air and water and by direct physical contact, and perceived by the chemical sense organs.

DIFFUSION

Diffusion is the spread of dissolved and gaseous molecules in a stationary medium (here water or air) as a result of random thermal movements. Diffusion is slow and cannot serve detection over long distances. Chemosensory cells, olfactory, vomeronasal, and taste receptors in both terrestrial and aquatic tetrapods are covered by thin layers of mucous. Within this thin 0.1 to 1 millimeter thick unstirred layer, molecules spread exclusively by diffusion (Dusenbery, 1992).

Diffusion is faster in air than in water (Table 2.1); if a micromole (6×10^{17} molecules) of a dissolved substance is released from a point source in still water, the first molecule reaches a point 1 centimeter from the source after approximately 10 minutes. In air, the first molecule reaches a point 1 meter from the source in 10 minutes. These observations can be quantified using diffusion constants, which are approximately 10^{-5} m^2/s and approximately 10^{-9} m^2/s, in air and water, respectively (Dusenbery, 1992). The diffusion constants are generally lower for large than for small molecules, and possible hydrophilic or amphoteric properties of the molecules aid their diffusion through mucous or fluid layers in the nasal and oral cavities (Bradbury and Vehrencamp, 1998). Still, although diffusion may serve communication between microorganisms, this mechanism is too slow to serve long-distance olfaction of

TABLE 2.1
Physical and Chemical Factors Relevant for Spread of Biological Signal Molecules

	IN AIR	IN WATER
Diffusion constant	Small, ca. 10^{-5} m^2/s	Very small, ca. 10^{-9} m^2/s
Odorant-carrying mechanism	Wind of varying velocity and direction	Water current, often predictable velocity and direction
Odorants carried by the medium	Small and medium-sized volatile alcohols, aldehydes, esters etc., with 15 to 20 carbon atoms	Hydrophilic amino acids, peptides, proteins, and nucleotides
Odorants in contact perception	Weakly volatile molecules, e.g., pheromones in sebum	Hydrophobic odorants embedded in slime

larger animals. Taste is different; masticated food in direct contact with the tongue is clearly within diffusion distance from the sensory cells of the taste buds.

During chemical perception in the nasal cavities of terrestrial tetrapods, airborne odorants diffuse both through a thin unstirred layer of air, and through a thin layer of mucous fluid covering the sensory epithelium (Dusenbery, 1992). Slow diffusion results in delays of up to 1 second, which is one factor making olfaction a relatively slow sensory modality compared with hearing and cone-driven vision.

ODORANTS AND THEIR PERCEPTION

The molecular characteristics of odorants determine their solubility in the medium and their volatility, and thus their release and transport toward the sensory epithelium. The spread of molecules is medium specific, and thus the evolution of efficient signal molecules is different in different habitats and for different types of spread: spread in air and water, or through direct physical contact (Dusenbery, 1992).

The physics of the chemical senses is ultimately concerned with interactions between biologically relevant molecules and the corresponding receptor proteins in sensory cells. However, it is unclear whether there are separate groups of olfactory receptor molecules for airborne odorants and for water-soluble substances (Eisthen, 1997; Freitag et al., 1998).

Independent of receptor proteins, aerial olfaction seems to be characterized by olfactory binding proteins in the mucus that covers the sensory epithelia of terrestrial tetrapods. Such binding proteins are also found in the air-exposed cavity of the *Xenopus* (African clawed frog) olfactory system. Olfactory binding proteins are not yet identified in fish, but in terrestrial tetrapods they are supposed to have a carrier role in transporting hydrophobic odorants through the liquid layers covering the olfactory epithelia (Millery et al., 2005).

Airborne odorants are small volatile molecules with few chemical bonds to the substrate. Thus they are easily released by modest thermal vibration, that is, when the thermal energy exceeds the chemical binding energy. Some smelly substances released from biological materials are very small molecules, such as hydrogen sulfide (H_2S) and ammonia (NH_3). These compounds warn at least some mammals of unhealthy food and water, but such small molecules are less suited for signaling between individuals, as important messengers must combine proper volatility with sufficient receptor specificity. Typical airborne pheromones are alcohols, aldehydes, ketones, fatty acids, esters, and steroles with 15 to 20 carbon atoms.

Alarm substances typically have only 6 to 15 carbon atoms and are thus less specific than pheromones. They serve instead as efficient (fast) warning signals (Dusenbery, 1992).

Waterborne odorants are hydrophilic substances ranging from small organic molecules to large proteins. Typical molecules with appetizing effect on aquatic predators are amino acids and nucleotides unintentionally released by prey animals. Waterborne pheromones are usually polar and thus hydrophilic peptides or proteins.

Chemical communication through direct physical contact includes both olfaction and taste; mammals and some other tetrapods transfer sexual pheromones by licking and touching. Direct contact is also involved in territorial markings. An interesting form of contact perception is the use of the snake tongue in following chemical trails (Bradbury and Vehrencamp, 1998). Especially in connection with territorial markings, the longevity of the signal is important, that is, the signal molecules should be released very slowly into the surrounding medium. In terrestrial tetrapods these substances must be weakly volatile pheromones in secreted sebum, while in aquatic tetrapods such as otters and seals they are hydrophobic odorants, sometimes embedded in slime.

STIMULUS STRENGTH AND DETECTION

Over distances longer than those served by diffusion, chemical stimuli are spread by movements of air and water. Essentially, flow in air is similar to flow in water: both media can be considered fluids for the purpose of the receiving sense organs. An important quantity characterizing the type of flow, laminar or turbulent, is the Reynolds number.

When a fluid flows slowly in a narrow channel, the flow is laminar: adjacent layers glide past each other without mixing. Increasing the fluid velocity or the width of the channel leads to turbulent flow. The velocity producing turbulent flow in a given flow field can be estimated by the dimensionless Reynolds number: $R = \rho L v / \mu$, where L is a chosen characteristic length of the flow field, v is a characteristic flow velocity, ρ is the density of the fluid, and μ is its dynamic viscosity (air: 18.3 μPa s; water: 1 mPa s). In all flow fields of similar shape (isometric flow fields) the laminar flow changes to turbulent at equal R values. For example, when a fluid flows in a tube and the tube diameter has been chosen as the characteristic size L, the flow turns turbulent when the value of R is approximately 2300. In narrow channels, flow is usually laminar. Open flow fields correspond to large L values, and in these flow is almost always turbulent, even at low velocities.

Turbulent flow results in efficient spread of odorants. When the source of an odorant is close to the signal-receiving animal, the main medium flow is produced by the animal itself, by its respiratory movements and active search, such as its breathing and sniffing. Differing from humans, many tetrapods have evolved highly specialized anatomical structures for actively collecting air samples for closer scrutiny. For example, among the functions of the elephant trunk is the collection of precisely localized samples of air, and the transport of these samples to a uniquely large and probably very sensitive olfactory organ (Boas and Paulli, 1925; Pihlström et al., 2005). Over longer distances, chemical stimuli are spread by wind and water currents. Water currents are more predictable than those in air.

In turbulent flows odorants do not spread homogeneously in the medium (Dusenbery, 1992). Thus, the rate at which odorant molecules reach an olfactory epithelium varies significantly. High peaks of odorant concentration may increase sensitivity, but the information obtained from such a noisy signal is limited. The type of odorant may be identified, but only a rough estimate of concentration is possible, and precise modulation in time is blurred by turbulence.

ACKNOWLEDGMENTS

We thank the editors, J. G. M. Thewissen and Sirpa Nummela, for inviting us to contribute to this volume.

We are grateful to Henry Pihlström for many helpful discussions concerning mammalian olfaction.

LITERATURE CITED

Bardach, J. E., J. H. Todd, and R. Crickmer. 1967. Orientation by taste in fish of the genus *Ictalurus*. Science 155:1276–1278.

Belluscio, L., C. Lodovichi, P. Feinstein, P. Mombaerts, and L. C. Katz. 2002. Odorant receptors instruct functional circuitry in the mouse olfactory bulb. Nature 419:296–300.

Boas, T. E., and S. Paulli. 1925. The Elephant's Head, Part 2. The Carlsberg Fund, Copenhagen.

Bradbury, J. W., and S. L. Vehrencamp. 1998. Principles of Animal Communication. Sinauer Associates, Sunderland, MA.

Breer, H., R. Hoppe, J. Kaluza, O. Levai, and J. Strotmann. 2005. Olfactory subsystems in mammals: specific roles in recognizing chemical signals? Chemical Senses 30 (Suppl. 1): 144–145.

Caprio, J. 1988. Peripheral filters and chemoreceptor cells in fishes; pp. 313–338 in J. Atema, R. R. Fay, A. N. Popper, and W. N. Tavolga (eds.), Sensory Biology of Aquatic Animals. Springer-Verlag, New York.

Dulac, C., and R. Axel. 1995. A novel family of genes encoding putative pheromone receptors in mammals. Cell 83:195–206.

Dusenbery, D. B. 1992. Sensory Ecology: How Organisms Acquire and Respond to Information. W. H. Freeman and Company, New York.

Eisthen, H. L. 1992. Phylogeny of the vomeronasal system and of receptor cell types in the olfactory and vomeronasal epithelia of vertebrates. Microscopy Research and Technique 23:1–21.

Eisthen, H. L. 1997. Evolution of vertebrate olfactory systems. Brain, Behavior, and Evolution 50:222–233.

Freitag, J., G. Ludwig, I. Andreini, P. Rössler, and H. Beer. 1998. Olfactory receptors in aquatic and terrestrial vertebrates. Journal of Comparative Physiology A 183:635–650.

Grus, W. E., P. Shi, Y. Zhang, and J. Zhang. 2005. Dramatic variation of the vomeronasal pheromone receptor gene repertoire among five orders of placental and marsupial mammals. Proceedings of the National Academy of Sciences USA 102:5767–5772.

Kuznetzov, V. B. 1990. Chemical sense of dolphins: quasi-olfaction; pp. 481–503 in J. A. Thomas and R. A. Kastelein (eds.), Sensory Abilities of Cetaceans: Laboratory and Field Evidence. Plenum Press, New York.

Millery, J., L. Briand, V. Bézirard, F. Blon, C. Fenech, L. Richard-Parpaillon, B. Quennedey, J.-C. Pernollet, and J. Gascuel. 2005. Specific expression of olfactory binding protein in the aerial olfactory cavity of adult and developing *Xenopus*. European Journal of Neuroscience 22:1389–1399.

Mombaerts, P. 2004. Genes and ligands for odorant, vomeronasal and taste receptors. Nature Reviews Neuroscience 5:263–278.

Pihlström, H., M. Fortelius, S. Hemilä, R. Forsman, and T. Reuter. 2005. Scaling of mammalian ethmoid bones can predict olfactory organ size and performance. Proceedings of the Royal Society of London B 272:957–962.

Schmidt, M., and A. Wöhrmann-Repenning. 2004. Suggestions for the functional relationship between differently located taste buds and vomeronasal olfaction in several mammals. Mammalian Biology 69:311–318.

Shepherd, G. M. 1994. Neurobiology. Oxford University Press, New York.

Smith, C. U. M. 2000. Biology of Sensory Systems. John Wiley and Sons, Chichester, UK.

Vassar, R., J. Ngai, and R. Axel. 1993. Spatial segregation of odorant receptor expression in the mammalian olfactory epithelium. Cell 74:309–318.

Zufall, F., and S. D. Munger. 2001. From odor and pheromone transduction to the organization of the sense of smell. Trends in Neurosciences 24:191–193.

3

The Chemical Stimulus
and Its Detection

Heather L. Eisthen and Kurt Schwenk

Volatility
Solubility
Polarity
Sensory Transduction
Concluding Comments

One of the most difficult aspects of studying the chemical senses is that researchers do not know which properties of the stimulus are relevant for detection and perception. In this article, we explain what we do and do not know about the features of chemical stimuli, the ways in which these features are detected and coded by chemosensory organs and the central nervous system of tetrapod vertebrates, and the characteristics of a stimulus that might make it likely to stimulate a particular chemosensory organ.

Before sensory stimuli can affect an animal's physiology or behavior (i.e., be "perceived"), they have to be transduced into electrical signals and processed in the central nervous system. In the visual system, the relevant parameters of light are its wavelength and intensity, which are transduced by specialized photoreceptor cells in the retina and, after processing in the central

nervous system, perceived as color and brightness, respectively. Many vertebrates also have two classes of photoreceptor cells that are differentially sensitive to light of different intensities, the rod and cone cells, enabling the visual system to operate in both bright and dim light. Furthermore, the photosensitive pigments (opsins) in these cells are sensitive to light within a limited range of wavelengths; an animal with two or more classes of photoreceptor cells with different opsins will be able to perceive color (Reuter and Peichl, chapter 10 in this volume). Water differentially attenuates light of different wavelengths, and these changes are reflected in the differences in spectral tuning of opsins seen in animals that live at different depths (e.g., Fasick and Robinson, 2000). Thus, the properties of light and the ways in which sensory specializations are tuned to these properties are well understood. (For a detailed discussion of the properties of visual stimuli and the ways in which they differ in air and water, see chapters 8 and 9 in this volume by Kröger and Kröger and Katzir).

In contrast to our detailed understanding of visual stimuli and the sensory ecology of the visual system, we know surprisingly little about chemosensory stimuli. Do chemical stimuli possess regular features, analogous to wavelength and intensity, that are coded by the nervous system? If so, how do changes in a given feature affect perception? Do animals have differentially sensitive chemosensory receptor cells, or chemosensory organs, for detecting stimuli with different features? Finally, how do air and water alter the features or availability of chemical stimuli?

Most tetrapods have three anatomically separate chemosensory organs: a gustatory (taste) organ, an olfactory (smell) organ, and a vomeronasal (accessory olfactory or smell) organ. The central projections of these three organs are anatomically distinct through several synapses, and this anatomical distinction is used to recognize and differentiate the chemosensory systems in diverse vertebrates. The organs may detect largely nonoverlapping sets of chemical stimuli, but information from the three organs is integrated within the central nervous system so that animals can make appropriate decisions about what to eat, flee from, fight with, or mate with.

In general, tetrapods employ the three chemosensory systems according to a rough temporal and informational hierarchy (e.g., Schwenk, 1995). The olfactory system is capable of detecting chemicals at a distance from their source. Once an animal has detected a potentially interesting odorant using its olfactory system, it may then investigate the source further using its close-range chemosensory organs, the tongue (taste buds) and vomeronasal organ. We still do not know why some chemicals preferentially stimulate the taste system instead of the vomeronasal organ, or why others stimulate the vomeronasal but not olfactory organ.

In general, the different chemosensory organs detect compounds with differing behavioral significance. The taste system is specialized for sampling possible food items before they are ingested and responds to compounds that are dissolved in fluids and do not become airborne to any significant degree: sugars, protons, salts, proteins, and amino acids, as well as diverse molecules that taste bitter and serve to warn of the presence of potential toxins. The olfactory and vomeronasal systems may serve less specialized functions. Although some researchers have suggested that the vomeronasal system could be specialized for detection of pheromones, the olfactory system mediates some pheromonal effects, and the vomeronasal system responds to nonpheromonal stimuli such as prey odorants and artificial chemicals with no inherent biological significance (reviewed in Baxi et al., 2006). The vomeronasal system also has been suggested to mediate unlearned responses to odorants (Meredith, 1986; Halpern, 1987), but this intriguing idea has not received much experimental attention.

If the olfactory system is specialized for detecting chemicals at a distance, what features of a chemical determine whether it will move from the source, or remain in place? In terrestrial environments, we might expect that a simple parameter of the chemicals themselves, such as volatility, might play a role. However, most biological sources of odorants contain water or lipids, and in all tetrapod chemosensory systems chemical stimuli must pass through a fluid covering the sensory epithelium before they are transduced by sensory neurons. Therefore, we might expect that other chemical parameters, such as solubility or polarity, might also be important in determining which organ detects a given stimulus. We consider each of these potentially important chemical qualities in turn.

VOLATILITY

The volatility of a given chemical depends on several factors, including its molecular weight. Within any chemical series, vapor pressure (and therefore volatility) decreases rapidly with increasing molecular weight. Early perfumers believed molecular weight to be the sole determinant of volatility, and that chemicals larger than about 500 daltons are too heavy to become airborne; nevertheless, volatility depends on several factors such as the length of side chains.

Volatility has been suggested to play a key role in determining which chemosensory organ detects a given stimulus. Most prominently, the olfactory epithelium has been hypothesized to detect volatile odorants, and the vomeronasal organ to detect nonvolatiles (for a thorough discussion, see work by Baxi et al. [2006]). This hypothesis is founded on the observation that some tetrapods have specialized behavioral or physiological mechanisms to promote access of nonvolatile molecules to the vomeronasal organs. For example, in squamate reptiles (lizards and snakes), the vomeronasal organs are sequestered from the nasal cavities, and chemical access is limited to two tiny openings in the anterior palate. Scent molecules are picked up by the mucus-covered surface of the tongue tip and physically delivered to the vomeronasal orifices in the mouth during a behavior known as tongue-flicking (reviewed in Halpern, 1992; Schwenk, 1995). The nasal epithelium, on the other hand, is stimulated by presumably volatile odorants that enter the nose relatively passively during breathing (Dial and Schwenk, 1996). In some rodents and other mammals, the vomeronasal organs are surrounded by vascular, contractile tissue that acts as a pump, drawing nonvolatile molecules into the organs along with the mucus surrounding their openings in the nose (Meredith, 1994). Nuzzling, nose-pressing, and licking are common mammalian behaviors that potentially introduce nonvolatile chemicals into the oral and nasal cavities. A peculiar lip-curling behavior known as *flehmen* is evident in artiodactyls and some carnivorans and is thought to facilitate suctioning of fluids from the mouth through palatal openings into the vomeronasal organ (reviewed in Wyatt, 2003). In contrast, the olfactory epithelium of most mammals is located in the posterodorsal region of the nasal cavity, a location suggesting that only volatile molecules would have access to the sensory receptor neurons.

Several problems complicate this tidy scenario. First, nonvolatile compounds, and even large particles such as viruses, can gain access to the olfactory epithelium (Mori et al., 2005). Conversely, there is no evidence suggesting that volatile compounds cannot gain access to the vomeronasal organ. The volatility hypothesis would be saved if one could demonstrate that "inappropriate" stimuli that gain access to the vomeronasal or olfactory organs are not transduced, but relevant data are lacking. Some studies demonstrate that volatile compounds can stimulate vomeronasal receptor neurons in vitro (e.g., Leinders-Zufall et al., 2000), but the results of in vivo studies are equivocal (e.g., Luo et al., 2003).

A second problem involves the presence of odorant binding proteins within the mucus in the nasal cavity (Baxi et al., 2006). In short, the importance of volatility may be significantly mitigated by the presence of large, nonvolatile carrier or transport proteins, particularly if odorants frequently or always arrive at the sensory epithelium in a bound state. Although some studies indicate that volatile compounds stimulate the vomeronasal system only in the presence of carrier proteins (e.g., Guo et al., 1997), other studies suggest that the volatile compounds and their carriers may be transduced by the vomeronasal neurons and then processed independently in the central nervous system (e.g., Brennan et al., 1999).

A final problem arises from the observation that some aquatic tetrapods, such as amphibians, possess a vomeronasal system (Reiss and Eisthen, chapter 4 in this volume), and volatility is irrelevant in water. Indeed, large molecules that are nonvolatile in air can be highly soluble in water if they are polar and, therefore, could be available to any of the three chemosensory systems (Wilson, 1970). As such, molecules of varying sizes and properties may have equal access to the olfactory and vomeronasal epithelia in aquatic tetrapods, such as amphibians, in which the vomeronasal epithelium is not sequestered (for discussion of this issue in marine turtles, see chapter 5 by Schwenk in this volume). Thus, the significance of volatility in determining which chemosensory organ detects a given stimulus is unclear, particularly for aquatic organisms.

SOLUBILITY

Chemical stimuli must cross a watery mucus layer before contacting receptor cells in taste buds, or in the olfactory or vomeronasal epithelium. Thus, one might expect that solubility plays a role in determining whether a chemical gains access to chemosensory receptor cells, and it may also determine which sensory organ is stimulated. As with volatility, the importance of solubility is complicated and poorly understood.

In some cases, odorant molecules probably diffuse across the mucus barrier directly to the sensory receptor cells. However, the mucus in the nasal cavity contains odorant binding proteins, which may function in transporting hydrophobic molecules across the mucus barrier to the sensory epithelium (Bignetti et al., 1987; Vogt, 1987). Nevertheless, the odorant binding proteins may bind odorants nonselectively, in which case hydrophobicity is irrelevant (Tegoni et al., 2000).

Although one might also expect that in aquatic environments chemical stimuli must be soluble to be detected, this does not appear to be the case. For example, the magnificent tree frog, *Litoria splendida*, produces a hydrophobic peptide pheromone. The molecule's hydrophobicity appears to be an important aspect of its function, as it causes the pheromone to disperse much more rapidly than would occur through simple diffusion (Apponyi and Bowie, 2005). Indeed, dispersal rate of chemical signals may be at a premium in aquatic environments owing to the diffusion kinetics of chemicals in water versus air. The time interval between chemical release and fade-out is 10,000 times greater in water; that is, a chemical signal takes much longer to fill the same volume of space in water compared to air (Wilson, 1970). For organisms inhabiting still waters, low rates of diffusion could pose a significant problem for chemical signaling. Therefore, the rapid spreading afforded by nonsoluble, hydrophobic chemical signals could be advantageous to aquatic organisms, a possibility that may strike some as nonintuitive.

POLARITY

Polarity is a virtually unexplored attribute of potential chemical signals in vertebrates. Its possible significance in the context of aquatic species is obvious in that it relates to chemical solubility: polar molecules are usually highly soluble in water and, therefore, potentially available in aquatic habitats. Indeed, as polarity increases, vapor pressure (volatility) decreases and solubility in water increases. High solubility owing to molecular polarity can offset the relatively slow diffusion rate in water versus air by greatly increasing the emission rate of a chemical source in water (Wilson, 1970). Further, although larger molecules are generally less volatile and less soluble than smaller molecules, even very large molecules, such as proteins, are soluble in water if they are polar. Thus, large, polar chemicals such as peptides that would make poor signals in terrestrial environments might provide an important source of chemical information to animals in aquatic habitats. This is known to be the case for some aquatic invertebrates and amphibians (Wilson, 1970; Kikuyama et al., 2002).

SENSORY TRANSDUCTION

The means by which odorants bind to and activate receptors, as well as the basis of receptor selectivity, are important unsolved problems in chemosensory research. One of the earliest models of olfactory transduction was Amoore's stereochemical model. Chemicals with similar odors, such as those that smell "fruity" or "floral," were postulated to have a similar shape and fit only into a subset of receptors, like a lock and key. The binding of a particularly shaped molecule to a receptor would then signal the presence of a particular class of molecule (Amoore, 1963). This idea stimulated much productive research, which ultimately proved the model false. Our current understanding is that two molecules that we perceive as similar in odor might bear no structural similarity. Nevertheless, both behavioral and electrophysiological work demonstrate that in some cases the processing of a stimulus depends on length of a side chain or other geometric features of a stimulus (Kent et al., 2003), suggesting that shape may not be completely irrelevant to perception.

A different model of olfactory processing suggests that odorant receptors detect molecular vibrations of stimulus compounds (Turin, 2002). The basic predictions of the theory do not stand up to experimental scrutiny (Keller and Vosshall, 2004).

CONCLUDING COMMENTS

This brief review demonstrates that we understand little about the parameters of chemical stimuli that are coded by chemosensory systems. No simple relationship predicts which chemosensory system will be stimulated, or in what fashion, by a particular stimulus. We therefore cannot make any broad generalizations about how the functional domains of the three tetrapod chemosensory systems are sorted according to the features of chemical stimuli, nor how these domains might differ between terrestrial and aquatic species.

To illustrate the difficulty we have in segregating the chemosensory systems, consider the apparently straightforward question of taste versus olfaction. We cannot define these systems based on the chemicals they respond to, as one might suppose; rather, we distinguish them on the basis of their anatomy and central nervous system coding. From empirical work, we know that taste receptor cells are involved in feeding behavior and usually respond to proximate

chemical cues released during physical contact with a chemical source, whereas olfactory receptors can respond to chemicals released by distant sources. Further, information from the two systems is combined to give rise to the perception of "flavor," which depends to a large degree on olfactory cues emanating from items in the oral cavity; thus, the same food item gives rise to both gustatory and olfactory stimuli. Overall, our understanding of the functional attributes of the system is based on empirical observation and not predictions based on first principles.

When it comes to distinctions between the olfactory and vomeronasal systems, we are at even more of loss, given the anatomical similarity between them. The functional dichotomy of sensitivity to low versus high molecular weight stimuli, respectively, may be true under some circumstances, but it is not a perfect distinction. Perhaps the more important question is one of accessibility. After all, one must consider not only whether a particular sensory receptor cell can respond to or transduce a particular class of chemicals, but the circumstances in which the cell will be exposed to those chemicals.

In mammals, squamate reptiles, and marine turtles, the vomeronasal epithelium is sequestered, and unlike the olfactory epithelium, its cells are not readily stimulated during normal breathing. Rather, stimulation requires a distinct action such as tongue-flicking in squamates, vomeronasal suction in rodents, flehmen in artiodactylans and carnivorans, or some other kind of physical contact with a chemical source. As such, the animal has the opportunity to sample a particular, localized source of chemicals, as well as to sample classes of chemicals not readily available in the surrounding air or water, such as nonvolatile or nonsoluble molecules. This is not to say that such molecules cannot also stimulate the olfactory epithelium or that volatile and/or soluble molecules cannot stimulate the vomeronasal receptors—only that the vomeronasal system may be used facultatively, actively, and selectively to add a new level to the sensory information an animal gathers about its environment. In amphibians and most turtles, in which the vomeronasal epithelium is broadly exposed within the nasal cavities (see Reiss and Eisthen, chapter 4 in this volume; Schwenk, chapter 5 in this volume), this hypothesis is harder to support, and the segregation of functional domains even more ambiguous.

A second question of accessibility relates to whether tetrapods flood their nasal cavities with water when submerged. For example, crocodylian and squamate reptiles evince adaptations for closing the nostrils during submersion, excluding water from the olfactory chamber, whereas aquatic turtles actively pump water into and out of the nasal cavities (Schwenk, chapter 5 in this volume). Thus, water-based olfaction is possible only in the latter taxon. Perhaps this behavioral difference relates to the fact that crocodylians lack a vomeronasal epithelium and in squamates it is inaccessible from the nasal cavity, whereas in turtles it is more or less exposed. Schwenk (chapter 5 in this volume) suggested that pulsing water through the nasal chambers in turtles represents vomeronasal stimulation rather than olfaction (a trapped air bubble, in fact, might prevent immersion of the dorsally located olfactory epithelium).

Finally, we note that the chemical parameters of volatility, solubility, and polarity to some extent determine the universe of potential chemosensory stimulants in any given environment, and that these chemical qualities suggest that quite different sets of biologically relevant chemicals may be available to a given species in air versus water. However, most aquatic tetrapods spend time in both terrestrial and aquatic environments, and aquatic specialization of the chemical senses, particularly the olfactory system, might come with insupportable fitness consequences. Furthermore, the long terrestrial history preceding even the most aquatic of extant tetrapods suggests a historical phenotypic burden that may be difficult to circumvent. The chemical senses of aquatic tetrapods might therefore exhibit far less adaptive specialization than we might expect.

In conclusion, the relevant properties of chemical stimuli seem straightforward, and one might assume that volatility, solubility, and polarity would play critical roles in determining which chemical stimuli are transduced, and by which chemosensory system. Furthermore, one might assume that these same properties would determine what kinds of compounds serve as chemical stimuli in air versus water. However, one only needs to consider that short-range pheromones used by brown algae tend to be nonpolar, volatile, hydrophobic hydrocarbons that become suspended in water but do not dissolve in it (Pohnert and Boland, 2002) to realize that we are a long way from predicting the relevant features of chemical stimuli in any environment or for any group of organisms.

ACKNOWLEDGMENTS

Thanks to Hans and Sirpa for inviting us to contribute to this book, and for asking many thought-provoking questions as we prepared the manuscript. This work was partly supported by a grant from the National Institutes of Health (DC05366 to HLE).

LITERATURE CITED

Amoore, J. E. 1963. Stereochemical theory of olfaction. Nature 198:271–272.

Apponyi, M. A., and J. H. Bowie. 2005. The discovery and characterisation of splendipherin, the first anuran sex pheromone; pp. 21–23 in R. Mason, M. LeMaster, and D. Müller-Schwarze (eds.), Chemical Signals in Vertebrates, 10. Springer-Verlag, New York.

Baxi, K. N., K. M. Dorries, and H. L. Eisthen. 2006. Is the vomeronasal system really specialized for detecting pheromones? Trends in Neurosciences 29:1–7.

Bignetti, E., G. Damiani, P. De Negri, R. Ramoni, F. Avanzini, G. Ferrari, and G. L. Rossi. 1987. Specificity of an immunoaffinity column for odorant-binding protein from bovine nasal mucosa. Chemical Senses 12:601–608.

Brennan, P. A., H. M. Schellinck, and E. B. Keverne. 1999. Patterns of expression of the immediate-early gene egr-1 in the accessory olfactory bulb of female mice exposed to pheromonal constituents of male urine. Neuroscience 90:1463–1470.

Dial, B. E., and K. Schwenk. 1996. Olfaction and predator detection in *Coleonyx brevis* (Squamata: Eublepharidae) with comments on the functional significance of buccal pulsing in geckos. Journal of Experimental Zoology 276:415–424.

Fasick, J. I., and P. R. Robinson. 2000. Spectral-tuning mechanisms of marine mammal rhodopsins and correlations with foraging depth. Visual Neuroscience 17:781–788.

Guo, J., A. Zhou, and R. L. Moss. 1997. Urine and urine-derived compounds induce c-*fos* mRNA expression in accessory olfactory bulb. NeuroReport 8:1679–1683.

Halpern, M. 1987. The organization and function of the vomeronasal system. Annual Review of Neuroscience 10:325–362.

Halpern, M. 1992. Nasal chemical senses in reptiles: structure and function; pp. 423–523 in C. Gans and D. Crews (eds.), Biology of the Reptilia, Vol. 18. University of Chicago Press, Chicago.

Keller, A., and L. B. Vosshall. 2004. A psychophysical test of the vibration theory of olfaction. Nature Neuroscience 7:337–338.

Kent, P. F., M. M. Mozell, S. L. Youngentob, and P. Yurco. 2003. Mucosal activity patterns as a basis for olfactory discrimination: comparing behavior and optical recordings. Brain Research 981:1–11.

Kikuyama, S., K. Yamamoto, T. Iwata, and F. Toyoda. 2002. Peptide and protein pheromones in amphibians. Comparative Biochemistry and Physiology B 132:69–74.

Leinders-Zufall, T., A. P. Lane, A. C. Puche, W. D. Ma, M. V. Novotny, M. T. Shipley, and F. Zufall. 2000. Ultrasensitive pheromone detection by mammalian vomeronasal neurons. Nature 405:792–796.

Luo, M., M. S. Fee, and L. C. Katz. 2003. Encoding pheromonal signals in the accessory olfactory bulb of behaving mice. Science 299:1196–1201.

Meredith, M. 1986. Vomeronasal organ removal before sexual experience impairs male hamster mating behavior. Physiology and Behavior 36:737–743.

Meredith, M. 1994. Chronic recording of vomeronasal pump activation in awake and behaving hamsters. Physiology and Behavior 56:345–354.

Mori, I., Y. Nishiyama, T. Yokochi, and Y. Kimura. 2005. Olfactory transmission of neurotropic viruses. Journal of NeuroVirology 11:129–137.

Pohnert, G., and W. Boland. 2002. The oxylipin chemistry of attraction and defense in brown algae and diatoms. Natural Products Reports 19:108–122.

Schwenk, K. 1995. Of tongues and noses: chemoreception in lizards and snakes. Trends in Ecology and Evolution 10:7–12.

Tegoni, M., P. Pelosi, F. Vincent, S. Spinelli, V. Campanacci, S. Grolli, R. Ramoni, and C. Cambillau. 2000. Mammalian odorant binding proteins. Biochimica Biophysica Acta 1482:229–240.

Turin, L. 2002. A method for the calculation of odor character from molecular structure. Journal of Theoretical Biology 216:367–385.

Vogt, R. G. 1987. The molecular basis of pheromone reception: its influence on behavior; pp. 385–431 in G. D. Prestwich and G. L. Blomquist (eds.), Pheromone Biochemistry. Academic Press, New York.

Wilson, E. O. 1970. Chemical communication within animal species; pp. 133–155 in E. Sondheimer and J. B. Simeone (eds.), Chemical Ecology. Academic Press, New York.

Wyatt, T. D. 2003. Pheromones and Animal Behaviour: Communication by Smell and Taste. Cambridge University Press, Cambridge.

4

Comparative Anatomy and Physiology of Chemical Senses in Amphibians

John O. Reiss and Heather L. Eisthen

In this chapter, we first introduce the chemosensory systems of tetrapods, then focus on their structure and function in amphibians.

Tetrapods possess three major chemosensory systems, the olfactory, vomeronasal, and gustatory (taste) systems. These are defined anatomically. The olfactory epithelium is found in the nasal cavity, and the axons of the olfactory receptor (OR) neurons project to the olfactory bulb at the rostral pole of the telencephalon. Most tetrapods also possess a vomeronasal system, an accessory olfactory system, the sensory epithelium of which is usually located in an organ that is distinct from the main nasal cavity. The axons of the vomeronasal receptor neurons project to the accessory olfactory bulb, a histologically distinct structure adjacent to the olfactory bulb. In contrast, the taste buds are found on the tongue, palate, and pharynx and are innervated by sensory neurons of cranial nerves VII, IX, X, which convey taste information to the hindbrain. In general, the olfactory system is involved in detecting chemicals emanating from distant sources, whereas the taste and vomeronasal systems are involved in detecting chemicals at close range (Eisthen and Schwenk, chapter 3 in this volume).

Tetrapods also possess chemosensors in the respiratory, circulatory, and digestive systems

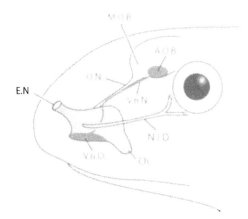

E.N

FIGURE 4.1. Schematic illustration of the olfactory organs and rostral telencephalon in a generalized terrestrial tetrapod. The vomeronasal organ is represented by a trough along the ventral side of the main olfactory organ, the condition seen in turtles and amphibian embryos. The olfactory nerve projects to the main olfactory bulb, and the vomeronasal nerve to the accessory olfactory bulb. Abbreviations: A.O.B., accessory olfactory bulb; Ch., choana; E.N., external naris; M.O.B., main olfactory bulb; N.l.D., nasolacrimal duct; O.N., olfactory nerve; V.n.O., vomeronasal organ; and V.n.N., vomeronasal nerve.

that detect gasses, ions, and nutrients. In general, these chemosensory systems consist of scattered, isolated sensory cells that project to the spinal cord and hindbrain. These systems are not considered further here.

OLFACTION

The peripheral olfactory system in tetrapods, including all amphibians, consists of paired nasal sacs (Fig. 4.1), each of which has an incurrent external naris and an excurrent internal naris, or choana, opening into the buccal cavity (Parsons, 1967; Bertmar, 1969). The lining of the nasal sac consists of one or more regions of olfactory epithelium, with the remainder respiratory epithelium. The olfactory sensory epithelium contains three main cell types: receptor cells; sustentacular (supporting) cells; and basal cells, the progenitors of the receptor and sustentacular cells. The receptor cells are bipolar neurons, with dendrites that extend to the

lumen of the nasal sac and axons that form an olfactory nerve projecting to the olfactory bulb at the rostral pole of the telencephalon. The odorant receptor proteins, members of a large family of G-protein-coupled receptors that possess seven membrane-spanning regions (Mombaerts, 2004), are localized to the dendrites of receptor cells, which terminate in cilia and/or microvilli, increasing the amount of membrane available for odorant transduction. It has recently been shown that vertebrate OR proteins form a number of well-distinguished phylogenetic groups, groups α through κ (Niimura and Nei, 2005, 2006). Two of these, groups α and γ (class II of Freitag et al. [1995, 1998]), are particularly well developed in tetrapods, and it has been suggested that they are specialized for olfaction in air. The cilia and microvilli carrying ORs are embedded in a layer of mucus that covers the surface of the sensory epithelium. This mucus is secreted by numerous simple Bowman's glands, and these glands are scattered throughout the epithelium. Accessory compound nasal glands are also frequently present. Both types of glands are lacking in fishes and in some aquatic amphibians. In many terrestrial taxa the nasal epithelia are borne on one or more conchae, which serve to increase surface area for olfaction and/or for humidifying the inspired air; in endotherms, the conchae may also warm the air.

VOMERONASAL SYSTEM

The vomeronasal system is a specialized olfactory subsystem found only in tetrapods. It has been suggested that it is specialized for detection of nonvolatile stimuli, including some pheromones, though this hypothesis does not account for all available data (Baxi et al., 2006). The relative functions of the vomeronasal and olfactory systems therefore remain unclear. In amphibians, the vomeronasal, or Jacobson's, organ consists of a diverticulum off the main nasal cavity, but in amniotes the vomeronasal organ can be directly connected to the nasal or oral cavity, or indirectly connected to both via a nasopalatine duct. The vomeronasal organ is secondarily absent in many taxa, including birds and crocodilians, many bats and cetaceans, and Old World primates, including humans. The vomeronasal sensory epithelium generally resembles the olfactory epithelium, although the dendrites of the vomeronasal receptor neurons terminate exclusively in microvilli. Axons from the vomeronasal receptor neurons project to an accessory olfactory bulb, which is distinct from the main olfactory bulb; higher-order projections also differ. Finally, vomeronasal receptor neurons use different transduction mechanisms than do OR neurons, including a different ion channel and different families of G-protein-coupled receptors (Liman et al., 1999; Mombaerts, 2004).

Supporting the idea that the vomeronasal system detects nonvolatile stimuli, odorant access to the vomeronasal organ in the terrestrial environment frequently involves direct transfer and/or fluid-pumping mechanisms. For example, in squamates the tongue is used to sample chemicals from the environment, which are then transferred to the vomeronasal organ. In many mammals the vomeronasal organ can acquire stimuli by means of the flehmen reflex, by vascular pumping, or due to contraction of the entire organ (Meredith, 1994).

The acquisition and loss of the vomeronasal system in tetrapods has long been suggested to be associated with transitions between aquatic and terrestrial habitats. Broman (1920) used his observation of fluid in the vomeronasal lumen to argue that the vomeronasal organ is specialized for smelling dissolved substances and represents a remnant of the piscine olfactory system, with the main olfactory system newly developed in tetrapods. In contrast, Bertmar (1981) suggested that the vomeronasal system arose in early tetrapods as an adaptation to terrestrial life. However, data from recent studies suggest that the "olfactory epithelium" of teleost fishes is probably a hybrid olfactory and vomeronasal epithelium: different morphologi-cal classes of receptor neurons express the olfactory and vomeronasal receptor genes and transduction elements, and these two populations of neurons project to different regions of the olfactory bulb (Cao et al., 1998; Naito et al., 1998; Hansen et al., 2003, 2004, 2005; Sato et al., 2005). The most likely scenario is that the vomeronasal system evolved in tetrapods by partitioning preexisting classes of ORs into distinct main olfactory and vomeronasal epithelia. The question then becomes whether this partitioning was associated with the assumption of terrestrial habits.

Although the vomeronasal system could be independently derived in amphibians and amniotes, this seems unlikely, as the major projections of the main and accessory olfactory bulbs are the same in both groups (Eisthen, 1997). Recent paleontological evidence suggests that amphibians and amniotes became terrestrial independently of each other, and that the last common ancestor was fully aquatic (Clack, 2002). In addition, the vomeronasal system is present in amphibians throughout life and does not arise at metamorphosis, as one might expect if the feature is an adaptation to terrestriality (Eisthen, 1997). Thus, the inter-mixed olfactory and vomeronasal systems in fishes must have become separated into distinct systems before tetrapods became terrestrial.

Interestingly, a distinct vomeronasal system has been secondarily lost in many amniotes that are aquatic or arboreal, as well as in the proteid family of salamanders, all of which are fully aquatic and paedomorphic as adults (Eisthen, 1997, 2000). Loss in arboreal animals is consistent with the hypothesis that the vomeronasal system is a specialization for detecting non-volatile stimuli, but loss in aquatic animals appears harder to explain on this basis.

TASTE

The sensory organs of the taste system are the taste buds, which in tetrapods are generally found on the tongue, palate, and pharynx

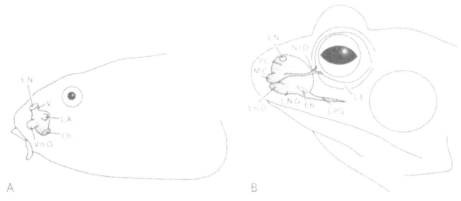

FIGURE 4.2. Schematic illustration of the olfactory organs and associated structures in an anuran tadpole (A) and adult (B), based largely on conditions in the midwife toad, *Alytes obstetricans* (Rowedder, 1937; Yvroud, 1966). Abbreviations: Ch., choana; E.N., external naris; L.A., lateral appendix; L.E., lower eyelid; L.N.G., lateral nasal groove (also called the lateral recess); L.P.G., lateral palatal groove; M.C., middle cavity; N.l.D., naso-lacrimal duct; P.C., principal cavity; V., vestibule; and V.n.O., vomeronasal organ (also called the medial recess).

(Northcutt, 2004). Each taste bud consists of taste receptor cells, supporting cells, and basal cells. At the molecular level, taste receptors are heterogeneous, including some that are simple ion channels and others that are members of distinct subfamilies of G-protein-coupled receptors (Bigiani et al., 2003; Mombaerts, 2004).

CHEMOSENSORY SYSTEMS IN AMPHIBIANS

Among tetrapods, amphibians are unique in primitively having an aquatic larval stage, followed by metamorphosis to a more terrestrial adult. Superimposed on this basic biphasic life-cycle, secondarily aquatic adults are found in each of the three groups of living amphibians. Amphibians can become secondarily aquatic in one of two ways (Duellman and Trueb, 1986). In many salamanders, the overall larval morphology is maintained while the gonads mature, an evolutionary process termed neoteny or paedomorphosis. On the other hand, in some frogs (e.g., pipids), salamanders (e.g., newts), and caecilians (e.g., typhlonectids) metamorphosis occurs but is not accompanied by a change in habitat (pipids, typhlonectids) or is followed by a second metamorphosis associated with reentry into the aquatic environment (newts). Intermediate conditions abound, with partially aquatic

adults in many taxa, in part because in most forms with aquatic larvae, adults return to water to breed. We first examine the changes that occur during metamorphosis from the aquatic to the terrestrial phase in taxa with a biphasic life history, then examine those evolutionary changes occurring in taxa in which metamorphosed adults have secondarily adopted an aquatic lifestyle.

OLFACTION AND VOMERONASAL CHEMORECEPTION

The morphology of the olfactory organ in adult amphibians has been reviewed by a number of authors (Seydel, 1895; Matthes, 1934; Jurgens, 1971; Saint Girons and Zylberberg, 1992a, 1992b), but the larval condition is less widely known. After a review of the morphology of the larval and adult olfactory organ in each group, we examine general issues of function in aquatic and terrestrial environments.

FROGS

In the larvae of frogs (Anura, Fig. 4.2A), the olfactory organ is highly specialized in association with the small mouth opening (Born, 1876; Hinsberg, 1901; Rowedder, 1937; Yvroud, 1966; Khalil, 1978; Jermakowicz et al., 2004). The nasal sac typically runs almost vertically from external naris to choana. The vomeronasal

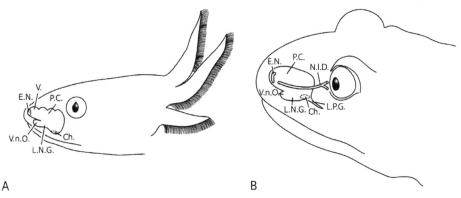

FIGURE 4.3. Schematic illustration of the olfactory organs and associated structures in a salamander larva (A) and adult (B), based largely on conditions in the Coastal Giant Salamander, *Dicamptodon tenebrosus* (Stuelpnagel and Reiss, 2005). Abbreviations: Ch., choana; E.N., external naris; L.N.G., lateral nasal groove; L.P.G., lateral palatal groove; N.l.D., nasolacrimal duct; P.C., principal cavity; V., vestibule; and V.n.O., vomeronasal organ.

organ is represented by a bean-shaped out-pocketing on the anterior (morphologically ventral) surface, while several other outpocketings typically occur, including a lateral appendix of unknown function. A nonsensory vestibule is present at the external naris, as is a valve guarding the choana (Gradwell, 1969). Of multicellular glands, only the vomeronasal gland is well developed. Much variation in form of the nose occurs among anuran tadpoles; for example, some have exposed areas of olfactory epithelium within the oral cavity, while others lack external nares (Altig and McDiarmid, 1999).

During metamorphosis, extreme remodeling occurs, as illustrated in Figure 4.2B: the vestibule is reduced or lost, the choana shifts posteriorly and loses its valve, and the vomeronasal organ (medial recess) connects to a lateral nasal groove (lateral recess) that runs posteriorly through the choana to continue as the lateral palatal groove (sulcus maxillopalatinus). The nasolacrimal duct develops and connects to the newly formed middle cavity of the nose. The postmetamorphic olfactory organ thus consists of three interconnected chambers: a principal (superior) cavity, a middle cavity, receiving the nasolacrimal duct; and an inferior cavity, with its lateral and medial recesses. Interestingly, the olfactory eminence

in the floor of the principal cavity is best developed in more terrestrial and especially fossorial forms (Jurgens, 1971). In most anurans, a special mechanism involving the lower jaw and the submentalis muscle develops to close the external nares during lung inflation (Gaupp et al., 1904; Nishikawa and Gans, 1996; Jorgensen, 2000). Bowman's glands usually appear only during metamorphosis, as do the rostral (internal oral) and lateral nasal glands.

SALAMANDERS

Among larval salamanders (Caudata), the olfactory organ typically consists of a tubular sac, extending from the external naris to the choana, as shown in Figure 4.3A (Seydel, 1895; Schuch, 1934; Stuelpnagel and Reiss, 2005). As in frogs, a nonsensory vestibule leads into the principal cavity of the olfactory organ, and a nonmuscular choanal valve is usually present at the medial border of the choana, presumably preventing reverse flow of water through the nose (Bruner, 1914a, 1914b). The olfactory epithelium is found in troughs separated by folds of nonsensory epithelium; this may be a primitive feature, because it resembles the condition seen in young lungfish larvae. The vomeronasal organ is usually a ventrolateral diverticulum of the main nasal sac (Seydel, 1895). Bowman's glands are present in most larval salamanders

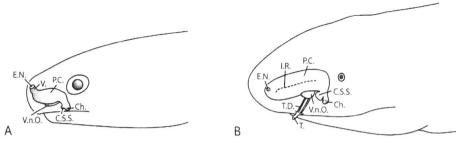

FIGURE 4.4. Schematic illustration of the olfactory organs and associated structures in a caecilian larva (A) and adult (B), based largely on conditions in the genus *Ichthyophis* (Sarasin and Sarasin, 1890; Badenhorst, 1978). Abbreviations: Ch., choana; C.S.S., choanal slime sac; E.N., external naris; I.R., internal ridge separating the medial and lateral portions of the principal cavity; P.C., principal cavity; T., tentacle; T.D., tentacle (naso-lacrimal) ducts; V., vestibule; and V.n.O., vomeronasal organ.

but are not abundant and may not be functional; by contrast, the vomeronasal gland is well developed.

During metamorphosis the vestibule is lost, a muscular mechanism for closing the external naris develops (Bruner, 1901; Nikitkin, 1986), the nasal sac widens, the nonsensory folds are greatly reduced or disappear completely, and the choana widens and loses its choanal valve (Fig. 4.3B). The responsiveness of the olfactory epithelium changes: sensitivity to dissolved odorants decreases, while that to volatile odorants increases (Arzt et al., 1986). Much as in anurans, the vomeronasal organ acquires a connection to the oral cavity by extension of the lateral nasal groove posteriorly along the lateral edge of the choana and into the mouth as the lateral palatal groove. The nasolacrimal duct develops, connecting the medial angle of the eye to the lateral nasal sac, just anterior to the vomeronasal organ. The Bowman's glands enlarge and proliferate, and in hynobiids and salamandrids the compound lateral nasal glands develop (Saint Girons and Zylberberg, 1992a). In plethodontid salamanders, the vomeronasal organ shifts forward to acquire a functional connection with the newly developed nasolabial groove (Wilder, 1925). Although some have suggested that ciliated and microvillar OR neurons are specialized for detecting odorants in air and water, respectively, both types of cells are found in salamanders at all stages of develop-

ment, including in neotenic adults (Eisthen, 1992, 2000).

CAECILIANS

In caecilians (Gymnophiona), the larval olfactory organ is a simple, triangular sac, as depicted in Figure 4.4A (Sarasin and Sarasin, 1890; Badenhorst, 1978). The vomeronasal organ is represented by a diverticulum of the nasal sac; it lies laterally for most of its length but shifts medially at the level of the choana. As in frogs and salamanders, a vestibule and (in at least some species) a choanal valve are present. During metamorphosis the vestibule is lost, and the principal cavity acquires a large ridge on its floor, somewhat resembling the olfactory eminence of anurans (Fig. 4.4B); this divides the nasal cavity into a medial, sensory part, and a lateral, respiratory part (Schmidt and Wake, 1990). The vomeronasal organ changes position, coming to lie transversely, and its distal tip connects with the newly formed nasolacrimal (tentacular) ducts. The tentacle, including the tentacular (Harderian) gland and the retractor muscle, forms, and its lumen connects with the tentacular ducts (Badenhorst, 1978; Billo and Wake, 1987). The choana widens, and the choanal slime sac greatly enlarges, but, unlike in frogs and salamanders, no lateral palatal groove forms. As in frogs and salamanders, the vomeronasal gland is present in larvae, but Bowman's glands and the lateral nasal gland form only during metamorphosis (Badenhorst, 1978).

Because of their large cells and simple nasal cavity architecture, amphibians have long been used as model animals for neurobiological research in olfaction, and much detailed information is available concerning mechanisms of olfactory system function at the level of both the sensory receptor cells and the olfactory bulb (Kauer, 2002). Nevertheless, studies of function of the chemical senses in amphibians are spotty, with much information available in some areas, while others are almost completely unexplored. In general, for animals that clearly must deal with both, surprisingly little attention has been paid to differences between function in aquatic and terrestrial environments.

In larvae and neotenes, the nasal sac is typically irrigated with water by the respiratory pump, aided by ciliary action. The flow in larvae is usually unidirectional inward, with backflow being prevented by the choanal valves. As in fishes, air used to inflate the lungs is gulped through the mouth, not taken through the nose. In neotenic salamanders the situation is more variable, with some taking only water through the nose, but others inspiring air as well; a choanal valve may be present or absent (Bruner, 1914b). In larval amphibians, the chemical senses have been shown to be behaviorally important for feeding, predator avoidance, and kin recognition (reviewed in Dawley, 1998), although many of these studies have not distinguished among the roles of the olfactory, vomeronasal, and taste systems. Interestingly, the ability to smell in the aquatic environment is retained after metamorphosis in at least some species. For example, blinded tiger salamanders (*Ambystoma tigrinum*) will preferentially nose tap and bite bags containing earthworms underwater; occlusion of the nares eliminates this response (Nicholas, 1922). Likewise, pheromones involved in aquatic reproduction are widely used in both salamanders (reviewed in Arnold, 1977; Dawley, 1998) and frogs (Wabnitz et al., 1999).

In the terrestrial environment, the olfactory system typically serves to sample air. In all three groups of amphibians, oscillations of the buccal floor with the mouth closed serve to continually bring fresh air into the nasal sac and oral cavity, and airflow is bidirectional (reviewed in Jorgensen, 2000). Lung inflation occurs intermittently by closing the external naris and compressing the floor of the buccal cavity. Sensing of airborne chemicals is important for feeding, homing behavior, and reproduction (reviewed in Dawley, 1998). In plethodontid salamanders, nonvolatile chemicals are transported along the nasolabial groove to the vomeronasal organ (Dawley and Bass, 1989), and in caecilians the tentacular ducts may serve a similar function (Schmidt and Wake, 1990; Himstedt and Simon, 1995). In frogs and salamanders, the vomeronasal organ may also serve to sample fluid transported from the lateral palatal groove forward through the choana (Seydel, 1895).

TASTE

The biology, anatomy, physiology, and development of the taste system of amphibians has recently been reviewed by Barlow (1998) and Zuwala and Jakubowski (2001b). In frogs, significant differences occur between larvae and metamorphosed adults (Fig. 4.5). In larvae, taste buds are present on papillae throughout the oral epithelium. During metamorphosis, the fleshy secondary tongue arises, and the larval taste buds are replaced by taste discs, which are found on the secondary tongue as well on as the oral and pharyngeal epithelium. Taste discs differ from taste buds in cellular composition as well as morphology. A similar change from taste buds to taste discs at metamorphosis has recently been shown to occur in salamanders as well (Takeuchi et al., 1997; Zuwala and Jakubowski, 2001a; Zuwala et al., 2002). In contrast, in caecilians taste buds have been described only in larvae and adults of the secondarily aquatic *Typhlonectes* and may be lost in most metamorphosed forms (Wake and Schwenk, 1986). While the morphological changes in the taste periphery at metamorphosis are profound, the functional significance of these changes remains completely unclear

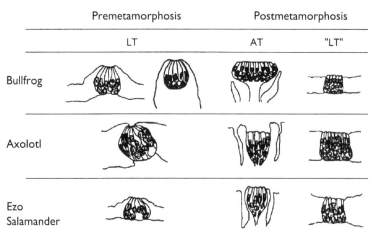

	Premetamorphosis	Postmetamorphosis	
	LT	AT	"LT"
Bullfrog			
Axolotl			
Ezo Salamander			

FIGURE 4.5. Overview of tastebud development in bullfrogs *(Rana catesbeiana)* and two species of salamanders, the axolotl *(Ambystoma mexicanum)* and the Ezo salamander *(Hynobius retardatus)*. Barrel-shaped taste buds on the larval tongue (LT) are replaced by fungiform taste disks on the adult tongue (AT), and by taste disks embedded in the oral epithelium on the more posterior "larval tongue remnant" ("LT"). Figure modified after Takeuchi et al. (1997) (permission granted by T. Nagai, Department of Biology, Keio University School of Medicine).

(Barlow, 1998). In particular, although responses to sweet, sour, salty, bitter, and amino acid tastes have been demonstrated electrophysiologically in larvae, neotenes, and adults, there is no evidence for any differences functionally correlated with feeding in aquatic versus terrestrial habitats. However, the results of a recent study (Nagai et al., 2001) suggest that salt taste is mediated by amiloride-sensitive sodium channels in metamorphosed salamanders but not in larvae or neotenes. This indicates that significant physiological differences may exist.

COMPARATIVE CHEMORECEPTION IN SECONDARILY AQUATIC AMPHIBIANS

Adult amphibians often cross the threshold between terrestrial and aquatic environments, but here we focus on three prominent cases in which metamorphosed adults have secondarily adapted to semipermanent or permanent residence in the aquatic environment: pipid frogs, newts (salamandrids), and typhlonectid caecilians. Because the taste system of amphibians has not been extensively explored, our examples focus on olfaction.

FROGS

Pipid frogs (family Pipidae) are the best-studied example of secondarily aquatic amphibians, largely due to the widespread use of the African clawed frog *Xenopus laevis* in laboratory research. Adult *Xenopus* are almost completely aquatic; excursions onto land appear to occur rarely (Tinsley et al., 1996), although terrestrial prey capture has been reported (Measey, 1998). As illustrated in Figure 4.6, adult *Xenopus* possess three main nasal cavities, as in typical terrestrial frogs. These are known as the principal cavity or medial diverticulum, middle cavity or lateral diverticulum, and vomeronasal organ, which comprises the entire inferior cavity (Föske, 1934; Paterson, 1939a, 1939b, 1951; Saint Girons and Zylberberg, 1992a, 1992b; Hansen et al., 1998). The principal cavity is tubular, with no sign of the olfactory eminence developed in terrestrial frogs. Moreover, while in typical frogs the middle cavity is nonsensory, in *Xenopus* the middle cavity contains a well-developed sensory epithelium. Studies by Altner (1962) provide evidence that the principal cavity is used to detect airborne chemical stimuli, while the blind-ended middle cavity is used to detect waterborne stimuli: a flap valve in the

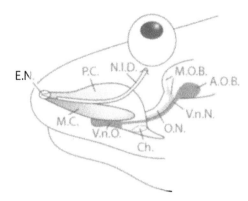

FIGURE 4.6. Schematic illustration of the nasal cavity of the African clawed frog *(Xenopus laevis)*. The axons of the receptor cells in the principal cavity ("air nose") project to the dorsal portion of the main olfactory bulb, and those of the middle cavity ("water nose") project to the ventral portion of the main olfactory bulb. Note the valve in the external naris separating the entrance to the middle and principal cavities, and the papilla under the eye bearing the opening of the nasolacrimal duct. Abbreviations: A.O.B., accessory olfactory bulb; Ch., choana; E.N., external naris; M.C., middle cavity; M.O.B., main olfactory bulb; N.l.D., nasolacrimal duct; O.N., olfactory nerve; P.C., principal cavity; V.n.O., vomeronasal organ; and V.n.N., vomeronasal nerve.

external naris allows access to one or the other chamber depending on the external medium. Water flow in and out of the middle cavity correlates with pulsations observed on the lateral side of the snout, apparently due to rhythmic contractions of the submentalis muscle, which helps close the naris in terrestrial frogs (Altner, 1962). As in typical frogs, the vomeronasal organ is connected to the oral cavity by the lateral nasal groove. A nasolacrimal duct is present, but its entrance is at the tip of a short papilla below the eye, rather than on the lower eyelid as in most frogs. The medial nasal (vomeronasal) gland is present, as is the rostral (internal oral) nasal gland, but the lateral nasal gland is absent.

The olfactory epithelia of the principal and middle cavities in *Xenopus* show profound differences at both morphological and molecular levels. Bowman's glands and associated olfactory binding proteins are present in the principal cavity, but lacking in the middle cavity (Millery et al., 2005). Receptor neurons of the principal cavity are ciliated, and supporting cells are secretory. By contrast, the middle cavity contains both microvillar and ciliated receptor cells, and both secretory and ciliated supporting cells (Weiss, 1986; Saint Girons and Zylberberg, 1992b; Hansen et al., 1998; Oikawa et al., 1998). In this respect the epithelium of the adult middle cavity closely resembles that of the larval principal cavity. Primary and secondary projections of the principal and middle cavity are also distinct (Weiss, 1986; Reiss and Burd, 1997; Gaudin and Gascuel, 2005). At the molecular level, middle cavity receptor cells express at least group ε (class I) odorant receptors, whereas principal cavity receptor cells express group γ (class II) odorant receptors; these have been suggested to be functionally correlated with olfaction in water versus air, respectively (Freitag et al., 1995, 1998; Mezler et al., 1999, 2001). Transduction mechanisms also differ (Mezler et al., 2001). Here too the adult middle cavity resembles the larval principal cavity, which (at least in early larval stages) is known to express only group ε receptors. By contrast, the vomeronasal epithelium resembles that of other anurans both morphologically and in containing microvillar receptor cells and ciliated supporting cells, and shows no striking changes during metamorphosis. It is worth noting that the "posterolateral epithelial area of the principal cavity" recently identified as expressing V2R vomeronasal receptor genes (Hagino-Yamagishi et al., 2004) is merely the posterior part of the vomeronasal organ itself, which is unusually extensive in *Xenopus* compared with terrestrial frogs.

The taste system of *Xenopus*, which lacks a tongue, has been examined by Toyoshima and Shimamura (1982) and Witt and Reutter (1994); as in terrestrial metamorphosed frogs, taste discs, rather than taste buds, are present in the oral epithelium. However these taste discs are not raised on fungiform papillae, as are the lingual taste buds of terrestrial species. Physiological studies have shown that the taste system of *Xenopus* is particularly sensitive to amino acids and bitter substances (Yoshii et al., 1982), and a recent genomics study (Shi and

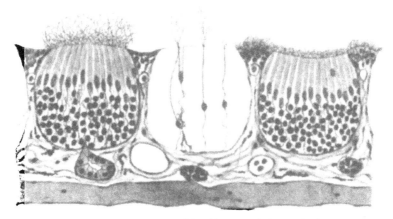

FIGURE 4.7. Diagrammatic comparison of the olfactory epithelium in land-phase *(left)* and water-phase *(right)* metamorphosed salamandrid, the newt *Triturus*. Note the much longer olfactory cilia in the land-phase animal, the elevation of the olfactory epithelium into a slight bulge, and the well-developed Bowman's gland at the base of the epithelium. By contrast, the water-phase animal shows much shorter cilia, depression of the olfactory epithelium into a groove, and ciliated respiratory epithelium between the "buds" of olfactory epithelium. The central region shows individual cells. From left to right these are an olfactory receptor cell with supporting cell from a land-phase animal, a receptor cell from a land-phase animal, and a receptor cell from a water-phase animal (from Matthes, 1927).

Zhang, 2006) has shown that *X. tropicalis* has 49 T2R (bitter) taste receptor genes but apparently no T1R (sweet/umami) receptors. The lack of comparative data makes the significance of these findings unclear.

SALAMANDERS

Among salamanders, newts (family Salamandridae) are notable for their frequent resumption of aquatic habits after metamorphosis. Studies of the smooth newt *(Triturus vulgaris)* and alpine newt *(T. alpestris)* provide the most complete information on changes in olfactory structure and function, due largely to the work of Matthes (1924a, 1924b, 1924c, 1926, 1927) and Schuch (1934). These European newts develop in ponds and have a typical metamorphosis to a terrestrial stage. After emerging in the fall, they return to the ponds to breed the following fall or spring and remain there for three months or more. In doing so, they undergo a secondary metamorphosis, marked most prominently by the development of a large tail fin, particularly in males. In the aquatic environment, olfaction functions both in food

localization (Matthes, 1924a, 1927) and in courtship behavior (Halliday, 1977). The olfactory organs of land- and water-phase newts are morphologically distinct, with the olfactory epithelium of water-phase newts having larger folds separating the grooves of olfactory epithelium, greatly reduced numbers of goblet cells in the respiratory epithelium, and much shorter cilia on the OR cells (Fig. 4.7) (Matthes, 1927). Blinded land-phase animals placed in the water can immediately find food, but blinded water-phase animals require several days before they can find food on land. These behavioral results correlate with morphology: olfactory cilia shorten immediately upon placing land-phase animals in water, perhaps due to osmotic effects, but the cilia of water-phase animals transported onto land take several days to lengthen. Matthes also showed that food-finding in water or on land did not depend on the vomeronasal organ, as blinded animals in which the vomeronasal nerve had been sectioned found food as easily as those in which it was intact. Unfortunately, these pioneering studies have not been confirmed or followed up by more recent workers. Nothing is known of the taste system in *Triturus*.

Among caecilians (Gymnophiona), the family Typhlonectidae has entirely aquatic adults and juveniles; typhlonectids have live birth, with no gilled larval stage. *Typhlonectes compressicauda* and *T. natans* have been common animals in the pet trade and thus have been the subject of a number of morphological and physiological investigations. Although aquatic, the species of *Typhlonectes* breathe air. Unlike more primitive caecilians, but like many other derived but terrestrial forms, *Typhlonectes* has a single, undivided nasal cavity (Schmidt and Wake, 1990). A short tentacle is present, and its duct communicates with an exceptionally well-developed vomeronasal organ (Schmidt and Wake, 1990). Interestingly, two distinct types of olfactory epithelium are present in the nasal cavity: the anteroventral region has a very thick epithelium containing both ciliated and microvillar receptor neurons and lacks Bowman's glands; and the posterodorsal region has a thin epithelium containing only ciliated receptor neurons (Saint Girons and Zylberberg, 1992b; although Schmidt and Wake [1990] reported this region as nonsensory). By analogy with *Xenopus*, it appears that the anteroventral region may be specialized for aquatic olfaction, and the posterodorsal for aerial olfaction. Unfortunately, no study has examined epithelial ultrastructure in a terrestrial caecilian, so no comparison can be made.

Both inspiration and expiration in *T. natans* occur through the nares. The choanae are protected by valves and are usually closed when the animals are filling the lungs or underwater (Prabha et al., 2000), but buccal floor oscillations still occur and increase in frequency when food is introduced to the aquarium, suggesting that pressure changes transmitted across the choanal valve are used to move water in and out of the nose (Wilkinson and Nussbaum, 1997). Recent work has shown that *T. natans* is able to use waterborne chemical cues to distinguish

sex and kinship relations of conspecifics (Warbeck and Parzefall, 2001). Finally, as noted above, *Typhlonectes* is the only adult caecilian in which taste buds have been reported (Wake and Schwenk, 1986); this may be correlated with its aquatic habits.

EVOLUTION OF THE CHEMICAL SENSES IN SECONDARILY AQUATIC AMPHIBIANS

For each of the species described in the previous section, incomplete data concerning both the exemplar species and the distribution of characters in other taxa make impossible a full characterization of the adaptations to a secondarily aquatic lifestyle. Nevertheless, some tantalizing hints allow us to piece together an evolutionary scenario for the origin and diversification of the chemical senses in the group to which each belongs (cf. Fig. 4.8).

PIPID FROGS

Pipids provide the best case to study evolution of chemosensation, as we know much about their morphology and physiology, as well as their evolutionary relationships. Current phylogenies (Cannatella and Trueb, 1988; Roelants and Bossuyt, 2005; Frost et al., 2006) agree that the sister group of the aquatic pipids is the Mexican burrowing toad, *Rhinophrynus dorsalis* (Rhinophrynidae) (Fig. 1.3 in this volume), which as an adult has typical anuran nasal cavities (Trueb and Cannatella, 1982). By contrast, the nasal cavities of the aquatic pipids have long been known to be strange and have been the subject of a number of investigations (reviewed in Helling, 1938; Paterson, 1951). *Xenopus* is the least specialized of the pipids: to derive its nose from a more typical anuran type, it would seem that the normally nonsensory middle cavity would only have to acquire a sensory epithelium of the larval type, and a valve separating it from the principal cavity. However, the homology of the "middle cavity" of *Xenopus* with that of other frogs is not entirely clear (Föske, 1934; Helling, 1938; Paterson, 1951). Helling (1938) argued that the "recessus olfactorius" (a small region of

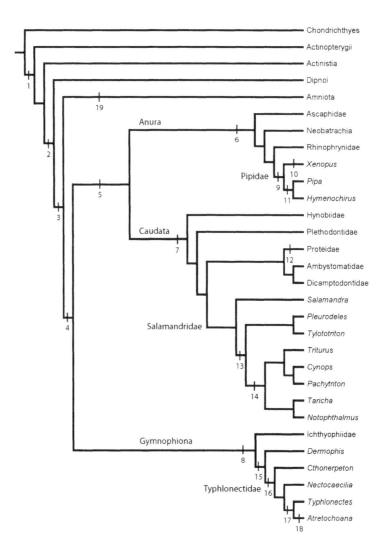

olfactory epithelium lacking Bowman's glands at the rostral end of the principal cavity in many frogs) is in fact the homolog of the middle cavity of *Xenopus*, and it has been suggested that in all frogs the anterior principal cavity may be exposed to waterborne odorants (Døving et al., 1993). Supporting this view, Benzekri (2006) has recently found that the primitive anuran *Ascaphus truei* has two distinct types of epithelium in the principal cavity, an anterolateral one resembling the middle cavity epithelium of *Xenopus*, and a posteromedial one resembling the principal cavity epithelium of other frogs. Whether the middle cavity of *Xenopus* derives from that of other frogs, from a region of the principal cavity, or both, the presence of anatom-

ically and functionally distinct nasal chambers, one used for aerial, the other for aquatic olfaction, is clearly associated with the assumption of a secondarily aquatic lifestyle.

Conditions in other pipids differ even further from the typical anuran type but are easily derived from that of *Xenopus* (Helling, 1938; Paterson, 1951). In *Pipa* and *Hymenochirus* the nasal sac consists of parallel medial and lateral nasal canals, connected by a narrow isthmus along their entire length. The medial nasal canal corresponds to the principal cavity of *Xenopus* and other anurans, while the lateral nasal canal corresponds to the middle cavity of *Xenopus* (Helling, 1938; Paterson, 1951; Meyer et al., 1997). A narial valve separating the two

FIGURE 4.8. A speculative phylogeny of the tetrapod olfactory system. Characters on branches are as follows: 1. Oste-ichthyes: Olfactory organ with both microvillar and ciliated receptor cells, secretory and ciliated supporting cells. V2R (and possibly V1R) vomeronasal receptor genes expressed on microvillar receptor cells, OR olfactory receptor genes on both microvillar and ciliated receptor cells. 2. Sarcopterygii: Expansion of groups α and γ (class II) OR genes. Olfactory lamel-lae bear sensory grooves containing receptor cells. 3. Tetrapoda: Further expansion of groups α and γ OR genes. Posterior nostril migrates into mouth to form choana. Receptor cells partitioned into vomeronasal organ expressing V2R and V1R genes, and main olfactory organ expressing OR genes. Vomeronasal gland forms. Larvae use water as olfactory medium, flow is unidirectional inward, have choanal valve. Adults use air (at least sometimes) as olfactory medium, flow is bidirec-tional. In adults, main olfactory organ divided into water-sensing and air-sensing areas. Bowman's glands form in air-sens-ing area. Multicellular accessory glands form. Nasolacrimal duct forms, drains lacrimal secretions into nasal sac. 4. Lis-samphibia: Adults undergo significant metamorphosis, lose choanal valve and vestibule, reduce sensory grooves in main olfactory organ, develop nasolacrimal duct, Bowman's and accessory glands. 5. Batrachia: Metamorphosed adults develop smooth muscular mechanism to close external nares, palatal groove connecting to vomeronasal organ. 6. Anura: Meta-morphosis extreme. Nasal sac approximately vertical in larvae, sensory grooves lost, lateral appendix forms. Adults develop nonsensory middle cavity, recessus olfactorius for detection of waterborne odors, olfactory eminence in principal cavity, mechanism to close nostrils by submentalis muscle. 7. Caudata: Assumed primitive for all features. 8. Gymnophiona: Sensory grooves lost, choanal slime sac forms. In adults, principal cavity divided into lateral (respiratory) and medial (olfactory) parts by ventral ridge, tentacle forms, connects to nasolacrimal duct, drains into vomeronasal organ, choanal slime sac enlarges. 9. Pipidae: Secondarily aquatic adult; overland excursions rare. In adult, great develop-ment of middle cavity/lateral diverticulum (probably derived from middle cavity and recessus olfactorius of other frogs) with sensory epithelium devoted to water-sensing. Middle cavity has both microvillar and ciliated receptors that express group δ, ε, and η OR genes. Principal cavity (medial diverticulum) specialized for air-sensing, has exclusively ciliated receptors expressing group α and γ OR genes. Vomeronasal organ and gland enlarged, lateral nasal gland lost (but rostral gland retained). 10. Xenopus: Valve in external naris separates middle cavity and principal cavity, middle cavity with enlarged nonsensory accessory sacs used for ventilation, nasolacrimal duct opens beneath eye at tip of tentacle. 11. Pipa plus Hymenochirus: Tubular vestibule leads to principal (medial nasal canal) and middle (lateral nasal canal) cavities. Lat-eral nasal canal formed by connection of lateral diverticulum with choana. Nasolacrimal duct lost. 12. Proteidae: Metamor-phosis lost; vomeronasal organ lost. 13. Newts: Secondarily aquatic adults, return to water for extended periods after meta-morphosis. Water moved through nasal cavity by buccal floor oscillation. 14. Derived newts: Ability to reversibly respond to aquatic environment by losing/gaining goblet cells, lengthening/shortening olfactory cilia. 15. Viviparous caecilians: Loss of aquatic larval stage, live birth. 16. Typhlonectidae: Adults partially aquatic. Main olfactory cavity single, not divided into medial and lateral parts. 17. Typhlonectes plus Atretochoana: Adults fully aquatic. Enlarged, subtriangular external nares, reduced, nonprotrusible tentacle. Main olfactory cavity divided into anterior and posterior parts, specialized for aquatic and aerial olfaction. Anterior part with ciliated and microvillar receptors, lacking Bowman's glands, posterior part with only ciliated receptors. Vomeronasal organ well developed, choanae large with well-developed, superficial choanal valve. Narial plugs well developed. 18. Atretochoana: Lungs lost. Choanae closed by membrane. External nares counter-sunk and extremely enlarged. 19. Amniota: Loss of aquatic larval stage. Loss of water-smelling ability in main olfactory organ, specialization for detection of volatile molecules. Further expansion of groups and OR genes, loss (or extreme reduction) of all others. Loss (or reduction) of microvillar receptor cells. Further specialization of vomeronasal organ (likely for detection of nonvolatile molecules). Placement of changes on the tree is unconstrained due to lack of informa-tion and reflects tentative judgment of the authors. Phylogeny is based on work by Frost et al. (2006) but follows Weis-rock et al. (2006) for salamandrids, and Canatella and Trueb (1988) for pipids.

chambers is lacking, however; instead a short (Hymenochirus) to long (Pipa) vestibule leads to both medial and lateral canals. A well-developed vomeronasal organ and associated gland are present in all of these forms, as is the rostral (internal oral) nasal gland, but the nasolacrimal duct is absent. Unlike Xenopus, both the medial and lateral nasal canals communicate with the oral cavity through the choana, and the vomeronasal organ opens into the lateral canal (middle cavity) rather than the medial canal (principal cavity). This suggests that water cur-rents might pass through the lateral nasal canal into the oral cavity. Unfortunately, little is known of the functional significance of olfac-tion in pipids, though anecdotal observations support an important role for olfaction in food localization and reproduction (reviewed in Elepfandt, 1996), and recent work has shown that female Hymenochirus are specifically attracted to pheromones from the breeding glands of males (Pearl et al., 2000).

The olfactory sense, particularly the "water nose," is clearly highly developed in Xenopus and other pipids compared with other frogs. This appears easily explained by habitat: these frogs typically inhabit, feed, and breed in rather muddy ponds, and olfaction is clearly a sense that would be emphasized here, whereas vision, the predominant sense used for terrestrial feeding, is less likely to be useful. What remains unclear is the functional distinction between the vomeronasal and main olfactory system in underwater smelling—do they have different classes of ligands and/or modulate distinct categories of behavior? On the other hand, it is surprising that all pipids appear to retain a well-developed "air nose," suggesting the continued importance of airborne odors, perhaps in locating bodies of water during overland migrations.

SALAMANDRID NEWTS

The situation in newts is less satisfactorily understood. The family Salamandridae is composed of two clades, the "true salamanders," *Salamandra* and its relatives, and the newts, including *Triturus*, *Pleurodeles*, and *Cynops* in the Old World, and *Taricha* and *Notophthalmus* in the New World (Titus and Larson, 1995; Weisrock et al., 2006). The true salamanders are largely terrestrial. By contrast, all of the newts are somewhat aquatic as adults, although the degree to which feeding occurs in the aquatic environment is quite variable (Özeti and Wake, 1969). Among the newts, there is a trend toward increased specialization for aquatic feeding from the basal *Pleurodeles* and *Tylototriton* through the highly derived *Pachytriton*, which is wholly aquatic as an adult. Unlike many terrestrial salamanders, newts characteristically ventilate the nose by buccal floor oscillations underwater (Joly and Caillere, 1983; reviewed in Jorgensen, 2000). Unfortunately, we simply do not have enough comparative information to determine whether the morphological changes in the olfactory organ upon entry into the aquatic environment described by Matthes and Schuch are characteristic only for the genus *Triturus*, for newts as a whole, or even for all salamanders. Saint Girons and Zylberberg (1992b) were unable to find any differences in the olfactory epithelium of land- and water-phase *Pleurodeles* at the light-microscopic level, suggesting that a comparative investigation of olfactory morphology and function in land- and water-phase newts could prove quite interesting.

Schmidt and colleagues (1988) investigated the central projections of the olfactory and vomeronasal organs in two species of salamandrids, including *T. alpestris*, and eight species of plethodontid salamanders. The authors report that the number of lobes in the accessory olfactory bulb is greater in adults of metamorphosing species than in direct developers, suggesting an association between vomeronasal function and the aquatic larval period, or possibly aquatic breeding. Species-specific female-attracting peptide pheromones, sodefrin and silefrin, have recently been isolated and characterized from male abdominal glands of the Japanese newts *Cynops pyrrhogaster* and *C. ensicauda* (Toyoda et al., 2004). Bilateral plugging of the nares and olfactory nerve transection demonstrate that the nasal chemosensory systems are necessary for this attraction response, and electro-olfactogram recordings further suggest that the vomeronasal system may be primarily responsible for mediating responses to these compounds (Toyoda et al., 2004).

Finally, as with the vomeronasal system, we have only tantalizing evidence for secondary adaptation of the taste system to aquatic life. In a recent study, Zuwala and Jakubowski (2001b) showed that the fire salamander *Salamandra salamandra* undergoes a morphological transition from taste buds to taste discs during metamorphosis. However, a previous report on the newt *Cynops pyrrhogaster* noted both "bud-shaped" and "barrel-shaped" taste buds (Toyoshima and Shimamura, 1987), suggesting that this species may retain larval-type taste structures, or that there may be a secondary metamorphosis from adult-type to larval-type structures with resumption of aquatic habits.

The situation in newts shows that not only is the commitment of the olfactory system to aquatic versus terrestrial olfaction able to change in evolutionary time, it is able to change in ontogeny as well. This argues for the importance of careful attention to husbandry conditions in morphological, physiological, and behavioral studies on the chemical senses of salamanders, and amphibians in general.

TYPHLONECTID CAECILIANS

Within the typhlonectids, an evolutionary series can be constructed from the partially aquatic *Chthonerpeton* and *Nectocaecilia* through the fully aquatic *Typhlonectes* to the fully aquatic, highly specialized, lungless *Atretochoana* (Nussbaum and Wilkinson, 1995; Wilkinson and Nussbaum, 1997, 1999). Along this series we see accentuation of features associated with the nasal cavities, including enlarged external nares, reduced tentacular aperture with nonprotrusible tentacle, enlarged choanae with superficial choanal valves, and enlarged narial plugs on the buccal floor. In *Atretochoana* the choanal valves have fused, so that the choanae are not patent. Wilkinson and Nussbaum (1997) have argued that this suite of features represents a transition to a nose adapted to smelling in water: they postulate that air is drawn into the lungs through the nasal sac and choanae, but water is moved in and out of the nasal sac by buccal floor oscillations with the choanal valve closed. With the loss of lungs in *Atretochoana*, it was possible to close this valve permanently. A more complete examination of olfactory structure and function in *Typhlonectes* could help to support this scenario; it would be of great interest to know whether air is retained in the nasal cavity in submerged animals.

CONCLUSIONS

The brief overview presented here is summarized in Figure 4.8. It is clear that we still have much to learn about the structure and function of the amphibian olfactory and taste systems in general, and their modification in secondarily aquatic amphibians in particular. In the olfactory system apparent morphological adaptations to a secondarily aquatic existence include a reduction in glands (aside from the vomeronasal gland) and the nasolacrimal duct, segregation of distinct epithelia for olfaction in water and air, and the provision of a special mechanism for moving water through the nasal cavities. Unsurprisingly, many of these evolutionary modifications involve the reversal of changes that usually occur in metamorphosis from aquatic larva to terrestrial adult. However, given that amphibians primitively return to water to breed, it may also be that some of these adaptations are more widespread among amphibians than we presently realize. For example, as noted above, peptide pheromones occur in the secondarily aquatic newt *Cynops* (Kikuyama et al., 1995; Toyoda et al., 2004), but also in the terrestrial neobatrachian treefrog *Litoria*, which returns to water to breed (Wabnitz et al., 1999). The possibility that the ability to smell in both water and air is primitive for amphibians is supported by the recent discovery of two distinct regions of olfactory epithelium, apparently associated with aerial versus aquatic olfaction, in the principal cavities of the salamander *Dicamptodon tenebrosus* (Stuelpnagel and Reiss, 2005) (Dicamptodontidae of Fig. 1.3 in this volume; Fig. 4.8) and the frog *Ascaphus truei* (Benzekri, 2006). In amphibians, secondarily aquatic forms have generally evolved from ancestors that were partially aquatic already. Only a broader understanding of the functional diversity of amphibian olfaction, and chemoreception in general, will enable us to place the modifications in secondarily aquatic forms in proper evolutionary context.

A key issue that has not been well examined in the previous literature is the relative functional role of the olfactory and vomeronasal systems in aquatic olfaction. Matthes (1927) showed that food localization in aquatic *Triturus* depends on an intact olfactory, but not vomeronasal, system, and recent work with peptide pheromones in newts (Toyoda et al., 2004) supports the role of the vomeronasal system in detecting these compounds. However, in axolotls *(Ambystoma mexicanum)* the olfactory and vomeronasal organs respond equally to chemical cues from conspecifics (Park et al., 2004). Moreover, in the terrestrial environment, Placyk and Graves (2002) found that prey detection in *Plethodon cinereus* is facilitated by the presence of an intact vomeronasal system. Thus, it is not yet possible to make any broad distinction between the function of the olfactory and vomeronasal systems in aquatic amphibians.

Data from teleosts are relevant here. The discovery that microvillar OR neurons in teleost fishes express class V2R vomeronasal receptor proteins, while ciliated olfactory receptor neurons express OR proteins (Hansen et al., 2003, 2004, 2005; Sato et al., 2005) suggests that the two cell types may be specialized for different functions. Sato et al. (2005) proposed that the microvillar receptor neurons may be specialized to detect polar molecules, such as amino acids and nucleic acids, while ciliated receptor neurons may be specialized to detect relatively nonpolar molecules, such as bile acids, steroids, and prostaglandins. They note, however, that ciliated receptor neurons have also been shown to respond to amino acids. In addition, studies using different techniques with other teleost species have produced wildly inconsistent results on the relative functions of ciliated and microvillar cells (reviewed in Eisthen, 2004). Regardless of their functions in teleosts, phylogenetic analysis indicates that one cannot deduce a simple relationship between the ciliated receptor cells of teleosts and the ciliated main OR cells of tetrapods, and between the microvillar receptor cells of teleosts and vomeronasal receptor cells of tetrapods (Eisthen, 2004). The evidence from *Xenopus*, the only amphibian for which comprehensive molecular and morphological data are available, makes it clear that microvillar receptor cells do not always express vomeronasal receptor proteins; instead, those located in the main olfactory cavity express group ε ORs (Freitag et al., 1995).

Integrating the evidence from teleost fishes and olfactory genomics (Niimura and Nei, 2005, 2006) with the data here reviewed for amphibians suggests the following, admittedly speculative, scenario for the origin and evolution of the tetrapod olfactory and vomeronasal systems (Fig. 4.8). The common ancestor of all bony fishes had at least three types of receptor cells in the olfactory epithelium: microvillar receptor neurons expressing V1R and V2R receptor proteins, and microvillar and ciliated receptor neurons expressing ORs. All OR groups were likely already present in this common ancestor (Niimura and Nei, 2005, 2006). In the sarcopterygian line leading to tetrapods, neurons expressing V1R and V2R receptor genes and bearing microvilli were segregated into a distinct vomeronasal organ, which maintained a fluid-filled lumen associated with copious mucus secretion from the newly evolved vomeronasal gland. As discussed above, paleontological data suggest that this likely occurred in the aquatic habitat, a hypothesis supported by the fact that both the organ and the gland are well developed in most larval, neotenic, and secondarily aquatic amphibians. The vomeronasal receptor neurons may have retained their ancestral sensitivity to amino acids and peptides. In the main olfactory organ, ciliated neurons expressing group γ (class II of Freitag et al. [1995, 1998]) ORs greatly proliferated, but microvillar and ciliated neurons expressing other groups of ORs were initially present as well. With the evolution of a terrestrial adult stage, a functional separation developed between a region of olfactory epithelium expressing group γ (and α?) ORs and specialized for olfaction in air, and a region expressing other groups of receptors, which continued to function in the aquatic environment. This separation necessarily was associated with the evolution of mechanisms separating water and air within the nasal cavity itself, and the evolution of accessory nasal glands (including Bowman's glands). In living amphibians, we see various evolutionary offshoots of this early specialization for olfaction in water and air, reaching their extreme in secondarily aquatic amphibians, such as the pipid frogs and typhlonectid caecilians discussed here, which have greatly developed the water nose, while maintaining the air nose. The vomeronasal organ and gland are typically well developed in secondarily aquatic species, although the neotenic and permanently aquatic proteid salamanders provide an interesting case in which the vomeronasal system has been secondarily lost, and the main olfactory epithelium is exclusively involved in aquatic olfaction. Finally, in the amniote line we see a further specialization with the loss of water-smelling ability in the main olfactory system (correlated with the loss of all but group α and γ ORs), and perhaps greater differentiation of function between the olfactory and vomeronasal systems. In amniotes, it is the vomeronasal system that becomes specialized for aquatic olfaction; in its absence, aquatic olfaction is apparently lacking (see Schwenk, chapter 5 in this volume). This scenario is testable by additional data on the diversity of olfactory structure and function in fishes (including lungfish), amphibians, and amniotes.

As in the olfactory system, profound morphological changes in the gustatory system at metamorphosis suggest that significant functional differences exist between taste in aquatic and terrestrial environments. Clearly, there is much yet to be learned about vertebrate chemoreception across the water-air threshold.

ACKNOWLEDGMENTS

We thank the editors for inviting us to contribute to this book, and for helpful comments on the manuscript. Thanks also to Gianluca Polese for assistance with drawings. We gratefully acknowledge the sources on which our Figures 4.5 and 4.7 are based (Matthes, 1927; Takeuchi et al., 1997). This work was partly supported by grants from the National Science Foundation (IBN-0092070 to JOR) and the National Institutes of Health (DC05366 to HLE).

LITERATURE CITED

Altig, R., and R. McDiarmid. 1999. Body plan: Development and morphology; pp. 24–51 in R. Altig and R. McDiarmid (eds.), Tadpoles: The Biology of Anuran Larvae. University of Chicago Press, Chicago.

Altner, H. 1962. Untersuchungen über Leistungen und Bau der Nase des südafrikanischen Krallenfrosches *Xenopus laevis* (Daudin, 1803). Zeitschrift für vergleichende Physiologie 45:272–306.

Arnold, S. J. 1977. The evolution of courtship behavior in new world salamanders with some comments on old world salamandrids; pp. 141–183 in D. H. Taylor and S. I. Guttman (eds.), The Reproductive Biology of Amphibians. Plenum, New York.

Arzt, A. H., W. L. Silver, J. R. Mason, and L. Clark. 1986. Olfactory responses of aquatic and terrestrial tiger salamanders to airborne and waterborne stimuli. Journal of Comparative Physiology A 158:479–487.

Badenhorst, A. 1978. The development and the phylogeny of the organ of Jacobson and the tentacular apparatus of *Ichthyophis glutinosus* (Linné). Annals of the University of Stellenbosch A2 1:1–26.

Barlow, L. 1998. The biology of amphibian taste; pp. 743–782 in H. Heatwole and E. Dawley (eds.), Amphibian Biology, Vol. 3, Sensory Perception. Surrey Beatty and Sons, Chipping Norton, NSW, Australia.

Baxi, K. N., K. M. Dorries, and H. L. Eisthen. 2006. Is the vomeronasal system really specialized for detecting pheromones? Trends in Neurosciences 29:1–7.

Benzekri, N. A. 2006. From ontogeny to phylogeny: the system of *Ascaphus truei*. M. A. thesis, Humboldt State University, Arcata, California.

Bertmar, G. 1969. The vertebrate nose, remark on its structural and functional adaptation and evolution. Evolution 23:131–152.

Bertmar, G. 1981. Evolution of vomeronasal organs in vertebrates. Evolution 35:359–366.

Bigiani, A., V. Ghiaroni, and F. Fieni. 2003. Channels as taste receptors in vertebrates. Progress in Biophysics and Molecular Biology 83:193–225.

Billo, R., and M. H. Wake. 1987. Tentacle development in *Dermophis mexicanus* (Amphibia, Gymnophiona) with an hypothesis of tentacle origin. Journal of Morphology 192:101–111.

Born, G. 1876. Ueber die Nasenhöhlen und den Thränennasengang der Amphibien. Morphologisches Jahrbuch 2:577–646.

Broman, I. 1920. Das Organon vomero-nasale Jacobsoni—ein Wassergeruchsorgan! Anatomische Hefte 58:143–191.

Bruner, H. 1901. The smooth facial muscles of Anura and Salamandrina. Morphologisches Jahrbuch 29:317–364.

Bruner, H. L. 1914a. Jacobson's organ and the respiratory mechanism of amphibians. Morphologisches Jahrbuch 48:157–165.

Bruner, H. L. 1914b. The mechanism of pulmonary respiration in amphibians with gill clefts. Morphologisches Jahrbuch 48:63–82.

Cannatella, D. C., and L. Trueb. 1988. Evolution of pipoid frogs: intergeneric relationships of the aquatic from family Pipidae (Anura). Zoological Journal of the Linnean Society 94:1–38.

Cao, Y., B. C. Oh, and L. Stryer. 1998. Cloning and localization of two multigene receptor families in goldfish olfactory epithelium. Proceedings of the National Academy of Sciences USA 95:11987–11992.

Clack, J. A. 2002. Gaining Ground: The Origin and Evolution of Tetrapods. Indiana University Press, Bloomington.

Dawley, E. M. 1998. Olfaction; pp. 713–742 in H. Heatwole (ed.), Amphibian Biology: Sensory Perception. Surrey Beatty and Sons, Chipping Norton, NSW, Australia.

Dawley, E. M., and A. H. Bass. 1989. Chemical access to the vomeronasal organs of a plethodontid salamander. Journal of Morphology 200:163–174.

Døving, K. B., D. Trotier, J.-F. Rosin, and A. Holley. 1993. Functional architecture of the vomeronasal organ of the frog (genus *Rana*). Acta Zoologica 74:173–180.

Duellman, W. E., and L. Trueb. 1986. Biology of Amphibians. MacMillian, New York.

Eisthen, H. L. 1992. Phylogeny of the vomeronasal system and of receptor cell types in the olfactory and vomeronasal epithelia of vertebrates. Microscopy Research and Technique 23:1–21.

Eisthen, H. L. 1997. Evolution of vertebrate olfactory systems. Brain, Behaviour, and Evolution 50: 222–233.

Eisthen, H. L. 2000. Presence of the vomeronasal system in aquatic salamanders. Philosophical Transactions of the Royal Society of London B 355:1209–1213.

Eisthen, H. L. 2004. The goldfish knows: olfactory receptor cell morphology predicts receptor gene expression. Journal of Comparative Neurology 477:341–346.

Elepfandt, A. 1996. Sensory perception and the lateral line system in the clawed frog, *Xenopus*; pp. 97–120 in R. Tinsley and H. Kobel (eds.), The Biology of *Xenopus*. Oxford University Press, Oxford.

Föske, H. 1934. Das Geruchsorgan von *Xenopus laevis*. Zeitschrift für Anatomie und Entwicklungsgeschichte 103:519–550.

Freitag, J., J. Krieger, J. Strotmann, and H. Breer. 1995. Two classes of olfactory receptors in *Xenopus laevis*. Neuron 15:1383–1392.

Freitag, J., G. Ludwig, I. Andreini, P. Rössler, and H. Breer. 1998. Olfactory receptors in aquatic and terrestrial vertebrates. Journal of Comparative Physiology A 183:635–650.

Frost, D., T. Grant, J. Faivovich, R. H. Bain, H. A, C. F. B. Haddad, R. O. De Sá, A. Channing, M. Wilkinson, S. C. Donnelan, C. J. Raxworthy, J. A. Campbell, B. L. Blotto, P. Moler, R. C. Drewes, R. A. Nussbaum, J. D. Lynch, D. M. Green, and W. C. Wheeler. 2006. The amphibian tree of life. Bulletin of the American Museum of Natural History 297:1–370.

Gaudin, A., and J. Gascuel. 2005. 3D atlas describing the ontogenetic evolution of the primary olfactory projections in the olfactory bulb of *Xenopus laevis*. Journal of Comparative Neurology 489:403–424.

Gaupp, E., A. Ecker, and R. Wiedersheim. 1904. Anatomie des Frosches. Friedrich Vieweg, Braunschweig.

Gradwell, N. 1969. The function of the internal nares of the bullfrog tadpole. Herpetologica 25:120–121.

Hagino-Yamagishi, K., K. Moriya, H. Kubo, Y. Wakabayashi, N. Isobe, S. Saito, M. Ichikawa, and K. Yazaki. 2004. Expression of vomeronasal receptor genes in *Xenopus laevis*. Journal of Comparative Neurology 472:246–256.

Halliday, T. R. 1977. The courtship of European newts: an evolutionary perspective; pp. 185–232 in D. H. Taylor and S. I. Guttman (eds.), The Reproductive Biology of Amphibians. Plenum, New York.

Hansen, A., J. O. Reiss, C. L. Gentry, and G. D. Burd. 1998. Ultrastructure of the olfactory organ in the clawed frog, *Xenopus laevis*, during larval development and metamorphosis. Journal of Comparative Neurology 398:273–288.

Hansen, A., S. H. Rolen, K. Anderson, Y. Morita, J. Caprio, and T. E. Finger. 2003. Correlation between olfactory receptor cell type and function in the channel catfish. Journal of Neuroscience 23:9328–9339.

Hansen, A., K. T. Anderson, and T. E. Finger. 2004. Differential distribution of olfactory receptor neurons in goldfish: structural and molecular correlates. Journal of Comparative Neurology 477: 347–359.

Hansen, A., S. Rolen, K. Anderson, Y. Morita, J. Caprio, and T. Finger. 2005. Olfactory receptor neurons in fish: structural, molecular, and functional correlates. Chemical Senses 30 (Suppl. 1): i311.

Helling, H. 1938. Das Geruchsorgan der Anuren, vergleichend-morphologisch betrachtet. Zeitschrift für Anatomie und Entwicklungsgeschichte 108:587–643.

Himstedt, W., and D. Simon. 1995. Sensory basis of foraging behaviour in caecilians. Herpetological Journal 5:266–270.

Hinsberg, V. 1901. Die Entwicklung der Nasenhöhle bei Amphibien. Archiv für mikroskopische Anatomie 58:411–482.

Jermakowicz, W. J., III, D. A. Dorsey, A. L. Brown, K. Wojciechowski, C. L. Giscombe, B. M. Graves, C. H. Summers, and G. R. T. Eyck. 2004. Development of the nasal chemosensory organs in two terrestrial anurans: the directly developing frog, *Eleutherodactylus coqui* (Anura: Leptodactylidae), and the metamorphosing toad, *Bufo americanus* (Anura: Bufonidae). Journal of Morphology 261:225–248.

Joly, P., and L. Caillere. 1983. Smelling behavior of urodele amphibians in an aquatic environment: study of *Pleurodeles waltl*. Acta Zoologica 64:169–175

Jorgensen, B. C. 2000. Amphibian respiration and olfaction and their relationships: from Robert Townson (1794) to the present. Biological Reviews 75:297–345.

Jurgens, J. D. 1971. The morphology of the nasal region of Amphibia and its bearing on the phylogeny of the group. Annals of the University of Stellenbosch 46A:1–146.

Kauer, J. S. 2002. On the scents of smell in the salamander. Nature 417:336–342.

Khalil, S. H. 1978. Development of the olfactory organ of the Egyptian Toad, *Bufo regularis* Reuss. Folia Morphologica 26:69–87.

Kikuyama, S., F. Toyoda, Y. Ohmiya, K. Matsuda, S. Tanaka, and H. Hayashi. 1995. Sodefrin: a female-attracting peptide pheromone in newt cloacal glands. Science 267:1643–1645.

Liman, E. R., D. P. Corey, and C. Dulac. 1999. TRP2: a candidate transduction channel for mammalian pheromone sensory signaling. Proceedings of the National Academy of Sciences USA 96:5791–5796.

Matthes, E. 1924a. Das Geruchsvermögen von *Triton* beim Aufenthalt unter Wasser. Zeitschrift für vergleichende Physiologie 1:57–83.

Matthes, E. 1924b. Das Geruchsvermögen von *Triton* beim Aufenthalt an Land. Zeitschrift für vergleichende Physiologie 1:590–606.

Matthes, E. 1924c. Die Rolle des Gesichts-, Geruchs- und Erschütterungssinnes für den Nahrungserwerb von *Triton*. Biologisches Zentralblatt 44:72–87.

Matthes, E. 1926. Die physiologische Doppelnatur des Geruchsorgans der Urodelen im Hinblick auf seine morphologische Zusammensetzung aus Haupthöhle und "Jacobsonschem Organe." Zeitschrift für vergleichende Physiologie 4:81–102.

Matthes, E. 1927. Der Einfluss des Mediumwechsels auf das Geruchsvermögen von *Triton*. Zeitschrift für vergleichende Physiologie 5:83–166.

Matthes, E. 1934. Geruchsorgan; pp. 879–948 in L. Bolk, E. Göppert, E. Kallius, and W. Lubosch (eds.), Handbuch der vergleichenden Anatomie der Wirbeltiere, Vol. II-2. Urban and Schwarzenberg, Berlin.

Measey, G. 1998. Terrestrial prey capture in *Xenopus laevis*. Copeia 1998:787–791.

Meredith, M. 1994. Chronic recording of vomeronasal pump activation in awake and behaving hamsters. Physiology and Behavior 56:345–354.

Meyer, D. L., I. R. Fackler, A. G. Jadhao, B. D'Aniello, and E. Kicliter. 1997. Differential labelling of primary olfactory system subcomponents by SBA (lectin) and NADPH-d histochemistry in the frog *Pipa*. Brain Research 762:275–280.

Mezler, M., S. Konzelman, J. Freitag, P. Rössler, and H. Breer. 1999. Expression of olfactory receptors during development in *Xenopus laevis*. Journal of Experimental Biology 202:365–376.

Mezler, M., J. Fleischer, and H. Breer. 2001. Characteristic features and ligand specificity of the two olfactory receptor classes from *Xenopus laevis*. Journal of Experimental Biology 204:2987–2997.

Millery, J., L. Briand, V. Bézirard, C. Blon, C. Fenech, L. Richard-Parpaillon, B. Quennedey, J.-C. Pernollet, and J. Gascuel. 2005. Specific expression of olfactory binding protein in the aerial olfactory cavity of adult and developing *Xenopus*. European Journal of Neuroscience 22:1389–1399.

Mombaerts, P. 2004. Genes and ligands for odorant, vomeronasal and taste receptors. Nature Reviews Neuroscience 5:263–278.

Nagai, T., D. Nii, and H. Takeuchi. 2001. Amiloride blocks salt taste transduction of the glossopharyngeal nerve in metamorphosed salamanders. Chemical Senses 26:965–969.

Naito, T., Y. Saito, J. Yamamoto, Y. Nozaki, K. Tomura, M. Hazama, S. Nakanishi, and S. Brenner. 1998. Putative pheromone receptors related to the Ca^{2+}-sensing receptor in *Fugu*. Proceedings of the National Academy of Sciences USA 95:5178–5181.

Nicholas, J. S. 1922. The reactions of *Amblystoma tigrinum* to olfactory stimuli. Journal of Experimental Zoology 35:257–281.

Niimura, Y., and M. Nei. 2005. Evolutionary dynamics of olfactory receptor genes in fishes and tetrapods. Proceedings of the National Academy of Sciences USA 102:6039–6044.

Niimura, Y., and M. Nei. 2006. Evolutionary dynamics of olfactory and other chemosensory receptor genes in vertebrates. Journal of Human Genetics 51:505–517.

Nikitkin, V. B. 1986. On the nasal muscles in Anura and Urodela; pp. 251–254 in Z. Roček (ed.), Studies in Herpetology. Charles University, Prague.

Nishikawa, K., and C. Gans. 1996. Mechanisms of tongue protraction and narial closure in the marine toad *Bufo marinus*. Journal of Experimental Biology 199:2511–2529.

Northcutt, R. G. 2004. Taste buds: development and evolution. Brain, Behaviour, and Evolution 64:198–206.

Nussbaum, R., and M. Wilkinson. 1995. A new genus of lungless tetrapod: a radically divergent caecilian (Amphibia: Gymnophiona). Proceedings of the Royal Society of London B 261:331–335.

Oikawa, T., K. Suzuki, T. R. Saito, K. W. Takahashi, and K. Taniguchi. 1998. Fine structure of three types of olfactory organs in *Xenopus laevis*. Anatomical Record 252:301–310.

Özeti, N., and D. Wake. 1969. The morphology and evolution of the tongue and associated structures in salamanders and newts. Copeia 1969:91–123.

Park, D., J. M. McGuire, A. L. Majchrzak, J. M. Ziobro, and H. L. Eisthen. 2004. Discrimination of conspecific sex and reproductive condition using chemical cues in axolotls *(Ambystoma mexicanum)*. Journal of Comparative Physiology A 190:415–427.

Parsons, T. S. 1967. Evolution of the nasal structure in the lower tetrapods. American Zoologist 7:397–413.

Paterson, N. F. 1939a. The head of *Xenopus laevis*. Quarterly Journal of Microscopical Science 81:161–233.

Paterson, N. F. 1939b. The olfactory organ and tentacles of *Xenopus laevis*. South African Journal of Science 36:390–404.

Paterson, N. F. 1951. The nasal cavities of the toad *Hemipipa carvalhoi* Mir.-Rib. and other Pipidae. Proceedings of the Zoological Society of London 121:381–415.

Pearl, C., M. Cervantes, M. Chan, U. Ho, R. Shoji, and E. Thomas. 2000. Evidence for a mate-attracting chemosignal in the dwarf African clawed frog *Hymenochirus*. Hormones and Behavior 38:67–74.

Placyk, J. S., Jr., and B. M. Graves. 2002. Prey detection by vomeronasal chemoreception in a plethodontid salamander. Journal of Chemical Ecology 28:1017–1036.

Prabha, K. C., D. G. Bernard, M. Gardner, and N. J. Smatresk. 2000. Ventilatory mechanics and the effects of water depth on breathing pattern in the aquatic caecilian *Typhlonectes natans*. Journal of Experimental Biology 203:263–272.

Reiss, J. O., and G. D. Burd. 1997. Metamorphic remodeling of the primary olfactory projection in *Xenopus*: developmental independence of projections from olfactory neuron subclasses. Journal of Neurobiology 32:213–222.

Roelants, K., and F. Bossuyt. 2005. Archaeobatrachian paraphyly and Pangaean diversification of crown-group frogs. Systematic Biology 54:111–126.

Rowedder, W. 1937. Die Entwicklung des Geruchsorgans bei *Alytes obstetricians* und *Bufo vulgaris*. Zeitschrift für Anatomie und Entwicklungsgeschichte 107:91–123.

Saint Girons, H., and L. Zylberberg. 1992a. Histologie comparée des glandes céphaliques exocrines et des fosses nasales des Lissamphibia. I. Anatomie générale et glandes céphaliques. Annales des Sciences Naturelles, Zoologie (Paris) 13:59–82.

Saint Girons, H., and L. Zylberberg. 1992b. Histologie comparée des glandes céphaliques exocrines et des fosses nasales des Lissamphibia. II. Épitheliums des fosses nasales. Annales des Sciences Naturelles, Zoologie (Paris) 13:121–145.

Sarasin, F., and P. Sarasin. 1890. Zur Entwicklungsgeschichte und Anatomie der Ceylonischen Blindwühle *Ichthyophis glutinosus*. Ergebnisse naturwissenschaftlicher Forschung en auf Ceylon in den Jahren 1884–1886 2:153–263.

Sato, Y., N. Miyasaka, and Y. Yoshihara. 2005. Mutually exclusive glomerular innervation by two distinct types of olfactory sensory neurons revealed in transgenic zebrafish. Journal of Neuroscience 25:4889–4897.

Schmidt, A., and M. H. Wake. 1990. Olfactory and vomeronasal systems of caecilians (Amphibia: Gymnophiona). Journal of Morphology 205:255–268.

Schmidt, A., C. Naujoks-Manteuffel, and G. Roth. 1988. Olfactory and vomeronasal projections and the pathway of the nervus terminalis in ten species of salamanders. Cell and Tissue Research 251:45–50.

Schuch, K. 1934. Das Geruchsorgan von *Triton alpestris*. Zoologisches Jahrbuch 59:69–134.

Seydel, O. 1895. Über die Nasenhöhle und das Jacobson'sche Organ der Amphibien: Eine vergleichend-anatomische Untersuchung. Morphologisches Jahrbuch 23:453–543.

Shi, P., and J. Zhang. 2006. Contrasting modes of evolution between vertebrate sweet/umami receptor genes and bitter receptor genes. Molecular Biology and Evolution 23:292–300.

Stuelpnagel, J. T., and J. O. Reiss. 2005. Olfactory metamorphosis in the coastal giant salamander *(Dicamptodon tenebrosus)*. Journal of Morphology 266:22–45.

Takeuchi, H., S. Ido, Y. Kaigawa, and T. Nagai. 1997. Taste disks are induced in the lingual epithelium of salamanders during metamorphosis. Chemical Senses 22:535–545.

Tinsley, R., C. Loumont, and H. Kobel. 1996. Geographical distribution and ecology; pp. 35–59 in R. Tinsley and H. Kobel (eds.), The Biology of *Xenopus*. Clarendon Press, Oxford.

Titus, T., and A. Larson. 1995. A molecular phylogenetic perspective on the evolutionary relationships of the salamander family Salamandridae. Systematic Biology 44:125–151.

Toyoda, T., K. K. Yamamoto, T. Iwata, I. Hasunuma, M. Cardinali, G. Mosconi, A. M. Polzonetti-Magnic, and S. Kikuyama. 2004. Peptide pheromones in newts. Peptides 25:1531–1536.

Toyoshima, K., and A. Shimamura. 1982. Comparative study of ultrastructures of the lateral-line organs and the palatal taste organs in the African clawed toad, *Xenopus laevis*. Anatomical Record 204:371–381.

Toyoshima, K., and A. Shimamura. 1987. Monoamine-containing basal cells in the taste buds of the newt *Triturus pyrrhogaster*. Archives of Oral Biology 32:619–621.

Trueb, L., and D. C. Cannatella. 1982. The cranial osteology and hyolaryngeal apparatus of *Rhinophrynus dorsalis* (Anura: Rhinophrynidae) with comparisons to recent pipid frogs. Journal of Morphology 171:11–40.

Wabnitz, P. A., J. H. Bowie, M. J. Tyler, J. C. Wallace, and B. P. Smith. 1999. Aquatic sex pheromone from a male tree frog. Nature 401:444–445.

Wake, M. H., and K. Schwenk. 1986. A preliminary report on the morphology and distribution of taste buds in gymnophiones, with comparison to other amphibians. Journal of Herpetology 20:254–256.

Warbeck, A., and J. Parzefall. 2001. Mate recognition via waterborne chemical cues in the viviparous caecilian *Typhlonectes natans* (Amphibia: Gymnophiona); pp. 263–268 in A. Marchlewska-Koj, J. J. Lepri, and D. Müller-Schwarze (eds.), Chemical Signals in Vertebrates, 9. Kluwer Academic/Plenum Publishers, New York.

Weisrock, D., T. Papenfuss, J. Macey, S. Litvinchuk, R. Polymeni, I. Ugartas, E. Zhao, H. Jowkar, and A. Larson. 2006. A molecular assessment of phylogenetic relationships and lineage accumulation rates within the family Salamandridae (Amphibia, Caudata). Molecular Phylogenetics and Evolution 41:368–383.

Weiss, G. 1986. Die Struktur des Geruchsorgans und des Telencephalons beim südafrikanischen Krallenfrosch *Xenopus laevis* (Daudin) und ihre Veränderungen während der Metamorphose. Doctoral Dissertation, University of Regensburg, Germany.

Wilder, I. W. 1925. The Morphology of Amphibian Metamorphosis. Smith College, Northampton, MA.

Wilkinson, M., and R. A. Nussbaum. 1997. Comparative morphology and evolution of the lungless caecilian *Atretochoana eiselti* (Taylor) (Amphibia: Gymnophiona: Typhlonectidae). Biological Journal of the Linnean Society 62:39–109.

Wilkinson, M., and R. A. Nussbaum. 1999. Evolutionary relationships of the lungless caecilian *Atretochoana eiselti* (Amphibia: Gymnophiona: Typhlonectidae). Zoological Journal of the Linnean Society 126:191–223.

Witt, M., and K. Reutter. 1994. Ultrastructure of the taste disk of the African clawed frog, *Xenopus laevis*. Chemical Senses 19:433.

Yoshii, K., C. Yoshii, Y. Kobatake, and K. Kurihara. 1982. High sensitivity of *Xenopus* gustatory receptors to amino acids and bitter substances. American Journal of Physiology 243:R42–R48.

Yvroud, M. 1966. Développement de l'organe olfactif et des glandes annexes chez *Alytes obstetricans* Laurenti au cours de la vie larvaire et de la métamorphose. Archives d'Anatomie Microscopique 55:387–410.

Zuwala, K., and M. Jakubowski. 2001a. Morphological differentiation of taste organs in the ontogeny of *Salamandra salamandra*. Anatomy and Embryology 204:413–420.

Zuwala, K., and M. Jakubowski. 2001b. Taste organs in lower vertebrates: morphology of the taste organs in Amphibia; pp. 221–239 in H. Dutta and J. D. Munshi (eds.), Vertebrate Functional Morphology: Horizon of Research in the Twenty-first Century. Science Publishers, Enfield, NH.

Zuwala, K., S. Kato, and M. Jakubowski. 2002. Two generations of the tongue and gustatory organs in the development of *Hynobius dunni* Tago. Journal of Anatomy 201:91–97.

5

Comparative Anatomy and Physiology of Chemical Senses in Nonavian Aquatic Reptiles

Kurt Schwenk

The nonavian reptiles are an exceptionally diverse group of tetrapods that have radiated into a variety of habitats, including marine and freshwater environments. While most aquatic reptiles retain strong ties to the land, some species are among the most fully aquatic of any tetrapod, including, for example, the marine turtles (Cheloniidae) and pelagic sea snakes (Elapidae). Indeed, some species of seasnake have forsaken their connection to the land entirely, not even returning to give birth. They thus achieve a level of adaptive commitment to the water comparable to those most aquatic of tetrapods, the Cetacea. Nonetheless, only about 8% (approximately 600 species) of living, nonavian reptiles are even partially aquatic.

Crown group (nonavian) reptiles comprise the turtles (Testudines), alligators, crocodiles and gharials (Archosauria, Crocodylia), and tuatara (Rhynchocephalia), lizards, and snakes (Squamata). Each of these major clades includes some partially or fully aquatic species (see the section by Schwenk and Thewissen in chapter 1 in this volume). Among extinct lineages

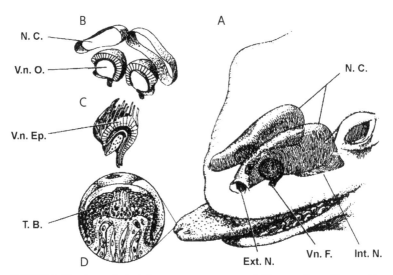

FIGURE 5.1. The three principal chemosensory systems available to reptiles as illustrated in a squamate. Gustation is mediated by taste buds in the mouth and pharynx. Olfaction is mediated by olfactory epithelium in the nasal cavities. Vomeronasal chemoreception is mediated by vomeronasal epithelium in the nasal cavities (turtles) or sequestered in separate vomeronasal organs (squamates). (A) Lizard head showing location of chemosensory organs; (B) transverse section showing the spatial relationship between the nasal cavities (N. C.) and vomeronasal organs (V.n. O.); (C) sagittal section of one vomeronasal organ showing the sensory epithelium (V.n. Ep.) and sensory nerves on its convex surface; (D) magnified image of a taste bud (T. B.) within the epithelium of the tongue tip. Other abbreviations: Int. N., internal naris (choana); Ext. N., external naris; and V.n. F., vomeronasal fenestra (opening into mouth through palate). Modified from Schwenk (1995).

for which we have limited information on the chemical senses are the mosasaurs, large marine lizards, the phytosaurs, aquatic archosaurs related to crocodylians, and the plesiosaurs, a euryapsid group probably related to placodonts and ichthyosaurs.

Given the phylogenetic breadth spanned by these taxa, the extreme morphological divergence among them, and the multiple origins of their secondarily aquatic habits, one might expect pronounced diversity both in the nature of their chemical senses and in any putative aquatic adaptations they exhibit, and indeed, this is the case. In this chapter, I review briefly what is known about the structure and function of the chemical senses in aquatic reptiles and identify, to the extent possible, evolutionary patterns. Sadly, this chapter serves mostly to highlight our ignorance about the chemical senses in these groups.

COMPARATIVE ANATOMY AND FUNCTION OF THE CHEMICAL SENSES

The chemical senses potentially available to tetrapods, including reptiles, are gustation, olfaction, and vomeronasal chemoreception (Fig. 5.1) (see Wyatt, 2003; Eisthen and Schwenk, chapter 3 in this volume; Reiss and Eisthen, chapter 4 in this volume, for summaries). Gustation is mediated by taste buds located within the oral and pharyngeal cavities. Taste buds are innervated by cranial nerves VII (facial: chorda tympani and palatine branches), IX (glossopharyngeal), and X (vagus) to the floor of the mouth, anterior tongue and palate, posterior tongue and pharynx, and larynx, respectively. Olfaction and vomeronasal chemoreception constitute the nasal chemical senses associated developmentally and evolutionarily with the nasal capsule. These are often described as the "main" and "accessory" olfactory senses, respectively, and are distinguished topologically, anatomically, neuroanatomically, and functionally. Sensory epithelia of both systems are innervated by cranial nerve I (olfactory), although the vomeronasal system is supplied by a separate branch called the vomeronasal nerve.

Below, each system is described separately. In addition to these three, widely recognized chemosensory systems, free nerve endings of cranial nerve V (trigeminal) are widely distributed in the mouth and nasal cavity of most tetrapods and are known to be generally sensitive to chemical (and mechanical) stimuli in turtles (e.g., Tucker, 1971; Scott, 1979). However, the role of the trigeminal system is poorly understood and is not considered further here.

COMPARATIVE ANATOMY OF THE NASAL CAVITIES: OLFACTORY AND VOMERONASAL SYSTEMS

The morphology of the nose in reptiles has been extensively reviewed by Parsons (1959a, 1959b, 1967, 1970a). Information reported here comes from these sources unless otherwise noted. Each paired nasal cavity comprises a tube that opens to the outside through the external nares (nostrils) and to the mouth through the internal nares (choanae). Three parts of this cavity are recognized in most reptiles: an anterior vestibulum a middle cavum and a posteroventral nasophyarngeal duct. Protrusions of the wall within the cavum of lepidosaurs and crocodylians are called conchae. Olfactory epithelium is located within the cavum. The shape and relative contribution to the nasal cavity of these chambers varies among taxa. They are most clearly evident in turtles, which possess a relatively simple nasal cavity with clearly differentiated regions (Fig. 5.2). *Sphenodon* (Rhynchocephalia) and most lizards lack a well-defined nasopharyngeal duct so that the cavum opens directly into the mouth. Crocodylians and some turtles possess secondary palates that are associated with an elongated nasopharyngeal duct (Fig. 5.3). However, some aquatic turtles lacking a secondary palate also have a relatively long nasopharyngeal duct.

TURTLES

In turtles the vestibulum is usually a short tube that leads directly into the expanded chamber of the cavum. In aquatic species with snorkel-like noses, the vestibulum is elongated (e.g., *Chelus*, *Carettochelys* and trionychids; Fig. 5.2). In sea turtles (Cheloniidae), the proximal part of the vestibulum is surrounded by vascular erectile tissue that constricts the nostrils when the turtles are resting while submerged (Walker, 1959; Parsons, 1970a). The nostrils remain open when the turtles are active, however (Walker, 1959; see below).

The cavum in turtles consists of a dorsal chamber containing the olfactory epithelium and a ventral "intermediate region." In most turtles the olfactory chamber is domed, but broadly open to the rest of the cavum below. In sea turtles, however, the olfactory region forms a partial sphere with a constricted opening at its base. The vomeronasal epithelium lies within shallow sulci on the floor of the intermediate region of most turtles, but in sea turtles it is sequestered within small dorsal and ventral diverticulae (Fig. 5.2). It has been suggested that in aquatic turtles the dorsal olfactory chamber retains an air bubble when the nasal cavity floods during submersion (Tucker, 1971). This would be most likely to occur in sea turtles given the form of the olfactory chamber. In contrast, the ventral position of the vomeronasal epithelium suggests that it is covered by water when the cavum is flooded (see below).

The turtle nasopharyngeal duct forms a tube that connects the cavum to the buccal cavity. It seems to be especially long in highly aquatic turtles (Fig. 5.2). Furthermore, the entire nasal tube (not including diverticulae) appears to be relatively straight (e.g., *Chelydra*, *Trionyx*, *Chelus*), whereas in terrestrial (*Testudo*, Testunidae) (Fig. 1.4. in this volume) and some semiaquatic turtles (Emydidae) it is kinked (Fig. 5.2).

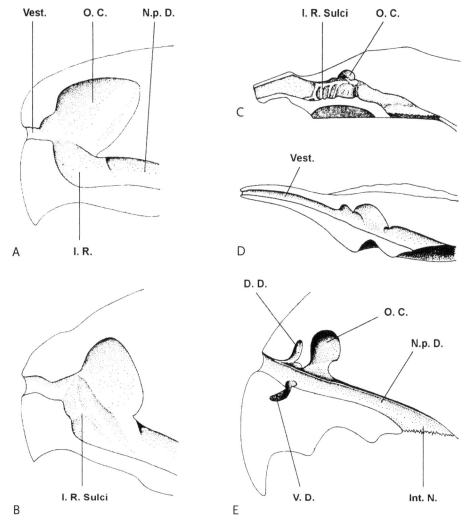

FIGURE 5.2. The nasal cavities in turtles shown in sagittal section. (A) Terrestrial *Testudo graeca* (Testudinidae); (B) semiaquatic *Emys orbicularis* (Emydidae); (C) aquatic *Apalone ferox* (Trionychidae); (D) aquatic *Chelus fimbriatus* (Chelidae); (E) marine *Chelonia mydas* (Cheloniidae). Note the relatively long, straight nasal cavities of the aquatic species. Both freshwater turtles (C and D) use the elongated snout and vestibulum (Vest.) to snorkel. In all turtles, vomeronasal epithelium is found on the floor of the cavum in the "intermediate region" (I. R.), often located in sulci (I. R. sulci) in semiaquatic and aquatic species. In sea turtles (E) the vomeronasal epithelium is sequestered within two separate pockets, the dorsal and ventral diverticulae (D. D. and V. D.). The bubblelike form of the olfactory chamber of the cavum in sea turtles might help to trap air during submersion. Other abbreviations: Int. N., internal naris (choana); N.p. D., nasopharyngeal duct; and O. C., olfactory chamber of cavum. All figures modified from Parsons (1970a).

These conformational differences might relate to the hydrodynamics of water flow through the nasal cavity and the mechanism of underwater olfaction (see below).

CROCODYLIANS

The nasal chamber of crocodylians is relatively uniform but considered to be the most complex of reptiles (Parsons, 1970a) (Fig. 5.3). This complexity relates to the presence in archosaurs of paranasal sinuses that penetrate the surrounding bone and soft tissues (Witmer, 1995) and of several well-developed conchae within the cavum (Parsons, 1970a; Witmer, 1995). In addition, the entire nasal cavity and the nasopharyngeal duct, in particular, is elongate owing to

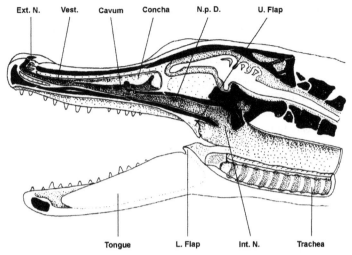

FIGURE 5.3. The head of a crocodylian *(Alligator mississippiensis)* in sagittal section, left lateral view. The nasal cavity is elongate in crocodilians owing to a long snout and a secondary palate that displaces the internal nares (Int. N.), or choanae, posteriorly into the pharynx. The cavum contains several conchae and openings into the paranasal sinuses that invade the surrounding tissues of the snout. Other abbreviations: Ext. N., external naris; L. Flap, lower flap of buccopharyngeal valve; N.p. D., nasopharyngeal duct; U. Flap, upper flap of buccopharyngeal valve; and Vest., vestibulum. Modified from Bellairs (1970).

the long snout and extensive secondary palate. The choanae lie posteriorly in the pharynx, behind a buccopharyngeal valve. This valve forms between a flap of soft tissue suspended from the back of the palate anterior to the choanae and a ventral flap extending dorsally from the floor of the pharynx just behind the tongue. The ventral flap is elevated along with the back of the tongue, thus forming a watertight seal at the back of the buccal cavity, anterior to the choanae. This permits retention of an open airway even when the mouth is flooded. Thus crocodylians can breathe while submerged as long as the nostrils remain above the surface (Negus, 1958; Bellairs, 1970). This behavior is facilitated by the dorsal placement of the nostrils on an elevated rostrum. The nasal protuberance can become extremely large in mature male *Gavialis* (Bellairs, 1970). The crescent-shaped nostrils are closed during submersion by vascular erectile tissue and smooth muscles within the walls of the vestibulum (Bellairs and Shute, 1953; Bellairs, 1970). Olfactory epithelium covers the dorsal surfaces of the cavum but does not extend into deep sinuses

(Saint Girons, 1976). There is no vomeronasal epithelium and no functional vomeronasal system in adult crocodylians (Parsons, 1970a; Saint Girons, 1976).

SQUAMATES

Squamates are morphologically diverse, and this is reflected in their nasal anatomy (Fig. 5.1). In most species the nostrils are positioned laterally on the snouts, but in many aquatic snakes (Kathariner, 1900; Gabe and Saint Girons, 1976; Green, 1997), the aquatic lizard *Lanthanotus* (McDowell and Bogert, 1954; Parsons, 1970a; personal observations), and some aquatic varanid lizards (Bellairs, 1970) they are medial and dorsal. The length and path of the vestibulum is highly variable. It is surrounded, in part, by cavernous erectile tissue that acts to close the nostrils in aquatic snakes (Kathariner, 1900; Santos-Costa and Hofstadler-Deiques, 2002) and sand-dwelling lizards (Stebbins, 1948); however, even terrestrial lizards have some cavernous tissue in the walls of the vestibulum (Oelrich, 1956). Kathariner (1900) also observed that in some aquatic snakes the vestibulum-cavum junction is constricted by a circular ridge. Conceivably, this constriction is also valvular, serving as an additional barrier to cavum flooding during submersion. According to Bellairs (1970), the nostrils of some aquatic monitor lizards *(Varanus)* are situated posteriorly near the eyes, and this condition is correlated with a convoluted vestibulum that might serve as an air supply during submersion. The highly aquatic homalopsine (Colubridae) snakes have slitlike, valvular nostrils and close fitting labial scale rows to seal the nose and mouth when underwater (Greene, 1997). Sea snakes (Elapidae) also have specialized structures to plug the small lingual orifice at the front of the mouth (Greene, 1997).

The squamate cavum is relatively simple, though variable. In most species there is a well-developed concha, but in some iguanians and aquatic snakes it is reduced or lost (Parsons, 1970a; Gabe and Saint Girons, 1976). The dorsal surfaces of the cavum and the concha are covered by olfactory epithelium (Gabe and Saint Girons, 1976). The olfactory region is particularly reduced in aquatic snakes (Kathariner, 1900; Gabe and Saint Girons, 1976), in contrast to the well-developed vomeronasal organ (see below). The cavum is lubricated by a large external nasal gland located at the cavum's junction with the vestibulum, but this is lost in aquatic snakes and the nasal cavity reduced to a simple, cylindrical tube (Gabe and Saint Girons, 1976). The posteroventral part of the cavum either opens directly into the mouth or is extended slightly by a short nasopharyngeal duct.

Uniquely among tetrapods, the vomeronasal epithelium in squamates is sequestered within discrete capsules lacking any connection to the nasal cavity (Fig. 5.1). However, these vomeronasal organs retain their ventral position relative to the nasal tube and open into the anterior end of the mouth via narrow ducts and tiny fenestrae. The vomeronasal organs are filled with fluid delivered from the orbital Harderian gland by the nasolacrimal duct (Bellairs and Boyd, 1950; Rehorek et al., 2000). Scent molecules are sampled from the external environment and delivered to the

vomeronasal organ by the tongue during "tongue-flicking" behavior (Halpern, 1992; Schwenk, 1995).

Tongue-flicking concentrates scent molecules in the fluid covering the tongue tips by molecular diffusion (Schwenk, 1996). Scent transfer from the tongue to the vomeronasal organs is poorly understood, but it is probably enhanced by vomeronasal suction (Broman, 1920; Young, 1993). Although there is substantial variation in the degree of vomeronasal organ development, it is always well developed in aquatic snakes, in contrast to the main olfactory system (Gabe and Saint Girons, 1976).

OLFACTION AND VOMERONASAL CHEMORECEPTION IN AQUATIC REPTILES

TURTLES

There is ample evidence that most turtles use chemical cues to direct some behaviors (Burghardt, 1970; Manton et al., 1972a, 1972b; Manton, 1979; Scott, 1979; Chelazzi and Delfino, 1986; Halpern, 1992; Quinn and Graves, 1998; Druzisky and Brainerd, 2001; Bartol and Musick, 2003; Constantino and Salmon, 2003; Muñoz, 2004). Turtles employ buccal oscillation, in which the buccal floor is rapidly depressed and elevated, as a kind of "sniffing" behavior during olfaction (McCutcheon, 1943; Root, 1949; Druzisky and Brainerd, 2001). Aquatic turtles continue buccal oscillation underwater with the nostrils open, thereby flooding the nasal cavities (Root, 1949; Walker, 1959; Manteifel et al., 1992; Druzisky and Brainerd, 2001). Therefore, aquatic olfaction is possible. Underwater nasal chemoreception is strongly implicated in general exploratory behavior (Root, 1949; Walker, 1959), food location and discrimination (Manton, 1979; Constantino and Salmon, 2003), and reproductive behavior (Manteifel et al., 1992; Muñoz, 2004). It may be a factor in long-range migratory homing in some sea turtles, which involves chemical "imprinting" of natal beaches (Koch et al., 1969; Grassman,

1993; Luschi et al., 1998). Aquatic turtles possess well-developed "musk" or Rathke's glands in axillary, inguinal, and/or inframarginal regions (Ehrenfeld and Ehrenfeld, 1973; Rostal et al., 1991). These glands appear to be an ancient chelonian characteristic (Weldon and Gaffney, 1998), but their function remains obscure (Manton, 1979). It is unknown whether underwater chemosensory behaviors are mediated by the main olfactory system or vomeronasal system, but for theoretical reasons, I suggest that the vomeronasal system is most likely (see below).

CROCODYLIANS

Crocodylians, like turtles, employ buccal oscillation (usually called "gular pumping") as an olfactory behavior (Naifeh et al., 1970; Gans and Clark, 1976; Weldon and Ferguson, 1993). Unlike turtles, however, adult crocodylians lack a vomeronasal system, therefore all olfactory behavior is attributable to the main olfactory system. Also unlike turtles, crocodylians close their nostrils and do not exhibit gular pumping when submerged, so underwater olfaction is unlikely (Bellairs and Shute, 1953; Weldon and Ferguson, 1993; Bellairs, 1971, in Weldon and Ferguson, 1993). Underwater food detection is presumably gustatory (below) and tactile (Davenport et al., 1990).

Aerial olfaction is well developed. Crocodylians locate carrion from great distances and are able to distinguish food sources without visual cues (Neill, 1971; Scott and Weldon, 1990; Weldon et al., 1990). They are also endowed with paired gular and paracloacal integumentary glands (Weldon and Sampson, 1988; Weldon and Wheeler, 2001) and, in some cases, dorsal glands (Richardson and Park, 2000). These glands are believed to function in pheromonal attraction of mates and/or nest-site marking (Weldon and Ferguson, 1993). The secretions are chemically complex, contain large volatile fractions, and exhibit taxonomic, intergland, sex, and age class differences (Whyte et al., 1999; Weldon and Wheeler, 2001; García-Rubio et al., 2002).

SQUAMATES

In general it is thought that the vomeronasal system is the predominant chemosensory system in squamates (Halpern, 1992; Ford and Burghardt, 1993; Schwenk, 1995). There is extensive literature on the squamate vomeronasal system, and this system is implicated in almost every important behavior. Only its relevance in aquatic behavior is discussed here. The role of the main olfactory system is poorly studied, but it is well developed in some species (Gabe and Saint Girons, 1976; Schwenk, 1993; Dial and Schwenk, 1996). However, highly aquatic snakes show marked regression of the main olfactory system without concomitant reduction of the vomeronasal system (Matthes, 1934; Gabe and Saint Girons, 1976). Furthermore, aquatic snakes exhibit putative adaptations to prevent flooding of the nose during submersion (see above). In contrast, aquatic snakes and lizards foraging underwater use typical tongue-flicking behavior, implicating vomeronasal system functioning (Rand, 1964; Heatwole, 1975; Hibbard, 1975; Heatwole and Cogger, 1993; Shine and Houston, 1993; Shine et al., 2003, 2004; Vincent et al., 2005). Together, these observations suggest that olfaction is unimportant to aquatic snakes and that the vomeronasal system alone functions during submersion in squamates. In seasnakes and filesnakes (Acrochordidae), underwater tongue-flicking occurs during exploratory behavior (Heatwole, 1975), foraging (Shine et al., 2003, 2004; Vincent et al., 2005) and reproductive behavior (R. Shine, pers. comm., June 2004). These behaviors are all typical of vomeronasal system function in terrestrial squamates.

COMPARATIVE ANATOMY OF THE ORAL CAVITY: GUSTATORY SYSTEM

Taste buds are small, flask-shaped organs comprising clusters of sensory and support cells embedded within an epithelium (Fig. 5.1). They are plesiomorphic for vertebrates, and therefore their wide occurrence among reptiles is unsurprising. Many fishes have taste buds scattered across their body surface, often in high densities (e.g., Bardach and Atema, 1971); however, with the exception of some larval amphibians, taste buds are restricted to the oral and pharyngeal mucosa in tetrapods, including fully aquatic species. Also unlike fish (and some amphibians), the sensory processes of the receptor neurons are rarely exposed on the epithelial surface. Rather, they are sunken to varying degrees, accessible through a mucus-filled pore or pit. Taste receptors are regarded as sensors of proximate chemical stimuli, typically food molecules that have been released by mechanical processing within the mouth (e.g., Atema, 1987).

TURTLES

Taste buds are usually found on turtle tongues (Tuckerman, 1892; Winokur, 1973; Pevzner and Tikhonova, 1979; Korte, 1980; Uchida, 1980; Spindel et al., 1987; Iwasaki et al., 1996b; Beisser et al., 1998, 2001), but not always (Iwasaki, 1992; Iwasaki et al., 1992, 1996a, 1996c, 1996d; Beisser et al., 1995, 2004; Lemell et al., 2000). It remains possible that taste buds are present in other parts of the mouth and pharynx. Too few taxa have been examined to reveal any phylogenetic or ecological patterns in taste bud distribution among turtles. The only two sea turtles examined (hawksbill and Pacific Ridley Turtle; Cheloniidae) appear to lack tastebuds (Iwasaki et al., 1996a, 1996c), whereas in freshwater (and terrestrial) species they are variably present, even within a single family such as the semi- to highly aquatic Emydidae. Although present in modest numbers in some species (e.g., *Chrysemys scripta*) (Korte, 1980), they never achieve the densities found in many squamate reptiles (Schwenk, 1985).

Virtually no study has attempted to distinguish the role of gustation in turtles independent of nasal olfaction. One study showed that taste, alone, was insufficient for chemosensory-based operant conditioning in green sea turtles (Manton et al., 1972b). Studies of food imprinting, however, suggest a role for gustation in turtle behavior (Burghardt and Hess, 1966; Burghardt, 1970). In any case, given the relatively large proportion of the peripheral and central nervous system devoted to the nasal chemical senses (Scott, 1979), it is likely that these are the predominant chemical senses in turtles, including aquatic species (see below).

CROCODYLIANS

Crocodylians possess modest numbers of typical taste buds on the mucosal surfaces of the tongue, mouth, palate, and pharynx (Bath, 1906; Ferguson, 1981; Shimada et al., 1990; Yoshie and Yokosuka, 2001). Small, dermal sense organs in the scales surrounding the mouth were considered possible chemoreceptors (Neill, 1971; Jackson et al., 1996), but recently they were shown to be exquisitely sensitive mechanoreceptors responsive to ripples at the water's surface (Soares, 2002). As with turtles, no studies have addressed specifically the role of gustation in crocodylian behavior, but circumstantial evidence suggests that taste may be important in underwater prey detection. American alligators exhibit increased rates of lateral head movements, mouth-opening, and snapping when food extracts are introduced into the water (Neill, 1971; Weldon et al., 1990; Banta et al., 1992). Since crocodylians do not employ olfaction underwater (see below), gustation is implicated in these behaviors. Limited evidence also suggests that taste is used to identify food and to reject unpalatable prey following contact with the mouth (Scott and Weldon, 1990; Weldon and McNease, 1991).

SQUAMATES

Taste buds are abundant in the mouth and pharynx of most lepidosaurs, reaching extremely high densities on the tongue tips of some lizards (Schwenk, 1985, 1986). Snakes lack lingual taste buds, but many, if not most, species have them in the palate (Burns, 1969; Kroll, 1973; Nishida et al., 2000; Berkhoudt et al., 2001; Atobe et al., 2004). In caenophidian snakes, including terrestrial and aquatic species of colubrids, elapids, and viperids, the taste buds are usually elevated on small projections or papillae that also contain a variety of proprioreceptive mechanoreceptors, forming a complex, compound sensory system putatively associated with the identification and manipulation of prey within the mouth (Nishida et al., 2000). As for other reptiles and tetrapods generally, gustation is implicated in the discrimination of palatable from unpalatable food once mouth or tongue contact occurs (Cooper and Pérez-Mellado, 2001; Stanger-Hall et al., 2001; Cooper et al., 2002a, 2002b); lizards and snakes are often observed to reject food items once they are held in the mouth (Burghardt, 1969; personal observations). Taste may underlie learned aversion to food following sickness (Burghardt et al., 1973; Boyden, 1976; Stanger-Hall et al., 2001; Paradis and Cabanac, 2004), although olfaction cannot be ruled out. Seasnakes are similar to other snakes in the presence of palatal taste buds (Burns, 1969; Hibbard, 1975). Drummond (1979, 1983) observed aquatic, natricine snakes (included in Colubridae) making open-mouthed searching movements in water containing prey extracts, possibly indicating gustatory behavior. Nevertheless, there is no indication that aquatic squamates, or aquatic reptiles generally, evince specializations of the gustatory system.

CHEMICAL SENSES IN FOSSIL AQUATIC REPTILES

MOSASAURS

Using phylogenetic bracketing and fossil evidence, Schulp et al. (2005) recently reconstructed tongue form in mosasaurs, ancient marine lizards related to living varanoids. They suggested that the tongue was probably deeply cleft, as in living Gila monsters and *Lanthanotus*, but not as deeply forked as in snakes and varanid lizards. Mosasaurs spent little, if any, time on land. It is likely that the highly protrusible tongue was used for tongue-flicking and vomeronasal chemoreception underwater (see below).

PHYTOSAURS

Phytosaur skulls exhibit impressions that have been attributed to an accessory olfactory bulb (putatively indicating a vomeronasal system), but Senter (2002) showed that the impressions resemble those left by the ophthalmic nerve in extant crocodylians. This and phylogenetic bracketing suggest that phytosaurs, like other archosaurs, lacked a vomeronasal system. Senter (2002) erred, however, in concluding that phytosaurs could not have used chemically mediated behavior or pheromonal communication. In living crocodylians, chemosensory behavior is mediated by olfaction (Weldon and Ferguson, 1993; see above), and in mammals it is well established that pheromonal communication can occur via the main olfactory system (reviewed in Baxi et al., 2006). The vomeronasal system is neither necessary for chemically mediated behavior, generally, nor pheromone reception, specifically.

PLESIOSAURS

It has been suggested that plesiosaurs used underwater olfaction (Cruickshank et al., 1991). In contrast to the usual tetrapod pattern, the internal nares are anterior to the external nares (Fig. 5.4). Grooves run in the palate from small maxillary diastemata to anteriorly directed, scooplike choanae. Aided by negative pressure generated by flow across the dorsal nostrils, water is presumed to have moved through the mouth, into the nasal cavities, and out through the external nares. Such a mechanism would permit continuous olfactory (or much more likely, vomeronasal) sampling of the water. The only modern analogue for such a system is found in some fishes that use ram-generated, continuous flow across the olfactory epithelium by means of incurrent and excurrent nares (Hara, 1997). However, in fishes there is no connection between olfactory and buccal cavities.

OTHER AQUATIC MESOZOIC REPTILES

The chemical senses of other highly aquatic, Mesozoic reptile groups remain unknown.

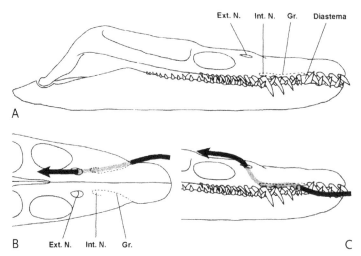

FIGURE 5.4. Reconstruction of the skull and narial anatomy in *Rhomaleosaurus*, a plesiosaur, in lateral (A, C) and doral (B) view. In contrast to other tetrapods, the internal nares (Int. N.) are anterior to the external nares (Ext. N.). Arrows show the putative path of water flow through the olfactory cavities as the animal swam. Water entered the mouth through a diastema between the teeth and was channeled along a groove (Gr.) in the palate (indicated with dotted lines) to scooplike internal nares. Water then passed through the nasal cavities, exiting posterodorsally through the external nares. Submersion and flushing of the nasal cavities and nasal epithelium presumably permitted underwater olfaction. Modified from Cruickshank et al. (1991).

However, the nostrils of many nothosaurs, placodonts, and ichthyosaurs are positioned posteriorly on the snout in front of the orbits, as seen in some other aquatic reptiles (see above; Carroll, 1988; McGowan, 1991).

FUNCTIONAL AND EVOLUTIONARY PATTERNS

Based on first principles, the chemosensory system most likely to adapt to secondarily aquatic habits is the vomeronasal system. The vomeronasal system has long been considered a "water sense" (the *Wassergeruchsorgan*, or water olfactory organ, of Broman [1920]) because its epithelium is virtually always flooded and because it is often responsive to high molecular weight, nonvolatile chemicals that do not stimulate the main olfactory system (Baxi et al., 2006). Indeed, association of the vomeronasal organ with fluidborne rather than airborne chemicals is so strong that homology of the vomeronasal organ to the main olfactory organ of fishes has been proposed, implying

that the tetrapod main olfactory system is neomorphic (Broman, 1920; Parsons, 1970b). Regardless, I suggest that the vomeronasal system of terrestrial reptiles is preadapted to aquatic conditions for three reasons: (1) mechanistically, it requires odorants to be in aqueous solution; (2) the set of chemicals to which it is responsive includes nonvolatile molecules; (3) the primitive mechanism for aerial acquisition of vomeronasal stimulants is compatible with aquatic acquisition. Next I evaluate the support for this hypothesis from the anatomy and function of specific aquatic reptile groups.

The strongest support comes from highly aquatic freshwater and marine snakes, which exhibit reduction of the main olfactory system while maintaining high levels of putative vomeronasal activity underwater (see above). Schwenk (1996) proposed that odorant molecules are collected on the tongue by diffusion into the fluid covering its surfaces. According to this model, tongue-flicking serves to decrease the thickness of the boundary layer at the tongue's surface, to increase the steepness of the velocity gradient between the tongue's surface and the surrounding medium, and to increase turbulence around the tongue in order to draw more odorant molecules into contact with its surface, all of which combine to increase the rate of diffusion. At first glance, it seems unlikely that this model could work for collecting odorant molecules underwater; however, all that is required is that a boundary is maintained between the fluid on the tongue and the surrounding water during a tongue-flick. Such an interface occurs, for example, between the mucus covering a fish's body scales and the surrounding water when it swims, and the mucus covering taste buds or olfactory epithelium in fishes.

Schwenk (1994) presented evidence supporting the notion that the squamate forked tongue morphology functions as a tropotactic mechanism used to detect chemical gradients. The length of the fork reflects the scale of the gradients sensed. In the case of terrestrial snakes, this is usually a relatively discrete chemical trail. In aquatic environments; however, odor plumes tend to mix and disperse rapidly owing to turbulence (Atema, 1987). This implies that, typically, the scale of biologically relevant chemical gradients will be larger in aquatic environments than in terrestrial. It is my impression, based on very few observations, that the tines of the forked tongue in aquatic snakes are, in fact, relatively longer than in terrestrial species. This hypothesis needs to be tested quantitatively and with phylogenetic controls. If true, it would represent a distinct vomeronasal specialization of aquatic snakes.

Simultaneous consideration of crocodylians and turtles provides additional support for the vomeronasal system preadaptation hypothesis. Both taxa are mostly aquatic and were probably ancestrally so. Both taxa primitively evolved a high-impedance, positive-displacement pump (buccal or gular oscillations) to pulse air through the nasal cavities for olfaction. Only turtles, however, continue the pumping behavior underwater. It is likely that this is because crocodylians lack a vomeronasal epithelium

within the nose. The main olfactory system, adapted as it is to aerial olfaction and the sensation of land-based, volatile chemicals, is nonfunctional when submerged. In contrast, turtles have a well-developed vomeronasal epithelium on the floor of the cavum that even in terrestrial species, is constantly flooded by external glands (Tucker, 1971). We know that the turtle vomeronasal epithelium is responsive to chemicals in aqueous solutions, including nonvolatiles (Tucker, 1971; Hatanaka and Matsuzaki, 1993; Fadool et al., 2001), and it is likely that the olfactory epithelium is protected from deluge by an air bubble (see above). Thus, putative underwater olfaction in turtles is almost certainly vomeronasal. It is conceivable that as water is pulsed through the nasal cavities, dissolved volatile chemicals escape into the air bubble surrounding the olfactory epithelium, thus providing olfactory access to waterborne chemicals, but this is purely speculative.

Speculating further, the nasal cavity of highly aquatic turtles, particular sea turtles, appears to be relatively longer, straighter, and more tubular than in other species, but this is based on the limited number of taxa figured by Parsons (1970a) and elsewhere (Fig. 5.2). Such a conformation might promote laminar flow through the nasal cavities, which would tend to reduce impedance and increase efficiency of the buccal pump. However, laminar flow creates boundary layers that limit exposure of the receptor cells to odorant molecules. Local, turbulent mixing generated by small protrusions in the mucosal surface, such as the sulcal ridges characteristic of the turtle vomeronasal region, could mitigate this problem. Receptor exposure might also be aided by the fact that fluid velocity within a pipe is slowest at the surfaces (Vogel, 1994). Finally, I note the unusual disposition of the vomeronasal epithelium in sea turtles, which is sequestered within dorsal and ventral diverticulae. The orientation of their openings suggests that the ventral diverticulum would be flooded during incurrent flow and the dorsal during excurrent flow (Fig. 5.2). Each diverticulum might then be emptied during the opposite cycle by negative pressures generated as water moves past its aperture. The utility of a system that sequentially floods and empties the vomeronasal chambers is obvious, but the advantage of separately sampling incurrent and excurrent streams (if this really happens) is not clear. The hypothesis could be tested with physical models.

Most aquatic reptiles are amphibious, spending time both in water and on land. Adaptive commitment of any sensory system solely to water would come with negative fitness consequences. It is therefore not surprising that few aquatic reptiles evince obvious chemosensory adaptations to their aquatic environments. Rather, highly aquatic species show refinements of the existing vomeronasal system, which was preadapted to aquatic chemoreception. Loss of the vomeronasal system in ancestral archosaurs, for whatever reasons, precluded the use of underwater nasal chemoreception in crocodylians, despite their possession of an appropriate pumping mechanism.

In addition to exploring some of the speculations presented here, future work considering the nature of the chemicals sampled by the main olfactory system and vomeronasal system is worthwhile (see Baxi et al., 2006; Eisthen and Schwenk, chapter 3 in this volume). The sets of chemicals to which each system responds are overlapping, but noncongruent in all tetrapods examined. One axis along which potential odorants vary is molecular weight. Although there is no crisp distinction between the nasal chemical senses in this regard, in general the vomeronasal system is more likely to respond to high molecular weight chemicals than the main olfactory system (but see cited references above). However, larger molecules are both less volatile and less soluble than smaller ones. In contrast, molecular polarity represents a clear trade-off in the properties of solubility and volatility: polar molecules are more soluble in water than nonpolar molecules, but they have lower vapor pressures and are therefore less volatile. The polarity axis could be another factor affecting the classes of chemicals to which each system responds and their ability to adapt to aquatic conditions (see Eisthen and Schwenk, chapter 3 in this volume, for discussion).

ACKNOWLEDGMENTS

I thank the editors, Hans Thewissen and Sirpa Nummela, for inviting me to contribute to this volume. I am grateful to Heather Eisthen and the editors for many helpful suggestions on an earlier draft, to Heather Eisthen for providing an advance copy of the article by Baxi et al. (2006), and to John Reiss and Heather Eisthen for providing an advance copy of their chapter. Work in my lab was supported by NSF IBN-9601173 and the University of Connecticut Research Foundation.

LITERATURE CITED

Atema, J. 1987. Aquatic and terrestrial chemoreceptor organs: morphological and physiological designs for interfacing with chemical stimuli; pp. 303–316 in P. Dejours, L. Bolis, C. R. Taylor, and E. R. Weibel (eds.), Comparative Physiology: Life in Water and on Land. Fidia Research Series. IX-Liviana Press, Padova.

Atobe, Y., M. Nakano, T. Kadota, T. Hisajima, R. C. Goris, and K. Funakoshi. 2004. Medullary efferent and afferent neurons of the facial nerve of the pit viper *Gloydius brevicaudus*. Journal of Comparative Neurology. 472:345–357.

Banta, M. R., T. Joanen and P. J. Weldon. 1992. Foraging responses by the American alligator to meat extracts; pp. 413–317 in R. L. Doty and D. Müller-Schwartze (eds.), Chemical Signals in Vertebrates, VI. Plenum Press, New York.

Bardach, J. E., and J. Atema. 1971. The sense of taste in fishes; pp. 293–336 in L. M. Beidler (ed.), Handbook of Sensory Physiology, Vol. IV, Chemical Senses, Part 2, Taste. Springer-Verlag, Berlin.

Bartol, S. M., and J. A. Musick, 2003. Sensory biology of sea turtles; pp. 79–102 in P. L. Lutz, J. A. Musick, and J. Wyneken (eds.), The Biology of Sea Turtles, Vol. II. CRC Press, Boca Raton, FL.

Bath, W. 1906. Die Geschmacksorgane der Vögel und Krokodile. Archiv für Biontologie 1:1–47.

Baxi, K. N., K. M. Dorries, and H. L. Eisthen. 2006. Is the vomeronasal system really specialized for detecting pheromones? Trends in Neurosciences 29:2–7.

Beisser, C. J., J. Weisgram, and H. Splechtna. 1995. Dorsal lingual epithelium of *Platemys pallidipectoris*

(Pleurodira, Chelidae). Journal of Morphology 226:267–276.

Beisser, C. J., J. Weisgram, H. Hilgers, and H. Splechtna. 1998. Fine structure of the dorsal lingual epithelium of *Trachemys scripta elegans* (Chelonia: Emydidae). Anatomical Record 250:127–135.

Beisser, C. J., P. Lemell, and J. Weisgram. 2001. Light and transmission electron microscopy of the dorsal lingual epithelium of *Pelusios castaneus* (Pleurodira, Chelidae) with special respect to its feeding mechanics. Tissue and Cell 33:63–71.

Beisser, C. J., P. Lemell, and J. Weisgram. 2004. The dorsal lingual epithelium of *Rhinoclemmys pulcherrima incisa* (Chelonia, Cryptodira). Anatomical Record A 277:227–235.

Bellairs, A. 1970. The Life of Reptiles, Vols. I and II. Universe Books, New York.

Bellairs, A. d'A., and J. D. Boyd. 1950. The lachrymal apparatus in lizards and snakes. II. The anterior part of the lachrymal duct and its relationship with the palate and with the nasal and vomeronasal organs. Proceedings of the Zoological Society of London 120:269–310.

Bellairs, A. d'A., and C. C. D. Shute. 1953. Observations on the narial musculature of Crocodilia and its innervation from the sympathetic system. Journal of Anatomy 87:367–378.

Berkhoudt, H., P. Wilson, and B. Young. 2001. Taste buds in the palatal mucosa of snakes. African Zoology 36:185–188.

Boyden, T. C. 1976. Butterfly palatability and mimicry: experiments with *Ameiva* lizards. Evolution 30:73–81.

Broman, I. 1920. Das Organon vomero-nasale Jacobsoni: ein Wassergeruchsorgan! Anatomische Hefte 58:143–191.

Burghardt, G. M. 1969. Comparative prey-attack studies in newborn snakes of the genus *Thamnophis*. Behaviour 33:77–114.

Burghardt, G. M. 1970. Chemical perception in reptiles; pp. 241–308 in J. W. Johnston, Jr., D. G. Moulton, and A. Turk (eds.), Chemical Communication by Chemical Signals. Appleton-Century-Crofts, New York.

Burghardt, G. M., and E. H. Hess. 1966. Food imprinting in the snapping turtle, *Chelydra serpentina*. Science 151:108–109.

Burghardt, G. M., H. C. Wilcoxon, and J. A. Czaplicki. 1973. Conditioning in garter snakes: aversion to palatable prey induced by delayed illness. Animal Learning and Behavior 1:317–320.

Burns, B. 1969. Oral sensory papillae in sea snakes. Copeia 1969:617–619.

Carroll, R. L. 1988. Vertebrate Paleontology and Evolution. W. H. Freeman, New York.

Chelazzi, G., and G. Delfino. 1986. A field test on the use of olfaction in homing by *Testudo hermanni* (Reptilia: Testudinidae). Journal of Herpetology 20:451–455.

Constantino, M. A., and M. Salmon. 2003. Role of chemical and visual cues in food recognition by leatherback posthatchlings (*Dermochelys coriacea* L). Zoology 106:173–181.

Cooper, W. E., Jr., and V. Pérez-Mellado. 2001. Chemosensory responses to sugar and fat by the omnivorous lizard *Gallotia caesaris* with behavioral evidence suggesting a role for gustation. Physiology Behavior 73:509–516.

Cooper, W. E., Jr., V. Pérez-Mellado, and L. J. Vitt. 2002a. Responses to major categories of food chemicals by the lizard *Podarcis lilfordi*. Journal of Chemical Ecology 28:709–720.

Cooper, W. E., Jr., V. Pérez-Mellado, L. J. Vitt, and B. Budzinsky. 2002b. Behavioral responses to plant toxins by two omnivorous lizard species. Physiology and Behavior 76:297–303.

Cruickshank, A. R. I., P. G. Smal, and M. A. Taylor. 1991. Dorsal nostrils and hydrodynamically driven underwater olfaction in plesiosaurs. Nature 352:62–64.

Davenport, J., D. J. Grove, J. Cannon, T. R. Ellis, and R. Stables. 1990. Food capture, appetite, digestion rate and efficiency in hatchling and juvenile *Crocodylus porosus*. Journal of Zoology (London) 220:569–592.

Dial, B. E., and K. Schwenk. 1996. Olfaction and predator detection in *Coleonyx brevis* (Squamata: Eublepharidae) with comments on the functional significance of buccal pulsing in geckos. Journal of Experimental Zoology 276:415–424.

Drummond, H. 1979. Stimulus control of amphibious predation in the northern water snake (*Nerodia s. sipedon*). Zeitschrift für Tierpsychologie 50:18–44.

Drummond, H. 1983. Aquatic foraging in garter snakes: a comparison of specialists and generalists. Behaviour 86:1–30.

Druzisky, K. A., and E. L. Brainerd. 2001. Buccal oscillation and lung ventilation in a semi-aquatic turtle, *Platysternon megacephalum*. Zoology 104:143–152.

Ehrenfeld, J. G., and D. W. Ehrenfeld. 1973. Externally secreting glands of freshwater and sea turtles. Copeia 1973:305–314.

Fadool, D. A., M. Wachiowiak, and J. H. Brann. 2001. Patch-clamp analysis of voltage-activated and chemically activated currents in the vomeronasal organ of *Sternotherus odoratus* (stinkpot/musk turtle). Journal of Experimental Biology 204:4199–4212.

Ferguson, M. W. J. 1981. The structure and development of the palate in *Alligator mississippiensis*. Archives of Oral Biology 26:427–443.

Ford, N. B., and G. M. Burghardt. 1993. Perceptual mechanisms and the behavioral ecology of snakes; pp. 117–164 in R. A. Seigel and J. T. Collins (eds.), Snakes: Ecology and Behavior. McGraw-Hill, New York.

Gabe, M., and H. Saint Girons. 1976. Contribution a la morphologie comparée des fosses nasales et de leurs annexes chez les lépidosoriens. Mémoires du Muséum National d'Histoire Naturelle (Paris) A 98:1–87.

Gans, C., and B. Clark, 1976. Studies on ventilation of *Caiman crocodilus* (Crocodilia: Reptilia). Respiration Physiology 26:285–301.

García-Rubio, S., A. B. Attygalle, P. J. Weldon, and J. Meinwald. 2002. Reptilian chemistry: volatile compounds from paracloacal glands of the American crocodile *(Crocodylus acutus)*. Journal of Chemical Ecology 28:769–781.

Grassman, M. 1993. Chemosensory orientation behavior in juvenile sea turtles. Brain, Behavior, and Evolution 41:224–228.

Greene, H. W. 1997. Snakes. The Evolution of Mystery in Nature. University of California Press, Berkeley.

Halpern, M. 1992. Nasal chemical senses in reptiles: structure and function; pp. 423–523 in C. Gans and D. Crews (eds.), Biology of the Reptilia, Vol. 18. University of Chicago Press, Chicago.

Hara, T. J. 1997. Chemoreception; pp. 191–218 in D. H. Evans (ed.), The Physiology of Fishes. CRC Press, Boca Raton, FL.

Hatanaka, T., and O. Matsuzaki. 1993. Odor responses of the vomeronasal system in Reeve's turtle, *Geoclemys reevesii*. Brain, Behavior, and Evolution 41:183–186.

Heatwole, H. 1975. Attacks by sea snakes on divers; pp. 503–516 in W. A. Dunson (ed.), The Biology of Sea Snakes. University Park Press, Baltimore.

Heatwole, H., and H. G. Cogger, 1993. Family Hydrophiidae; pp. 310–318 in C. J. Glasby G. J. B. Ross, and P. L. Beesley (eds.), Fauna of Australia, Vol. 2A, Amphibia and Reptilia. Australian Government Publishing Service, Canberra.

Hibbard, E. 1975. Eyes and other sense organs of sea snakes; pp. 355–382 in W. A. Dunson (ed.), The Biology of Sea Snakes. University Park Press, Baltimore.

Iwasaki, S.-I. 1992. Fine structure of the dorsal epithelium of the tongue of the freshwater turtle, *Geoclemyus reevesii* (Chelonia, Emydinae). Journal of Morphology 211:125–135.

Iwasaki, S.-I., T. Asami, Y. Asami, and K. Kobayashi. 1992. Fine structure of the dorsal epithelium of the tongue of the Japanese terrapin, *Clemmys japonica* (Cheloia [sic], Emydinae). Archives of Histology and Cytology 55:295–305.

Iwasaki, S.-I., T. Asami, and C. Wanichanon. 1996a. Fine structure of the dorsal lingual epithelium of the juvenile hawksbill turtle, *Eretmochelys imbricata bissa*. Anatomical Record 244:437–443.

Iwasaki, S.-I., T. Asami, and C. Wanichanon. 1996b. Ultrastructural study of the dorsal lingual epithelium of the soft-shell turtle, *Trionyx cartilagineus* (Chelonia, Trionychidae). Anatomical Record 246:305–316.

Iwasaki, S.-I., C. Wanichanon, and T. Asami. 1996c. Histological and ultrastructural study of the lingual epithelium of the juvenile Pacific ridley turtle, *Lepidochelys olivacea* (Chelonia, Cheloniidae). Annals of Anatomy 178:243–250.

Iwasaki, S.-I., C. Wanichanon, and T. Asami. 1996d. Ultrastructural study of the dorsal lingual epithelium of the Asian snail-eating turtle, *Malayemys subtrijuga*. Annals of Anatomy 178:145–152.

Jackson, K., D. G. Butler, and J. H. Youson. 1996. Morphology and ultrastructure of possible integumentary sense organs in the estuarine crocodile *(Crocodylus porosus)*. Journal of Morphology 229:315–324.

Kathariner, L. 1900. Die Nase der im Wasser lebenden Schlangen als Luftweg und Geruchsorgan. Zoologisches Jahrbuch Abteilung Systematik 13:415–442.

Koch, A. L., A. Carr, and D. W. Ehrenfeld. 1969. The problem of open-sea navigation: the migration of the green turtle *Chelonia mydas* to Ascension Island. Journal of Theoretical Biology 22:163–179.

Korte, G. E. 1980. Ultrastructure of the tastebuds of the red-eared turtle, *Chrysemys scripta elegans*. Journal of Morphology 163:231–252.

Kroll, J. C. 1973. Taste buds in the oral epithelium of the blind snake, *Leptotyphlops dulcis* (Reptilia: Leptotyphlopidae). Southwestern Naturalist 17:365–370.

Lemell, P., C. J. Beisser, and J. Weisgram. 2000. Morphology and function of the feeding apparatus of *Pelusios castaneus* (Chelonia; Pleurodira). Journal of Morphology 244:127–135.

Luschi, P., G. C. Hays, C. Del Seppia, R. Marsh, and F. Papi. 1998. The navigational feats of green sea turtles migrating from Ascension Island investigated by satellite telemetry. Proceedings of the Royal Society of London B 265:2279–2284.

Manteifel, Y., N. Goncharova, and V. Boyko. 1992. Chemotesting movements and chemosensory sensitivity to amino acids in the European pond turtle, *Emys orbicularis* L; pp. 397–401 in R. L. Doty and D. Müller-Schwartze (eds.), Chemical

Signals in Vertebrates, VI. Plenum Press, New York.

Manton, M. L. 1979. Olfaction and behavior; pp. 289–301 in M. Harless and H. Morlock (eds.), Turtles: Perspectives and Research. J. Wiley and Sons, New York.

Manton, M. L., A. Karr, and D. W. Ehrenfeld. 1972a. An operant method for the study of chemoreception in the green turtle, *Chelonia mydas*. Brain, Behavior, and Evolution 5:188–201.

Manton, M. L., A. Karr, and D. W. Ehrenfeld. 1972b. Chemoreception in the migratory sea turtle, *Chelonia mydas*. Biological Bulletin 143:184–195.

Matthes, E. 1934. Geruchsorgan; pp. 879–948 in L. Bolk, E. Göppert, E. Kallius, and W. Lubosch (eds.), Handbuch der vergleichenden Anatomie der Wirbeltiere. Urban and Schwarzenberg, Berlin [reprinted 1967, A. Asher and Co., Amsterdam].

McCutcheon, F. H. 1943. The respiratory mechanism in turtles. Physiological Zoology 16:255–269.

McDowell, S. B., Jr., and C. M. Bogert. 1954. The systematic position of *Lanthanotus* and the affinities of the anguinomorphan lizards. Bulletin of the American Museum of Natural History 105:1–142.

McGowan, C. 1991. Dinosaurs, Spitfires and Sea Dragons. Harvard University Press, Cambridge, MA.

Muñoz, A. 2004. Chemo-orientation using conspecific chemical cues in the stripe-necked terrapin *(Mauremys leprosa)*. Journal of Chemical Ecology 30:519–530.

Naifeh, K. H., S. E. Huggins, H. E. Hoff, T. W. Hugg, and R. E. Norton. 1970. Respiratory patterns in crocodilian reptiles. Respiratory Physiology 9:31–42.

Negus, V. 1958. The Comparative Anatomy and Physiology of the Nose and Paranasal Sinuses. E. and S. Livingstone, Edinburgh.

Neill, W. T. 1971. The Last of the Ruling Reptiles: Alligators, Crocodiles, and Their Kin. Columbia University Press, New York.

Nishida, Y., S. Yoshie, and T. Fujita. 2000. Oral sensory papillae, chemo- and mechano-receptors, in the snake, *Elaphe quadrivirgata*. A light and electron microscopic study. Archives of Histology and Cytology 63:55–70.

Oelrich, T. M. 1956. The anatomy of the head of *Ctenosaura pectinata* (Iguanidae). Miscellaneous Publications of the Museum of Zoology, University of Michigan 94:1–167.

Paradis, S., and M. Cabanac. 2004. Flavor aversion learning induced by lithium chloride in reptiles but not amphibians. Behavioral Processes 67:11–18.

Parsons, T. S. 1959a. Nasal anatomy and the phylogeny of reptiles. Evolution 13:175–187.

Parsons, T. S. 1959b. Studies on the comparative embryology of the reptilian nose. Bulletin of the Museum of Comparative Zoology (Harvard University) 120:101–277.

Parsons, T. S. 1967. Evolution of the nasal structure in the lower tetrapods. American Zoologist 7:397–413.

Parsons, T. S. 1970a. The nose and Jacobson's organ; pp. 99–191 in C. Gans and T. S. Parsons (eds.), Biology of the Reptilia, Vol. 2. Academic Press, New York.

Parsons, T. S. 1970b. The origin of Jacobson's organ. Forma et Functio 3:105–111.

Pevzner, R. A., and N. A. Tikhonova. 1979. [Fine structure of the taste buds of Reptilia. I. Chelonia]. Tsitologiya 21:132–138 [Russian, English Summary; English translation of the Discussion by E. A. Unumb, provided by P. J. Weldon].

Quinn, V. S., and B. M. Graves. 1998. Home pond discrimination using chemical cues in *Chrysemys picta*. Journal of Herpetology 32:457–461.

Rand, A. S. 1964. An observation on *Dracaena guianensis* foraging underwater. Herpetologica 20:207.

Rehorek, S. J., W. J. Hillenius, W. Quan, and M. Halpern. 2000. Passage of Harderian gland secretions to the vomeronasal organ of *Thamnophis sirtalis* (Serpentes: Colubridae). Canadian Journal of Zoology 78:1284–1288.

Richardson, K. C., and J. Y. Park. 2000. The histology of the dorsal integumentary glands in embryonic and young estuarine crocodiles *Crocodylus porosus* and Australian freshwater crocodiles *Crocodylus johnstoni*; pp. 180–187 in G. C. Grigg, F. Seebacher, and C. E. Franklin (eds.), Crocodilian Biology and Evolution. Surrey Beatty and Sons, Chipping Norton, NSW, Australia.

Root, R. W. 1949. Aquatic respiration in the musk turtle. Physiological Zoology 22:172–178.

Rostal, D. C., J. A. Williams, and P. J. Weldon. 1991. Rathke's gland secretion by loggerhead *(Caretta caretta)* and Kemp's ridley *(Lepidochelys kempi)* sea turtles. Copeia 1991:1129–1132.

Saint Girons, H. 1976. Données histologiques sur les fosses nasales et leurs annexes chez *Crocodylus niloticus* Laurenti et *Caiman crocodilus* (Linnaeus) (Reptilia, Crocodylidae). Zoomorphologie 84:301–318.

Santos-Costa, M. dos, and C. Hofstadler-Deiques. 2002. The ethmoidal region and cranial adaptations of the neotropical aquatic snake *Helicops infrataeniatus* Jan, 1865 (Serpentes, Colubridae). Amphibia-Reptilia 23:83–91.

Schulp, A. S., E. W. A. Mulder, and K. Schwenk. 2005. Did mosasaurs have forked tongues? Netherlands Journal of Geosciences 84:359–371.

Schwenk, K. 1985. Occurrence, distribution and functional significance of taste buds in lizards. Copeia 1985:91–101.

Schwenk, K. 1986. Morphology of the tongue in the tuatara, *Sphenodon punctatus* (Reptilia: Lepidosauria), with comments on function and phylogeny. Journal of Morphology 188:129–156.

Schwenk, K. 1993. Are geckos olfactory specialists? Journal of Zoology (London) 229:289–302.

Schwenk, K. 1994. Why snakes have forked tongues. Science 263:1573–1577.

Schwenk, K. 1995. Of tongues and noses: chemoreception in lizards and snakes. Trends in Ecology and Evolution 10:7–12.

Schwenk, K. 1996. Why snakes flick their tongues. American Zoologist 36:84A.

Scott, T. R., Jr. 1979. The chemical senses; pp. 267–287 in M. Harless and H. Morlock (eds.), Turtles: Perspectives and Research. J. Wiley and Sons, New York.

Scott, T. P., and P. J. Weldon. 1990. Chemoreception in the feeding behaviour of adult American alligators, *Alligator mississippiensis*. Animal Behaviour 39:398–405.

Senter, P. 2002. Lack of a pheromonal sense in phytosaurs and other archosaurs, and its implications for reproductive communication. Paleobiology 28:544–550.

Shimada, K., I. Sato, A. Yokoi, T. Kitagawa, M. Tezuka, and T. Ishi. 1990. The fine structure and elemental analysis of keratinized epithelium of the filiform papillae analysis on the dorsal tongue in the American alligator *(Alligator mississippiensis)*. Okajimas Folia Anatomica Japan 66:375–392.

Shine, R., and D. Houston. 1993. Family Acrochordidae; pp. 322–324 in C. J. Glasby, G. J. B. Ross, and P. L. Beesley (eds.), Fauna of Australia, Vol. 2A, Amphibia and Reptilia. Australian Government Publishing Service, Canberra.

Shine, R., T. Shine, and B. Shine. 2003. Intraspecific habitat partitioning by the sea snake *Emydocephalus annulatus* (Serpentes, Hydrophiidae): the effects of sex, body size, and colour pattern. Biological Journal of the Linnean Society 80:1–10.

Shine, R., X. Bonnet, M. J. Elphick, and E. G. Barrott. 2004. A novel foraging mode in snakes: browsing by the sea snake *Emydocephalus annulatus* (Serpentes, Hydrophiidae). Functional Ecology 18:16–24.

Soares, D. 2002. An ancient sensory organ in crocodilians. Nature 417:241–242.

Spindel, E. L., J. L. Dobie, and D. F. Buxton. 1987. Functional mechanisms and histologic composition of the lingual appendage in the alligator snapping turtle, *Macroclemys temmincki* (Troost) (Testudines: Chelydridae). Journal of Morphology 194:287–301.

Stanger-Hall, K. F., D. A. Zelmer, C. Bergren, and S. A. Burns. 2001. Taste discrimination in a lizard (*Anolis carolinensis*, polychrotidae). Copeia 2001:490–498.

Stebbins, R. C. 1948. Nasal structure in lizards with reference to olfaction and conditioning of the inspired air. American Journal of Anatomy 83:183–221.

Tucker, D. 1971. Nonolfactory responses from the nasal cavity: Jacobson's organ and the trigeminal system; pp. 151–181 in L. M. Beidler (ed.), Handbook of Sensory Physiology, Vol. IV, Chemical Senses, Part 1, Olfaction. Springer-Verlag, Berlin.

Tuckerman, F. 1892. On the terminations of the nerves in the lingual papillae of the Chelonia. Internationales Monatschrift für Anatomie und Physiologie. 9:1–5.

Uchida, T. 1980. Ultrastructural and histochemical studies on the taste buds in some reptiles. Archivum Histologicum Japonicum 43:459–478.

Vincent, S. E., R. Shine, and G. P. Brown. 2005. Does foraging mode influence sensory modalities for prey detection in male and female filesnakes, *Acrochordus arafurae*? Animal Behaviour 70:715–721.

Vogel, S. 1994. Life in Moving Fluids. Princeton University Press, Princeton, NJ.

Walker, W. F., Jr. 1959. Closure of the nostrils in the Atlantic loggerhead and other sea turtles. Copeia 1959:257–259.

Weldon, P. J., and M. W. J. Ferguson. 1993. Chemoreception in crocodilians: anatomy, natural history, and empirical results. Brain, Behavior, and Evolution 41:239–245.

Weldon, P. J., and E. S. Gaffney. 1998. An ancient integumentary gland in turtles. Naturwissenschaften 85:556–557.

Weldon, P. J., and L. McNease. 1991. Does the American alligator discriminate between venomous and nonvenomous snake prey? Herpetologica 47:403–406.

Weldon, P. J., and H. W. Sampson. 1988. The gular glands of *Alligator mississippiensis*: histology and preliminary analysis of lipoidal secretions. Copeia 1988:80–86.

Weldon, P. J., and J. W. Wheeler. 2001. The chemistry of crocodilian skin glands; pp. 286–296 in G. C. Grigg, F. Seebacher, and C. E. Franklin (eds.), Crocodilian Biology and Evolution. Surrey

Beatty and Sons, Chipping Norton, NSW, Australia.

Weldon, P. J., D. J. Swenson, J. K. Olson, and W. G. Brinkmeier. 1990. The American alligator detects food chemicals in aquatic and terrestrial environments. Ethology 85:191–198.

Whyte, A., Z.-C. Yang, K. Tiyanont, P. J. Weldon, T. Eisner, and J. Meinwald. 1999. Reptilian chemistry: characterization of dianeackerone, a secretory product from a crocodile. Proceedings of the National Academy of Sciences USA 96: 12246–12250.

Winokur, R. M. 1973. Adaptive modifications of the buccal mucosae in turtles. American Zoologist 13: 1347–1348A.

Witmer, L. M. 1995. Homology of facial structures in extant archosaurs (birds and crocodilians), with special reference to paranasal pneumaticity and nasal conchae. Journal of Morphology 225:269–327.

Wyatt, T. D. 2003. Pheromones and Animal Behaviour: Communication by Smell and Taste. Cambridge University Press, Cambridge.

Yoshie, S., and H. Yokosuka, 2001. Chemo- and mechano-receptors in the tongue of the Surinam caiman, *Caiman crocodilus crocodiles* [abstract]. Journal of Morphology 248:303.

Young, B. A. 1993. Evaluating hypotheses for the transfer of stimulus particles to Jacobson's organ in snakes. Brain, Behavior, and Evolution 41: 203–209.

6

Comparative Anatomy and Physiology of Chemical Senses in Aquatic Birds

Tobin L. Hieronymus

Birds are incredibly speciose in the extant time plane, comprising nearly two-thirds of all living sauropsid species. Although not as morphologically diverse as their combined reptilian relatives, they are ecologically diverse, occupying habitats ranging from polar oceans to equatorial deserts. Within this ecological diversity, a number of avian taxa swim or forage facultatively in shallow-water settings and are thus aquatic in some sense. However, one of the most concerted bodies of data on avian chemical sensation pertains to the procellariiform seabirds (petrels, albatross, and shearwaters; see Fig. 1.6 and Table 1.3 in this volume for a brief discussion of bird diversity), and the most illuminating comparisons of aquatic versus terrestrial chemical sensation in birds thus arise in contrasting these and other obligate deepwater (generally marine) taxa with the rest of birds as a whole. For the purposes of this chapter, "aquatic birds" are limited to birds that (1) swim as an obligate part of feeding behavior, or (2) forage on the outer shelf or beyond (sensu Shealer, 2002). This definition includes most extant species of Pelecaniformes, Procellariiformes, Sphenisciformes, Gaviiformes, and Podicipediformes, as well as larid and alcid Charadriiformes and many Anseriformes.

Of the four major components of chemical sensation discussed in previous chapters, three are present in birds: (1) taste (gustation), (2) smell (olfaction), and (3) chemesthesis, or noxious stimulation in response to air- or water-borne chemicals. Birds lack vomeronasal organs and the accompanying sensitivity to nonvolatile chemicals that they provide (Mason and Clark, 2000). Since so little is known about physiology and behavioral response to chemical stimulants in birds as a whole, much of this chapter points to gaps in our understanding, and as such will provide for discussion of areas for further research instead of a comprehensive review.

COMPARATIVE ANATOMY OF AVIAN CHEMICAL SENSES

Morphological relationships described in this section are after Bubień-Waluszewska (1981), Baumel and Witmer (1993), Dubbeldam (1993), and Evans and Martin (1993), unless otherwise noted. More detailed and complete reviews of the nasal cavity and organs of olfaction are provided by Bang (1971) and Bang and Wenzel (1985). The organs of chemical sense in birds are situated within the prominent beak. A secondary hard palate separates the nasal cavum, which houses the organs of olfaction, from the oropharynx. These two spaces communicate through a narrow slitlike internal naris or choana. In most birds, the nasal cavum communicates rostrally with the vestibule, which opens through the beak in two external nares. The external nares are generally placed rostrally with respect to the cavum, although in some birds (e.g., hornbills, toucans) the external nares are displaced caudally (Bang, 1971). The horny skin of the beak (rhamphotheca) generally covers the caudal portion of the bony naris, often forming a raised operculum (Lucas and Stettenheim, 1972; Clark, 1993), such that only a small rostral portion of the bony external naris is open for airflow (Witmer, 2001). In some pelecaniforms (boobies, gannets, cormorants, and anhingas), the bony external naris

is greatly reduced or closed; in these taxa and all penguins except *Spheniscus* spp. and *Eudyptula minor* (Sphenisciformes), the external bony naris is occluded by the rhamphotheca, preventing effective airflow (Bang, 1971; Clark, 1993; Bertelli and Giannini, 2005). The jointed jugal operculum on the upper beaks of derived pelecaniforms is thought to function as a secondary external naris, allowing breathing while the rest of the beak is closed (MacDonald, 1960). In these taxa, inspired air passes directly into the oropharynx, and any odorants must then pass up through the choana to reach olfactory epithelium.

The nasal cavum contains three paired cartilaginous or bony protrusions from the lateral wall: the rostral, medial, and caudal nasal conchae. Rostral or medial conchae have been secondarily lost in some pelecaniforms (e.g., sulids, phalacrocoracids) (Bang, 1971). Procellariiforms develop additional paired conchae that protrude from the nasal septum (septal conchae). Most of the epithelium lining the nasal cavum is respiratory epithelium; olfactory epithelium is found only on the caudal nasal concha in most birds. Procellariiforms show olfactory epithelium on the caudal septum and septal conchae as well (Portmann, 1961; Bang, 1966, 1971).

At least five bird lineages have independently acquired an epithelial nasal valve between the nasal septum and the middle nasal conchae that blocks foreign substances from contact with the caudal nasal conchae (Fig. 6.1). In most taxa its occurrence can be readily related to aquatic foraging or diving behavior (Bang, 1971; Bang and Wenzel, 1985). The independent losses of the nasal valve in derived pelecaniforms, pelicans, and penguins are most likely related to reduction or secondary closure of the external nares.

Birds lack a vomeronasal organ and accompanying terminal nerve. The roots of the olfactory nerves (cranial nerve I) gather into separate dorsal and ventral rami, which then merge in the nasal cavity into a single nerve for each side. These nerves pass through the

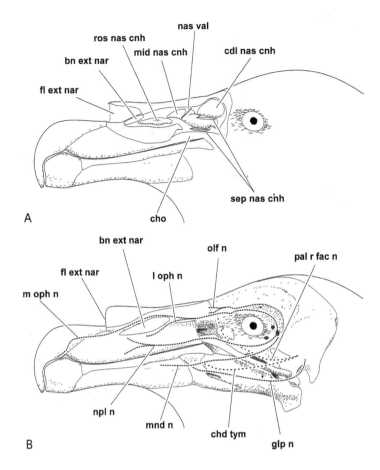

FIGURE 6.1. Nasal cavity (A) and superficial dissection (B) view of the face of a procellariiform bird. Abbreviations: bn ext nar, bony external naris; cdl nas cnh, caudal nasal concha; chd tym, chorda tympani; cho, choana; fl ext nar, flange of the external naris; glp n, glenopalatine nerve; l oph n, inferior branch of ophthalmic nerve; m oph n, medial ophthalmic nerve; mid nas cnh, middle nasal concha; mnd n, mandibular nerve; nas val, nasal valve; npl n, nasopalatine nerve; olf n, olfactory nerve; pal r fac n, palatine ramus of the facial nerve; ros nas cnh; rostral nasal concha; and sep nas cnh, septum of the nasal conchae.

orbitonasal foramen into the orbit, where they run medial to the eye in the olfactory groove of the mesethmoid bone, before entering the cranial cavity via the olfactory foramina. The diameter of the olfactory nerve along this route shows a positive relationship to the size of the olfactory bulb and thence, presumably, to olfactory acuity (Strong, 1911; Cobb, 1960a, 1960b). The ratio of greatest diameter of the olfactory bulbs to greatest diameter of the cerebrum (olfactory ratio) shows a clear relationship with olfactory homing and foraging behaviors (Bang, 1971; Bang and Wenzel, 1985). However, some of the most well-known examples of olfactorily mediated behaviors are present in taxa with intermediate olfactory ratios, as is seen in homing pigeons (Bang, 1971; Wenzel, 1992); and taxa with very low olfactory ratios, such as songbirds and hummingbirds, still respond to olfactory cues (Buitron and Nuechterlein, 1985; Harriman and Berger, 1986; Ioalè and Papi, 1989).

Free nerve endings throughout the nasal and oral cavities are associated with trigeminal nociception (Mason and Clark, 2000). General somatic afferent fibers from nociceptors in the rostromedial nasal cavum and palate are included in the medial and lateral rami of the ophthalmic nerve (cranial nerve V_1); fibers from nociceptors in the caudolateral nasal cavum and palate are included in the nasopalatine ramus of the maxillary nerve (V_2); and fibers from nociceptors on the floor of the oropharynx are included in the ramus of the angle of the mouth and sublingual ramus of the mandibular nerve (cranial nerve V_3). Although these nerve branches do leave characteristic osteological correlates, their combined modalities (including general somatic afferent, special visceral afferent, and general visceral efferent) preclude any gross assessment of function similar to the olfactory ratio for the olfactory nerve.

Birds generally have fewer taste buds than other vertebrates, and most of these are not clustered on the dorsal surface of the tongue but are instead distributed throughout the oral cavity, with concentrations on the palate and in the soft mucosa lateral and rostral to the tongue (Botezat, 1904; Bath, 1906; Berkhoudt, 1977, 1985, 1992). The tongue itself is a predominantly bony structure that is extensively cornified in some birds (e.g., penguins) and is mostly used in food handling (Berkhoudt, 1992).

Most of the taste buds on the ventral surface of the oropharynx are linked to the glossopharyngeal nerve (cranial nerve IX), the major carrier of taste sensation in birds, although the palatine ramus and chorda tympani of the facial nerve (cranial nerve VII) are involved in transmitting taste sensation as well. Special visceral afferent fibers from individual taste buds are first collected into adjacent branches of the medial ophthalmic (part of cranial V_1), nasopalatine (part of cranial nerve V_2), sublingual (part of cranial nerve V_3), and lingual (part of cranial nerve V_3) nerves (Berkhoudt, 1977; Kroll and Dubbeldam, 1979). Fibers in the medial ophthalmic and nasopalatine nerves pass into the dorsal and ventral rami of the palatine nerve (part of cranial nerve VII), respectively, through connections between the trigeminal branches and the ethmoid and sphenopalatine ganglia. Fibers in the sublingual nerve pass into the chorda tympani (cranial nerve VII). Fibers in the lingual nerve join with the other branches of the glossopharyngeal nerve before entering the cranial cavity in the parabasal fossa.

COMPARATIVE PHYSIOLOGY AND BEHAVIOR OF AVIAN CHEMICAL SENSES

AVIAN OLFACTION

In line with the predictions of Cobb (1960a, 1960b), most birds that have been examined to date show physiological and behavioral responses to olfactory cues (Tucker, 1965; Henton et al., 1966; Wenzel, 1968, 1971, 1992; Archer and Glen, 1969; Wenzel and Sieck, 1972; Bang and Wenzel, 1985). These responses are present even in Passeriformes, which were previously thought to be anosmic or microsmic (Buitron and Nuechterlein, 1985; Harriman and Berger, 1986; Wenzel, 1992). Based on the distribution of these examples in living birds and other sauropsids (Schwenk, chapter 5 in this volume), olfactorily mediated behavior in birds is most parsimoniously interpreted as a rule rather than an exceptional state. Birds do have a smaller number of genes in the olfactory receptor family compared to mammals (Nef et al., 1996). Because olfactory receptor gene counts are not available for other sauropsids, it is difficult to determine whether this difference represents a reduction in birds, a hypertrophy in mammals, or both.

HOMING BY OLFACTORY CUES

Most studies on olfactorily mediated homing in terrestrial birds have focused on homing pigeons (Wallraff, 2004), in which olfaction is thought to play a role in the initial "map" phase of homing (Benvenuti et al., 1992). Pigeons associate windborne olfactory cues detected at the roost with wind direction and use local olfactory cues upon release to set a "compass" course back to the roost (Wiltschko and Wiltschko, 1992). Anosmic pigeons show significantly impaired homing abilities (Wiltschko and Wiltschko, 1992; Benvenuti and Ranvaud, 2004).

Among the aquatic birds, only procellariiforms are known to use olfactory cues in homing (Grubb, 1973, 1974, 1979; Benvenuti et al., 1993; Minguez, 1997; Bonadonna et al., 2001, 2003a, 2003b, 2004; Bonadonna and Bretagnolle, 2002; De León et al., 2003). The degree of reliance on olfactory cues is strongest in nocturnal burrow-nesting taxa (Bonadonna and Bretagnolle, 2002). Returning to the nest at night has been hypothesized as a means for smaller seabird taxa to avoid predation (Bonadonna et al., 2004). Whether similar olfactory homing occurs in other burrow-nesting seabirds (e.g., *Synthliborhamphus*, *Cerorhinca*, and *Fratercula* in Charadriiformes, *Eudyptula* and *Spheniscus* in Sphenisciformes) remains an open question (Schreiber and Burger, 2002).

FORAGING BY OLFACTORY CUES

Foraging by olfaction has been reported for a number of birds; oft-cited terrestrial examples include kiwis (Wenzel, 1968, 1971) and turkey vultures (Stager, 1964). These taxa have obvious morphological adaptations for increased olfactory sensitivity (enlarged olfactory lobes in both, distally placed external nares in kiwis). However, other taxa that forage by olfactory cues, such as honeyguides (Archer and Glen, 1969), do not show marked departures from generalized avian morphology.

Foraging by olfactory cues is well documented in procellariiform seabirds (Grubb, 1972; Hutchison and Wenzel, 1980; Lequette et al., 1989; Verheyden and Jouventin, 1994). Different mechanisms of chemoattraction in these taxa have been linked to dimethyl sulfide (DMS) and pyrazines (Clark and Shah, 1992; Nevitt et al., 1995, 2004; Nevitt, 1999a, 1999b, 2000; Cunningham et al., 2003; Nevitt and Haberman, 2003), two odorants commonly found in plankton "hot spots" (Dacey and Wakeham, 1986). Storm-petrels and prions are more responsive to DMS, while petrels and albatross show greater responses to pyrazines (Nevitt, 1999a; Nevitt et al., 2004). This phylogenetically incongruent pattern suggests multiple changes in odorant sensitivity within Procellariiformes. There is also some evidence that Humboldt penguins *(Spheniscus humboldti)* can use DMS as a behavioral cue to track upwind plankton blooms over long distances (Culik, 2001).

SOCIAL BEHAVIOR AND OLFACTORY CUES

There are few studies on the role of olfaction in sociality and reproduction in birds. Seasonal variations in odorant detection threshold in European starlings *(Sturnus vulgaris)* are likely to be linked to reproductive behaviors (Clark and Smeraski, 1990), but the behavioral manifestations of increased sensitivity are unclear. Olfactorily mediated mate recognition has been demonstrated in Antarctic prions *(Pachyptila desolata)* (Bonadonna and Nevitt, 2004), and olfactory cues are known to be linked to reproductive effort in male mallard ducks (Balthazart and Schoffeniels, 1979); the distinctive odors of other anseriform taxa (Bang, 1971) may provide cues for reproductive behaviors as well. The distinctive tangerine odor of crested auklets *(Aethia cristatella)* has been hypothesized to play a social and reproductive role (Hagelin et al., 2003), although the overlapping hypothesis that it functions as an arthropod parasite repellent (Douglas et al., 2001, 2005) must also be considered.

AVIAN CHEMESTHESIS

Although there are a number of studies that examine trigeminally mediated chemesthesis in birds (summarized in Mason and Clark, 2000), most research has focused on economically important terrestrial taxa, and little information is available for aquatic taxa. Although birds do not respond to some common mammalian trigeminal irritants (e.g., capsaicin, the chemical responsible for the "burn" of hot peppers [Mason and Maruniak, 1983]), they are chemesthetically sensitive to other compounds that do not provoke strong responses in mammals (e.g., methyl anthranilate [Clark et al., 1991; Shah et al., 1992]).

AVIAN GUSTATION

Most birds examined to date, including some aquatic taxa such as grebes, show their highest concentrations of taste buds on the oral mucosa lateral to the tongue (Botezat, 1904; Bath, 1906; Berkhoudt, 1985). This arrangement corresponds to generalized "pecking" feeding behaviors (Berkhoudt, 1992). Other taxa, such as mallard ducks and flamingos, show tracts of taste buds rostral to the tongue and on the rostral palate (Bath, 1906; Berkhoudt, 1977), in line with the path of food particles during filter-feeding (Berkhoudt, 1992).

Whether other filter-feeding aquatic taxa (e.g., prions) show similar taste bud distributions is currently unknown.

EVOLUTIONARY CHANGE IN AVIAN CHEMICAL SENSES

FOSSIL DATA

The fossil record provides a transformational context for extant clades of aquatic birds, as well as some independently derived aquatic lineages that are now extinct (Olson, 1985; Warheit, 2002). While few fossil taxa include the cranial and rostral material necessary to asses the osteological correlates of olfactory acuity, the phylogenetic context and autecological reconstructions proposed for these taxa offer interesting perspectives on the context of chemical sense evolution in aquatic birds.

BASAL ORNITHURINES

Although braincase material is available for the toothed aquatic basal ornithurines *Hesperornis, Parahesperornis,* and *Ichthyornis* (Marsh, 1872, 1880; Martin, 1984; Bühler et al., 1988), there are no published accounts of olfactory nerve anatomy or olfactory ratios in these taxa. Marsh's (1872) paleoneurological reconstructions depict elongate olfactory lobes, but this representation is based on typological conceptions of primitive bird brains rather than morphological data (Edinger, 1951). The bony external nares of hesperornithids are morphologically similar to those of some penguins and may have likewise been functionally occluded by the rhamphotheca in life. Recent discoveries of new fossil enantiornithine and basal ornithurine taxa (Chiappe and Walker, 2002; Hope, 2002) provide baseline comparative material for studying the morphological correlates of chemosensation in hesperornithiform and ichthyornithiform birds.

EXTINCT PELECANIFORMES

There are two extinct radiations within Pelecaniformes that may also show interesting morphological correlates of chemosensation. Plotopteridae were flightless wing-propelled divers, similar to penguins and the recently extinct auk genera *Pinguinus* and *Mancalla*, and are represented by 5 fossil taxa from the North Pacific (Howard, 1969; Olson and Hasegawa, 1979, 1996; Olson, 1985; Warheit, 2002). Pelagornithidae were giant (>5 meter wingspan) volant birds that have been reconstructed as pelagic foragers specializing on squid, similar to albatross, and are represented by 17 fossil taxa worldwide (Howard and Warter, 1969; Harrison and Walker, 1976; Olson, 1985; Warheit, 2002). As Pelecaniformes, these two groups are part of a lineage that shows a reduction in the morphological correlates of olfactory acuity among its extant members, but the extant ecological analogs of these fossil taxa are known to use olfactory cues in foraging. This contrast of phylogenetic and ecological influence may provide insight into patterns and processes of the evolution of chemosensation in aquatic birds.

COMPARATIVE MODERN DATA

ANSERIFORMES

The anseriforms sampled by Bang (1971) do have greater olfactory ratios than representatives of their sister clade Galliformes (Fig. 6.2), but the difference is not nearly as striking as the contrasts between *Apteryx* and other paleognaths, *Cathartes* and other Falconiformes, or Procellariiformes and their sister taxa (Sphenisciformes and Gaviiformes). However, accounts of chemosensation in social and feeding behaviors in anseriforms (Balthazart and Schoffeniels, 1979; Berkhoudt, 1992; Wenzel, 1992), coupled with neurological data on olfaction (Ebinger et al., 1991), point to derived states that may be linked to an aquatic lifestyle. Anseriforms present a potential transformational sequence between terrestrial and aquatic taxa (e.g., screamers to eiders, scoters, and steamer ducks, but see work by Olson and Feduccia [1980]) that is closely related to one of the most commonly used model taxa in olfaction studies, the domestic fowl *Gallus gallus* (Jones and Roper, 1997). New information regarding olfactory and

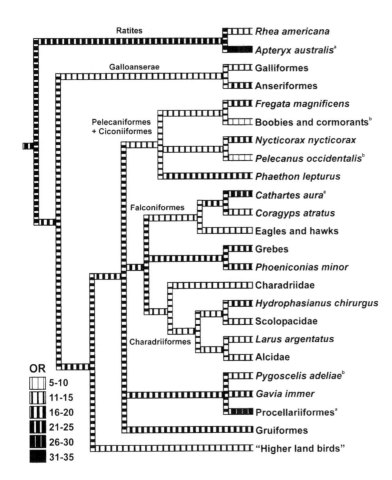

FIGURE 6.2. Phylogram showing an ancestral character state reconstruction by squared-change parsimony of Bang's (1971) olfactory ratios (OR) for basal birds based on representative taxa. Birds with olfactorily mediated foraging behaviors (indicated by superscript a) tend to show increased OR relative to outgroups, while closure of the external naris (indicated by superscript b) is generally associated with a decrease in OR.

gustatory behaviors in anseriform taxa other than mallard ducks may thus shed light on changes in avian chemosensation in aquatic environments.

PELECANIFORMES

Comparative analysis of extant Pelecaniformes (Fig. 6.2) suggests that the reduction in olfactory ratios seen in this group is directly related to the reduction or functional closure of their external nares (Bang, 1971). Many of the pelecaniform birds are nearshore foragers, but some taxa (e.g., phaethontids and sulids) travel far offshore to feed, effectively ruling out acute olfaction as a necessary adaptation for pelagic foraging in seabirds.

PROCELLARIIFORMES, GAVIIFORMES, AND SPHENISCIFORMES

Many procellariiforms, like other birds that display olfactory specializations, nest in burrows or cavities that enclose small pockets of stable air (Bang, 1971), although the validity of this relationship as a general rule has been called into question (Healy and Guilford, 1990). Squared-change parsimony ancestral character state reconstructions (Maddison and Maddison, 2004a) of Bang's (1971) olfactory lobe ratios (Fig. 6.2) show an expected increase of olfactory ratio at the base of Procellariiformes. A maximum-likelihood ancestral character state reconstruction (Maddison and Maddison, 2004a, 2004b) for discrete categories of nesting habit shows a significant probability that immediate descendants of that same node were burrow-nesters (Fig. 6.3). It is thus plausible (though as of yet untested) that the long-distance homing and foraging behaviors seen in Procellariiformes are exaptations of olfactory acuity that initially developed in stem taxa as an adaptation for burrow location. It is also interesting,

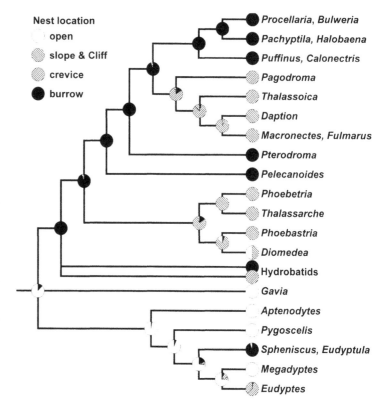

Nest location
- open
- slope & Cliff
- crevice
- burrow

Procellaria, Bulweria
Pachyptila, Halobaena
Puffinus, Calonectris
Pagodroma
Thalassoica
Daption
Macronectes, Fulmarus
Pterodroma
Pelecanoides
Phoebetria
Thalassarche
Phoebastria
Diomedea
Hydrobatids
Gavia
Aptenodytes
Pygoscelis
Spheniscus, Eudyptula
Megadyptes
Eudyptes

FIGURE 6.3. Phylogram showing a maximum likelihood ancestral character state reconstruction of nesting habit in tubenose (procellariiform), loon (gaviiform), and penguin (sphenisciform) genera. It is very probable that a change from open nesting to burrow nesting occurred within stem Procellariiformes, alongside the change in olfactory ratios described in the text. *Eudyptula* and *Spheniscus* spp. are the only penguin taxa that have patent external nares, and this change may have co-occurred with a transition from open nesting to burrow nesting as well.

though perhaps coincidental, that the only penguin taxa that nest in burrows or crevices (*Eudyptula minor* and *Spheniscus* spp.) (Schreiber and Burger, 2002) are also the only penguin taxa with patent external nares (Bertelli and Giannini, 2005).

The data on penguin foraging and behavioral response to DMS presented by Culik (2001) raise the possibility that DMS-mediated foraging behavior originated deep within the Sphenisciformes + Gaviiformes + Procellariiformes clade. Alternatively, this sensitivity may be convergent; either outcome provides an interesting perspective on the relative importance of foraging and diet versus nesting in olfactory adaptation (Bang, 1971; Bang and Wenzel, 1985). Data on DMS sensitivity in other penguin taxa, whether from adults (after Culik, 2001) or nestlings (after Porter et al., 1999; Cunningham et al., 2003), may help resolve the origins of this olfactory adaptation to aquatic feeding.

Since birds, to a greater degree than other vertebrates, are inhabitants of the atmospheric boundary layer above the air-water or air-land interface, the behavior of this body of air is likely to have a pronounced impact on their autecology. The offshore marine boundary layer is generally more stable than its terrestrial counterpart. This stability makes it much easier to trace the source of an odorant for long distances (Moore et al., 1992). The stability of the air masses that are encountered by foraging or homing birds is rarely considered (but see Clark and Shah, 1992), and may be another important environmental factor that relates to heightened olfactory acuity and the olfactory landscape proposed by Nevitt (1999b, 2000) for procellariiforms.

CHARADRIIFORMES

Very little comparative data is available to relate chemosensation in the deepwater larids and alcids to their shorebird relatives. Ecological differences between these groups do not appear to be accompanied by a concerted change in olfactory ratios (Fig. 6.2), although this does not automatically rule out the presence of differing adaptive responses to olfactory cues.

PODICIPEDIFORMES

There are limited data regarding chemosensation in grebes, but their relatively high olfactory ratios (Fig. 6.2) fit the general pattern for aquatic birds proposed by Bang (1971). Although their tentatively assigned sister taxon, flamingos (Cracraft et al., 2004), does not fall under this chapter's restricted definition of aquatic, the relatively high olfactory ratio of the pygmy flamingo (Fig. 6.2) also supports this pattern. Difficulty in resolving the relationships of these two groups to Falconiformes and Charadriiformes hampers drawing more detailed conclusions from comparative analysis.

CHEMOSENSATION AND AQUATIC HABITS IN BIRDS

The sharp contrast between olfactory reduction in Pelecaniformes and olfactory specialization in Procellariiformes suggests that there is not a single suite of stereotypical chemosensitive adaptations to an aquatic lifestyle in birds. Instead, the aquatic environment presents both opportunities and challenges for chemical sensation, and there are a number of possible solutions. On one hand, chemosensation provides cues for orientation and foraging in an otherwise featureless environment. Procellariiformes have exploited this opportunity and are currently among the most successful pelagic birds. On the other hand, birds are constrained by changes to their chemosensitive systems that occurred in their terrestrial archosaurian ancestors, such as loss of the vomeronasal organ and the hard keratinization of the margins of the external naris. Without the benefits of nonvolatile olfaction in an aquatic environment, occlusion of the external nares in diving taxa, such as some Pelecaniformes and Sphenisciformes, appears to have been a favorable trade-off.

Heightened olfaction in Procellariiformes is often upheld as the exception to the rule of microsmia in birds. However, given the evidence for olfaction in other bird groups (e.g., pigeons, penguins), chemosensation in birds may yet show systematic changes in response to the return to an aquatic milieu. Understanding chemosensitive physiology and behavior in outgroups to well-studied clades of aquatic birds is the necessary first step in uncovering such changes.

LITERATURE CITED

Archer, A. L., and R. M. Glen. 1969. Observations on the behavior of two species of honey-guides *Indicator variegatus* (Lesson) and *Indicator exilis* (Cassin). Los Angeles County Museum Contributions to Science 160:1–6.

Balthazart, J., and E. Schoffeniels. 1979. Pheromones are involved in the control of sexual behaviour in birds. Naturwissenschaften 66:55–56.

Bang, B. G. 1966. The olfactory apparatus of tubenosed birds (Procellariiformes). Acta Anatomica 65:391–415.

Bang, B. G. 1971. Functional anatomy of the olfactory system in 23 orders of birds. Acta Anatomica 79 (Suppl.): 1–76.

Bang, B. G., and B. M. Wenzel. 1985. Nasal cavity and olfactory system; pp. 195–225 in A. S. King and J. McLelland (eds.), Form and Function in Birds, Vol. 3. Academic Press, London.

Bath, W. 1906. Die Geschmacksorgane der Vögel und Krokodile. Archiv für Biontologie 1:5–74.

Baumel, J. J., and L. M. Witmer. 1993. Osteologia; pp. 45–132 in J. J. Baumel (ed.), Handbook of Avian Anatomy: Nomina Anatomica Avium. Nuttall Ornithological Club, Cambridge.

Benvenuti, S., and R. Ranvaud. 2004. Olfaction and homing ability of pigeons raised in a tropical area of Brazil. Journal of Experimental Zoology 301A:961–970.

Benvenuti, S., P. Ioalè, and F. Papi. 1992. The olfactory map of homing pigeons; pp. 429–434 in R. L. Doty and D. Müller-Schwarze (eds.), Chemical Signals in Vertebrates, 6. Plenum Press, New York.

Benvenuti, S., P. Ioalè, and B. Massa. 1993. Olfactory experiments on Cory's shearwater *(Calonectris diomedea)*: the effect of intranasal zinc sulphate treatment on short-range homing behaviour. Bolletino di Zoologia 60:207–210.

Berkhoudt, H. 1977. Taste buds in the bill of the mallard (*Anas platyrhynchos* L.): Their morphology, distribution, and functional significance. Netherlands Journal of Zoology 27:310–337.

Berkhoudt, H. 1985. Structure and function of avian taste receptors; pp. 463–496 in A. S. King and J. McLelland (eds.), Form and Function in Birds, Vol. 3. Academic Press, London.

Berkhoudt, H. 1992. Avian taste buds: topography, structure and function; pp. 15–20 in R. L. Doty and D. Müller-Schwarze (eds.), Chemical Signals in Vertebrates, 6. Plenum Press, New York.

Bertelli, S., and N. P. Giannini. 2005. A phylogeny of extant penguins (Aves: Sphenisciformes) combining morphology and mitochondrial sequences. Cladistics 21:209–239.

Bonadonna, F., and V. Bretagnolle. 2002. Smelling home: a good solution for burrow-finding in nocturnal petrels? Journal of Experimental Biology 205:2519–2523.

Bonadonna F., and G. A. Nevitt. 2004. Partner-specific odor recognition in an Antarctic seabird. Science 306:835.

Bonadonna, F., J. Spaggiari, and H. Weimerskirch. 2001. Could osmotaxis explain the ability of blue petrels to return to their burrows at night? Journal of Experimental Biology 204:1485–1489.

Bonadonna, F., G. B. Cunningham, P. Jouventin, F. Hesters, and G. A. Nevitt. 2003a. Evidence for nest-odour recognition in two species of diving petrel. Journal of Experimental Biology 206: 3719–3722.

Bonadonna, F., F. Hesters, and P. Jouventin. 2003b. Scent of a nest: discrimination of own-nest odours in Antarctic prions, Pachyptila desolata. Behavioral Ecology and Sociobiology 54:174–178.

Bonadonna, F., M. Villafane, C. Bajzak, and P. Jouventin. 2004. Recognition of burrow's olfactory signature in blue petrels, *Halobaena caerulea:* an efficient discrimination mechanism in the dark. Animal Behaviour 67:893–898.

Botezat, E. 1904. Geschmacksorgane und andere nervöse Endapparate im Schnabel der Vögel (vorläufige Mitteilung). Biologisches Zentralblatt 24:722–736.

Bubień-Waluszewska, A. 1981. The cranial nerves; pp. 385–438 in A. S. King and J. McLelland (eds.), Form and Function in Birds, Vol. 2. Academic Press, London.

Bühler, P. L. D. Martin, and L. M. Witmer. 1988. Cranial kinesis in the Late Cretaceous birds *Hesperornis* and *Parahesperornis.* Auk 105:111–112.

Buitron, D., and G. L. Nuechterlein. 1985. Experiments on olfactory detection of food caches by Black-Billed Magpies. Condor 87:92–95.

Chiappe, L. M., and C. A. Walker. 2002. Skeletal morphology and systematics of the Cretaceous Euenantiornithes (Ornithothoraces: Enantiornithes); pp. 240–267 in L. M. Chiappe and L. M.

Witmer (eds.), Mesozoic Birds: Above the Heads of Dinosaurs. University of California Press, Berkeley.

Clark, G. A. 1993. Integumentum commune; pp. 17–44 in J. J. Baumel (ed.), Handbook of Avian Anatomy: Nomina Anatomica Avium. Nuttall Ornithological Club, Cambridge.

Clark, L., and P. S. Shah. 1992. Information content of prey odor plumes: what do foraging Leach's storm petrels know? pp. 421–427 in R. L. Doty and D. Müller-Schwarze (eds.), Chemical Signals in Vertebrates, 6. Plenum Press, New York.

Clark, L., and C. A. Smeraski. 1990. Seasonal shifts in odor acuity by starlings. Journal of Experimental Zoology 255:22–29.

Clark, L., P. S. Shah, and J. R. Mason. 1991. Chemical repellency in birds: relationship between structure of anthranilate and benzoic derivatives and avoidance response. Journal of Experimental Zoology 260:310–322.

Cobb, S. 1960a. A note of the size of the avian olfactory bulbs. Epilepsia 1:394–402.

Cobb, S. 1960b. Observations on the comparative anatomy of the avian brain. Perspectives in Biology and Medicine 3:383–408.

Cracraft, J., F. K. Barker, M. Braun, J. Harshman, G. J. Dyke, J. Feinstein, S. Stanley, A. Cibois, P. Schikler, P. Beresford, J. García-Moreno, M. D. Sorenson, T. Yuri, and D. P. Mindell. 2004. Phylogenetic relationships among modern birds (Neornithes): toward an avian tree of life; pp. 468–489 in J. Cracraft and M. J. Donoghue (eds.), Assembling the Tree of Life. Oxford University Press, Oxford.

Culik, B. 2001. Finding food in the open ocean: foraging strategies in Humboldt penguins. Zoology 104:327–338.

Cunningham, G. B., R. W. Van Buskirk, F. Bonadonna, H. Weimerskirch, and G. A. Nevitt. 2003. A comparison of the olfactory abilities of three species of procellariiform chicks. Journal of Experimental Biology 206:1615–1620.

Dacey, J. W. H., and S. G. Wakeham. 1986. Oceanic dimethylsulphide: production during zooplankton grazing on phytoplankton. Science 233:1314–1316.

De León, A., E. Mínguez, and B. Belliure. 2003. Self-odour recognition in European storm-petrel chicks. Behaviour 140:925–933.

Douglas, H. D., III, J. E. Co, T. H. Jones, and W. E. Conner. 2001. Heteropteran chemical repellents identified in the citrus odor of a seabird (crested auklet: *Aethia cristatella*): evolutionary convergence in chemical ecology. Naturwissenschaften 88:330–332.

Douglas, H. D., III, J. E. Co, T. H. Jones, W. E. Conner, and J. F. Day. 2005. Chemical odorant of colonial seabird repels mosquitoes. Journal of Medical Entomology 42:647–651.

Dubbeldam, J. L. 1993. Systema nervosum periphericum; pp. 555–584 in J. J. Baumel (ed.), Handbook of Avian Anatomy: Nomina Anatomica Avium. Nuttall Ornithological Club, Cambridge.

Ebinger, P., G. Rehkämper, and H. Schröder. 1991. Forebrain specialization and the olfactory system in anseriform birds: An architectonic and tracing study. Cell and Tissue Research 268:81–90.

Edinger, T. 1951. The brains of the Odontognathae. Evolution 5:6–24.

Evans, H. E., and G. R. Martin. 1993. Organa sensum (organa sensoria); pp. 585–611 in J. J. Baumel (ed.), Handbook of Avian Anatomy: Nomina Anatomica Avium. Nuttall Ornithological Club, Cambridge.

Grubb, T. C., Jr. 1972. Smell and foraging in shearwaters and petrels. Nature 237:404–405.

Grubb, T. C., Jr. 1973. Colony location by Leach's petrel. Auk 90:78–82.

Grubb, T. C., Jr. 1974. Olfactory navigation to the nesting burrow in Leach's petrel (Oceanodroma leucorrhoa). Animal Behaviour 22:192–202.

Grubb, T. C., Jr. 1979. Olfactory guidance of Leach's storm petrel to the breeding island. Wilson Bulletin 91:141–143.

Hagelin, J. C., I. L. Jones, and L. E. L. Rasmussen. 2003. A tangerine-scented social odour in a monogamous seabird. Proceedings of the Royal Society of London B 270:1323–1329.

Harriman, A. E., and R. H. Berger. 1986. Olfactory acuity in the common raven (Corvus corax). Physiology and Behavior 36:257–262.

Harrison, C. J. O., and C. A. Walker. 1976. A review of the bony-toothed birds (Odontopterygiformes): with a description of some new species. Tertiary Research Special Papers 2:1–62.

Healy, S., and T. Guilford. 1990. Olfactory-bulb size and nocturnality in birds. Evolution 44:339–346.

Henton, W. W., J. C. Smith, and D. Tucker. 1966. Odor discrimination in pigeons. Science 153:1138–1139.

Hope, S. 2002. The Mesozoic radiation of Neornithes; pp. 339–388 in L. M. Chiappe and L. M. Witmer (eds.), Mesozoic Birds: Above the Heads of Dinosaurs. University of California Press, Berkeley.

Howard, H. 1969. A new avian from Kern County, California. Condor 71:68–69.

Howard, H., and S. L. Warter. 1969. A new species of bony-toothed bird (family Pseudodontornithidae) from the Tertiary of New Zealand. Records of the Canterbury Museum 8:345–357.

Hutchison, L. V., and B. M. Wenzel. 1980. Olfactory guidance in foraging by procellariiforms. Condor 82:314–319.

Ioalè, P., and F. Papi. 1989. Olfactory bulb size, odor discrimination and magnetic insensitivity in hummingbirds. Physiology and Behavior 45:955–959.

Jones, R. B., and T. J. Roper. 1997. Olfaction in the domestic fowl: a critical review. Physiology and Behavior 62:1009–1018.

Kroll, C. P. M., and J. L. Dubbeldam. 1979. On the innervation of taste buds by the n. facialis in the mallard (Anas platyrhynchos L.). Netherlands Journal of Zoology 29:267–274.

Lequette, B., C. Verheyden, and P. Jouventin. 1989. Olfaction in subantarctic seabirds: its phylogenetic and ecological significance. Condor 91:732–735.

Lucas, A. M., and P. R. Stettenheim. 1972. Avian Anatomy: Integument. Agriculture Handbook 362. United States Department of Agriculture, Washington, DC.

MacDonald, J. D. 1960. Secondary external nares of the gannet. Proceedings of the Zoological Society of London 135:357–363.

Maddison, W. P., and D. R. Maddison. 2004a. Mesquite: A Modular System for Evolutionary Analysis. Version 1.05. http://mesquiteproject.org.

Maddison, W. P., and D. R. Maddison. 2004b. StochChar: A package of Mesquite Modules for Stochastic Models of Character Evolution. Version 1.05. http://mesquiteproject.org.

Marsh, O. C. 1872. Notice of a new and remarkable fossil bird. American Journal of Science, 3rd Series 4:344.

Marsh, O. C. 1880. Odontornithes: a monograph on the extinct toothed birds of North America. Memoirs of the Peabody Museum of Natural History 1:1–201.

Martin, L. D. 1984. A new hesperornithid and the relationships of the Mesozoic birds. Transactions of the Kansas Academy of Science 87:141–150.

Mason, J. R., and L. Clark. 2000. The chemical senses in birds; pp. 39–56 in G. C. Wittow (ed.), Sturkie's Avian Physiology. Academic Press, San Diego.

Mason, J. R., and J. A. Maruniak. 1983. Behavioral and physiological effects of capsaicin in red-winged blackbirds. Pharmacology Biochemistry and Behavior 19:857–862.

Minguez, E. 1997. Olfactory recognition by British storm-petrel chicks. Animal Behaviour 53:701–707.

Moore, P. A., J. Atema, and G. A. Gerhardt. 1992. The structure of environmental odor signals: from turbulent dispersion to movement through boundary layers and mucus; pp. 79–83 in R. L. Doty and D. Müller-Schwarze (eds.), Chemical Signals in Vertebrates, 6. Plenum Press, New York.

Nef, S., I. Allaman, H. Fiumell, E. De Castro, and P. Nef. 1996. Olfaction in birds: differential embryonic expression of nine putative odorant receptor genes in the avian olfactory system. Mechanisms of Development 55:65–77.

Nevitt, G. A. 1999a. Olfactory foraging in Antarctic seabirds: a species-specific attraction to krill odors. Marine Ecology Progress Series 177:235–241.

Nevitt, G. A. 1999b. Foraging by seabirds on an olfactory landscape. American Scientist 87:46–53.

Nevitt, G. A. 2000. Olfactory foraging by Antarctic procellariiform seabirds: life at high Reynolds numbers. Biological Bulletin 196:245–253.

Nevitt, G. A., and K. Haberman. 2003. Behavioral attraction of Leach's storm-petrels (Oceanodroma leucorrhoa) to dimethyl sulfide. Journal of Experimental Biology 206:1497–1501.

Nevitt, G. A., R. R. Veit, and P. Kareiva. 1995. Dimethyl sulphide as a foraging cue for Antarctic procellariiform seabirds. Nature 376:680–682.

Nevitt, G. A, K. Reid, and P. Trathan. 2004. Testing olfactory foraging strategies in an Antarctic seabird assemblage. Journal of Experimental Biology 207:3537–3544.

Olson, S. L. 1985. The fossil record of birds; pp. 79–238 in D. S. Farner, J. R. King, and K. C. Parkes (eds.), Avian Biology, Vol. 8. Academic Press, New York.

Olson, S. L., and A. Feduccia. 1980. Presbyornis and the origin of the Anseriformes (Aves: Charadriomorphae). Smithsonian Contributions to Zoology 323:1–24.

Olson, S. L., and Y. Hasegawa. 1979. Fossil counterparts of Giant Penguins from the North Pacific. Science 206:688–689.

Olson, S. L., and Y. Hasegawa. 1996. A new genus and two new species of gigantic Plotopteridae from Japan (Aves: Pelecaniformes). Journal of Vertebrate Paleontology 16:742–751.

Porter, R. H., P. G. Hepper, C. Bouchot, and M. Picard. 1999. A simple method for testing odor detection and discrimination in chicks. Physiology and Behavior 67:457–462.

Portmann, A., 1961. Sensory organs: skin, taste and olfaction; pp. 37–48 in A. J. Marshall (ed.), Biology and Comparative Physiology of Birds, Vol. II. Academic Press, New York.

Schreiber, E. A., and J. Burger. 2002. Appendix 2: data on life-history characteristics, breeding range, size, and survival for seabird species; pp. 665–685 in E. A. Schreiber and J. Burger (eds.), Biology of Marine Birds. CRC Press, Boca Raton, FL.

Shah, P. S., J. R. Mason, and L. Clark. 1992. Avian chemical repellency: a structure-activity approach and implications; pp. 291–296 in R. L. Doty and D. Müller-Schwarze (eds.), Chemical Signals in Vertebrates, 6. Plenum Press, New York.

Shealer, D. A. 2002. Foraging behavior and food of seabirds; pp. 137–177 in E. A. Schreiber and J. Burger (eds.), Biology of Marine Birds. CRC Press, Boca Raton, FL.

Stager, K. E. 1964. The role of olfaction in food location by the turkey vulture (Cathartes aura). Los Angeles County Museum Contributions to Science 81:1–63.

Strong, R. W. 1911. On the olfactory organs and the sense of smell in birds. Journal of Morphology 22:619–661.

Tucker, D. 1965. Electrophysiological evidence for olfactory function in birds. Nature 207:34–36.

Verheyden, C., and P. Jouventin. 1994. Olfactory behavior of foraging procellariiforms. Auk 111:285–291.

Wallraff, H. G. 2004. Avian olfactory navigation: its empirical foundation and conceptual state. Animal Behaviour 67:189–204.

Warheit, K. I. 2002. The seabird fossil record and the role of paleontology in understanding seabird community structure; pp. 17–55 in E. A. Schreiber and J. Burger (eds.), Biology of Marine Birds. CRC Press, Boca Raton, FL.

Wenzel, B. M. 1968. Olfactory prowess of the kiwi. Nature 220:1133–1134.

Wenzel, B. M. 1971. Olfactory sensation in the kiwi and other birds. Annals of the New York Academy of Sciences 188:183–193.

Wenzel, B. M. 1992. The puzzle of olfactory sensitivity in birds; pp. 443–448 in R. L. Doty and D. Müller-Schwarze (eds.), Chemical Signals in Vertebrates, 6. Plenum Press, New York.

Wenzel, B. M., and M. H. Sieck. 1972. Olfactory perception and bulbar electrical activity in several avian species. Physiology and Behavior 9:287–293.

Wiltschko, W., and R. Wiltschko. 1992. Pigeon homing: the effect of temporary anosmia on orientation behavior; pp. 435–442 in R. L. Doty and D. Müller-Schwarze (eds.), Chemical Signals in Vertebrates, 6. Plenum Press, New York.

Witmer, L. M. 2001. Nostril position in dinosaurs and other vertebrates and its significance for nasal function. Science 293:850–853.

7

Comparative Anatomy and Physiology of Chemical Senses in Aquatic Mammals

Henry Pihlström

Among Recent mammals, about 200 species can reasonably be considered aquatic or semi-aquatic, that is, they regularly forage and/or seek refuge in water and have some specific morphological or physiological adaptations for aquatic living. From a sensory biology point of view, life in water places certain restrictions on mammals. Vision, hearing, and the tactile senses all function somewhat differently in water than they do in air, and the same is true

for the three main chemical senses: olfaction, vomeronasal sense, and taste.

CHEMICAL SENSES IN GENERALIZED MAMMALS

The olfactory receptor gene family is the largest identified gene family in mammals (Sullivan, 2002). This indicates the importance of olfaction in the lives of most mammals, although there are some species, such as humans, in which the majority of olfactory receptor genes are nonfunctional (Gilad et al., 2003). At some point during early tetrapod evolution the original olfactory receptor gene family, still present in fish and amphibians, was supplemented with a new set of olfactory genes, coding for olfactory receptor molecules adapted to detecting airborne odors; this suggests that mammals and other amniotes are incapable of detecting scents underwater (Freitag et al., 1995, 1998).

This is in contrast to mammalian gustation, which is based on a small number of sensory receptor types. The number of olfactory cells, with different receptor proteins, probably

95

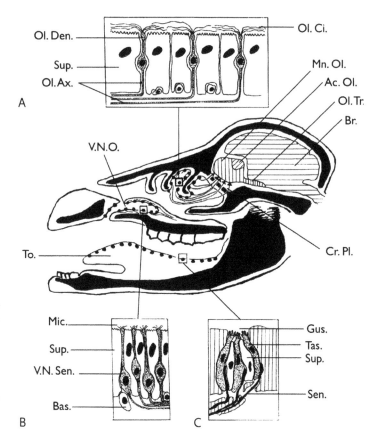

FIGURE 7.1. Diagram of bisected skull of an artiodactyl mammal showing the main components of the chemosensory organs. Insets show details of a section of the olfactory epithelium (A), a section of the vomeronasal epithelium (B), and a taste bud (C). Abbreviations: Ac. Ol., accessory olfactory bulb; Bas., basal cell; Br., brain; Cr. Pl., cribriform plate; Gus., gustatory pore; Mic., microvilli; Mn. Ol., main olfactory bulb; Ol. Ax., olfactory axons; Ol. Ci., olfactory cilia; Ol. Den., olfactory dendrite; Ol. Tr., olfactory tract; Sen., sensory nerve fibre; Sup., supporting cells; Tas., taste receptor cell; To., tongue; V.N.O., vomeronasal organ; and V.N. Sen., vomeronasal sensory neuron. Modified from Brodal (1992) and Smith (2000).

exceeds several hundred (Buck and Axel, 1991; Buck, 2000; Mombaerts, 2004).

The "sense of taste," as used in vernacular English, is actually a combination of different sensory systems: taste in the strict sense, olfaction, and somatosensory perception (i.e., sensory cells recording the texture, composition, and temperature of food items) (Mombaerts, 2004). The sense of taste proper (gustation) is mediated by taste receptor cells (Fig. 7.1). The output of these receptor cells is transmitted via cranial nerves VII, IX, and X (Herness and Gilbertson, 1999). The microvilli-carrying taste receptor cells are located in the taste buds, which are situated in the taste papillae on the upper side of the tongue and in other parts of the oral cavity (Herness and Gilbertson, 1999; Sullivan, 2002; Mombaerts, 2004). Humans, and presumably many other mammals, are capable of recognizing a vast array of different tastants; however, our species apparently can discern only five basic taste modalities, sweet, salt, sour, bitter, and umami. Additionally, water is sometimes considered a sixth basic taste (Lindemann, 1996, 2001; Herness and Gilbertson, 1999; Sullivan, 2002; Mombaerts, 2004).

OLFACTION

The various distal, morphological structures related to olfaction (Fig. 7.1) can be considered a functional whole—an olfactory organ (Negus, 1958; Pihlström et al., 2005). However, few previous investigations have taken such a comprehensive view and instead focused on the olfactory organ's individual components.

The olfactory bulb is a protruding part of the forebrain, the target of the olfactory nerves, and the main information processing center of the olfactory organ (Mori et al., 1999; Doty, 2001). It is situated in a distinctive fossa in the anterior part of the cranial cavity, which is often visible in cranial endocasts of both modern and fossil mammals (Radinsky, 1979, and references therein). This distinctiveness makes it relatively easy to measure olfactory bulb volume relative to the volume of the rest of the brain, and, consequently, the olfactory bulb is probably the most widely studied anatomical component of the olfactory organ.

The cribriform plate of the ethmoid bone is a bony plate separating the cranial and the nasal cavities. It is perforated with small holes through which the olfactory nerves project onto the olfactory bulb. A perforated cribriform plate occurs in mammals and possibly also in some of their therapsid ancestors (Crompton, 1958). A few studies have used the size of the ethmoid bone as a measure of olfactory organ size (Bhatnagar and Kallen, 1974; Pihlström et al., 2005).

The nasal turbinates are thin, highly convoluted sheets of bone and cartilage situated in the nasal cavity. They are covered with olfactory or respiratory epithelia and reach their greatest complexity in mammals (Hillenius, 1992, 1994). The respiratory epithelia of the more distal turbinates may regulate the temperature and moisture of the inhaled air. The olfactory receptors are located at the surface of the olfactory epithelium that covers the more proximal turbinates. The size and number of nasal turbinates varies in different mammals, with as many as 12 in some species (Negus, 1958). Due to the convoluted shape of the turbinates, precise histological measurement of the area of the olfactory epithelium is difficult, and few scientists have made systematic attempts to measure its surface (for a summary, see Pihlström et al., 2005).

The paired vertebrate nasal cavity ends in the left and the right nostrils, which are situated in the nose or snout. The nose has two main functions: it catches airborne odor molecules, and it is involved in respiration. In modern cetaceans, the nose has essentially been reduced to a pair of nostrils situated at the top of the head, thus facilitating respiration from a submerged position. In odontocete cetaceans, the echolocation organ is located in this region, and nasal structures are deeply involved in the generation of echolocation sound pulses. In many other mammals, the nose is a rather prominent organ and may have other uses than olfaction and respiration; for example, in the males of the pinniped genera *Mirounga* and *Cystophora*, the nose functions as an impressive display organ (Witmer, 2001).

VOMERONASAL SENSE

The second major chemosensory system, present in most mammals, is the accessory olfactory system, or the vomeronasal system. It is composed of the vomeronasal, or Jacobson's, organ, of the vomeronasal nerve fibers and of the accessory olfactory bulb. The vomeronasal organ of mammals is a paired, tubelike structure situated near the bases of both the left and the right sides of the nasal cavity, and communicating with both the oral and the nasal cavities (Fig. 7.1). In mammals, vomeronasal receptor cells generally carry microvilli, although some studies suggest the presence of ciliated receptor cells as well (Eisthen, 1992). The destination of the vomeronasal nerve bundles is not the main olfactory bulb, but a separate, although contiguous, part of it, the accessory olfactory bulb; the outputs of the main olfactory bulb and the accessory olfactory bulb are both carried by cranial nerve I (Scalia and Winans, 1975; Keverne, 1982, 1999; Døving and Trotier, 1998; Meisami and Bhatnagar, 1998; Doty, 2001; Sullivan, 2002; Halpern and Martínez-Marcos, 2003).

The functional details of the vomeronasal organ are still somewhat disputed. Before the discovery of vertebrate pheromones in the mid-twentieth century, researchers frequently did not make a clear distinction between the main and the accessory olfactory organ and treated the latter simply as an extension of the former. Subsequently, the opinion shifted to considering the vomeronasal system as an entirely separate sensory complex, the main function of which was to detect pheromones and thus play an important part in the reproductive biology of mammals (e.g., Estes, 1972). However, the view of the vomeronasal organ functioning exclusively as a detector of sexual signals has recently been challenged (Baxi et al., 2006; Eisthen and Schwenk, chapter 4 in this volume). It has been suggested that the vomeronasal organ has other functions in addition: it may play a role in intragender aggressive behavior (Leypold et al., 2002; Stowers et al., 2002). Also, several studies suggest, in at least a few mammal species, a functional overlap between the main olfactory organ and the vomeronasal organ (Sam et al., 2001; Wakabayashi et al., 2002; Trinh and Storm, 2003, 2004). Thus, the lack of a vomeronasal organ in any given mammal species may not imply a lack of pheromone detection (Brennan and Keverne, 2004).

TASTE

Most of the basic knowledge of mammalian gustation stems from studies using the common experimental species: the laboratory rat and mouse and the primates. Very few gustatory experiments have ever been made with living, aquatic or semiaquatic mammal species. However, the gross morphology of the mammalian tongue has been known since Charles F. Sonntag described the comparative anatomy of the tongue in virtually all major groups of living mammals, including several aquatic and semiaquatic taxa (Sonntag, 1922, 1923a, 1923b, 1924a, 1924b).

CHEMICAL SENSES IN AQUATIC MAMMALS

In general, there is very little quantitative data on the chemosensory organs of any group of mammals, and this is particularly true of aquatic and semiaquatic mammals. Inevitably, most available information comes from Recent species. The mammalian fossil record contains members of extant aquatic mammal taxa, specifically, early whales (for a review, see Thewissen and Williams, 2002), sirenians, pinnipeds, otters,

beavers, hippopotami, and platypus (Musser and Archer, 1998); aquatic members of taxa that now have only terrestrial species, namely, xenarthrans (Muizon and McDonald, 1995), rhinoceroses (Wall and Heinbaugh, 1999), and possibly proboscideans (Janis, 1988; Gaeth et al., 1999); and, finally, wholly extinct taxa, namely, certain Cretaceous triconodonts (Slaughter, 1969), desmostylians, and amynodontid perissodactyls (Wall and Heinbaugh, 1999). Unfortunately, with the partial exception of early whales, not much is known about the neuroanatomy of fossil aquatic and semiaquatic mammals, and thus the emphasis here is on living taxa.

MONOTREMATA

The only living semiaquatic monotreme species is the Australian platypus (*Ornithorhynchus anatinus*, Ornithorhynchidae of Fig. 1.7 and Table 1.4 in this volume). The various parts of the main olfactory organ of the platypus are generally weakly developed, with the most noteworthy feature being the lack of a perforated ethmoid bone in the adult platypus (Elliot Smith, 1895; Paulli, 1900; Beer and Fell, 1936). Such a structure is present only for a brief period of time during the ontogeny of the platypus, and the cribriform plate does not become ossified but remains cartilaginous (Zeller, 1988, 1989). The olfactory bulb volume of the platypus is relatively smaller than that of its relative, the terrestrial short-beaked echidna (*Tachyglossus aculeatus*) (Pirlot and Nelson, 1978), and the number of olfactory turbinates in the nasal cavity is only two in *Ornithorhynchus* (Zeller, 1989) but as many as seven in *Tachyglossus* (Kuhn, 1971). The vomeronasal organ is present in the platypus (Elliot Smith, 1895).

In spite of being only semiaquatic, the olfactory organ of the platypus is thus reduced to a degree surpassed only by the fully aquatic whales, which are the only other mammals that lack a perforated cribriform plate. The closest living relative of the platypus, the echidna, has a large olfactory organ, including a well-developed cribriform plate (Kuhn, 1971). The fossil record indicates that the platypus is anatomically closer to the last common ancestor of the modern monotremes than are echidnas (Musser, 2003). This raises the interesting possibility that echidnas have evolved from platypus-like ancestors and are thus secondarily terrestrial (Gregory, 1947; Musser, 2003). It is not at present known whether the last common ancestor of modern monotremes had a fully functional, perforated cribriform plate. Therefore, it is not certain whether this structure has subsequently been lost in the platypus, or evolved independently in the echidna and in therian mammals. The latter alternative seems less likely but is not entirely implausible, as the recent discovery of a parallel evolution of middle ear bones in monotremes and therians has shown (Rich et al., 2005).

DIDELPHIMORPHIA

There is only one Recent, semiaquatic marsupial species: the yapok, or water opossum (*Chironectes minimus*, Didelphidae of Fig. 1.7 and Table 1.4 in this volume), of South America. However, adaptation to life in water has not resulted in any pronounced differences in the skull morphology of the yapok in comparison with terrestrial opossum species (Sánchez-Villagra and Asher, 2002), and like all marsupials studied to date, the yapok possesses a vomeronasal organ (Sánchez-Villagra, 2001).

INSECTIVORES

Insectivorous placental mammals of various taxonomic groups (Chrysochloridae, Erinaceidae, Solenodontidae, and Tenrecidae of Fig. 1.7, and Afrosoricida and Eulipotyphla of Table 1.4 in this volume) have been the subjects of a considerable number of comparative studies on different brain components. These include studies of the volume and/or the relative size of the main olfactory bulb (Stephan and Andy, 1964; Bauchot and Stephan, 1966; Stephan, 1967; Stephan et al., 1981; Baron et al., 1983; Barton et al., 1995; Catania et al., 1999), as well as of the accessory olfactory bulb (Stephan, 1965). These studies focus on different members of the order Insectivora rather than the semiaquatic species in particular, but others

(Stephan and Bauchot, 1959; Bauchot and Stephan, 1968; Söllner and Kraft, 1980; Stephan and Kuhn, 1982; Stephan et al., 1986; Sánchez-Villagra and Asher, 2002) deal specifically with semiaquatic insectivores. The main finding of these studies is that semiaquatic tenrecids, soricids, and talpids do indeed have relatively smaller olfactory organs than their terrestrial relatives.

CARNIVORA

The olfactory organs of most aquatic and semiaquatic carnivores (Fig. 1.7 and Table 1.4 in this volume) remain insufficiently studied. Radinsky suggested olfactory bulb size reduction in otters (1968) and in semiaquatic viverrids and herpestids (1975). Gittleman (1991) provided the first truly comprehensive overview of olfactory bulb size in Recent carnivores, and his results support Radinsky's regarding the reduction of the olfactory bulb in otters. Unfortunately, Gittleman did not include seals, sea lions, or walruses in his study. The olfactory organ of pinnipeds has thus far received surprisingly little attention, and few studies have taken an explicitly evolutionary approach to this question. Noteworthy exceptions are the articles by Mitchell and Tedford (1973) and Repenning (1976) on early eared seals and their relatives. Repenning, in particular, emphasized the reduction of the olfactory organ as a main feature of otariid and odobenid evolution. However, he refrained from concluding whether this reduction has been primarily an adaptation to aquatic conditions or a trade-off resulting from an increase of eye size in pinnipeds. Modern pinnipeds are more accessible for study, but even here our knowledge is still very incomplete. Since pinnipeds spend a significant part of their life on land, most notably during mating and reproduction, they could be expected to rely on olfactory cues to some extent, and there are indeed observations to support this (e.g., Burton et al., 1975; Dobson and Jouventin, 2003; Phillips, 2003). Regarding the basic morphology of the pinniped olfactory organ, it has long been known that both the volume of the olfactory bulb (Spitzka, 1890) and the size and number of the nasal turbinates

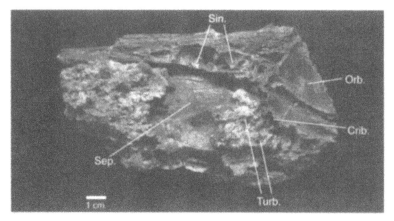

FIGURE 7.2. Left lateral view of the nasal cavity of the Eocene whale *Andrewsiphius*, with the left maxilla removed. This specimen retains a number of olfactory turbinates (Turb.), and these are covered by gypsum in this specimen. The specimen is in the collection of the Indian Institute of Technology, Roorkee (IITR-SB 2006-054). Other abbreviations: Crib., cribriform plate; Orb., medial orbital wall; Sep., nasal septum; and Sin., paranasal sinuses. Photo courtesy of Sunil Bajpai and J. G. M. Thewissen.

(Anthony and Iliesco, 1926) in seals and sea lions are reduced. However, more recently, Pihlström et al. (2005) noted that the cribriform plate of ethmoid in phocids and otariids does not markedly differ in size from that of terrestrial carnivores. That the pinniped sense of smell is, in fact, quite acute was corroborated by Kowalewsky et al. (2005); they showed that phocid seals are able to detect very low concentrations of dimethyl sulfide. The pinniped vomeronasal system has not been much investigated, but otariid seals, at least, possess an accessory olfactory bulb (Switzer et al., 1980).

We still have much to learn about olfaction in both terrestrial and semiaquatic carnivores, and a major comparative study of the morphology and evolution of the main olfactory organ in Carnivora is needed. The gustation abilities of pinnipeds and other semiaquatic carnivores have hardly been studied (but see Kvitek et al., 1991). Sonntag (1923a) considered the pinniped tongue anatomy intermediate between that of cetaceans and terrestrial carnivores. Kubota (1968) largely confirmed Sonntag's anatomical descriptions.

SIRENIA

The olfactory capacities of the only fully aquatic herbivorous mammals, the sirenians, have so far received relatively little attention, but it is evident that a significant reduction of the size of the main olfactory organ has taken place (Genschow, 1934; Pirlot and Kamiya, 1985). The vomeronasal organ in sirenians is lacking altogether (Genschow, 1934; Mackay-Sim et al., 1985). The gustatory capabilities of sirenians are poorly known. Anatomically the sirenian tongue is intermediate between that of cetaceans and terrestrial ungulates (Sonntag, 1922; Levin and Pfeiffer, 2002).

CETACEA

Whales, dolphins, and porpoises are the most fully aquatic of all mammals, and with regards to their olfactory abilities, the cetaceans are also thought to be the least well endowed among mammals. Endocasts of fossil skulls of Eocene archaeocete whales show that these forms still possessed relatively large olfactory bulbs (Dart, 1923; Kellogg, 1928; Edinger, 1955) and olfactory turbinates are present in the nasal cavity of some (Fig. 7.2), but in extant cetaceans the olfactory bulbs, as well as the ethmoid bone, the ethmoturbinals, and the vomeronasal organ, are either completely or almost completely rudimentary. As is the case with the platypus, embryonic whales possess better-developed

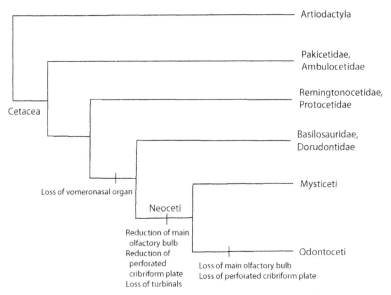

FIGURE 7.3. Hypothesis of chemical sense evolution in cetaceans. Loss of the vomeronasal organ is inferred from the absence of anterior palatine foramina, which are present in early archaeocetes (Nummela et al., 2006) but absent in later archaeocetes (Uhen, 2004). These foramina are a necessary conduit for the vomeronasal duct, although presence of the foramina does not imply that the organ is present. Character state transitions at the indicated nodes are consistent with available data but are not the only possible interpretation of transitions.

olfactory organs than do adults (Oelschläger and Buhl, 1985; Oelschläger et al., 1987; Oelschläger, 1989; Marino et al., 2001). Interestingly, the reduction of the olfactory organ is more complete in toothed than in baleen whales (Fig. 7.3) (Gruhl, 1911; Edinger, 1955; Breathnach, 1960; Morgane and Jacobs, 1972; Cave, 1988). If and to what extent baleen whales use olfaction under natural conditions is at present unknown.

Regarding taste, cetacean tongues are simpler in structure than those of terrestrial mammals (Sonntag, 1922), the taste buds and papillae being either very few or absent altogether. Sonntag's early observations have been corroborated by more recent light and electron microscopic investigations on the tongue structure of whales (Yamasaki et al., 1978; Pfeiffer et al., 2001). Nevertheless, there are experimental data suggesting that at least some odontocetes possess a rather well developed sense of taste. For example, Nachtigall and Hall (1984) showed that the threshold for detecting citric acid and quinine sulfate in the bottlenose dolphin (*Tursiops truncatus*) is almost as low as it is in humans.

ARTIODACTYLA

Many species of suids, cervids, and bovids prefer wet habitats and are excellent swimmers, but the only truly semiaquatic artiodactyls are the hippopotami (Hippopotamidae). The amount of published literature on the chemosensory organs of hippopotami is small and usually qualitative in nature (e.g., Pilleri, 1962).

RODENTIA

Most of what is known about the vomeronasal and the main olfactory organs of rodents is based on studies of the laboratory rat and mouse, but some research has been conducted on semiaquatic rodents. Comparative studies on the brain components of several semiaquatic rodent species led Pilleri (1959, 1983) to conclude that the main olfactory bulbs of semiaquatic rodents are not reduced and are indeed particularly large in the beaver (*Castor canadensis*).

TABLE 7.1

Presence of Olfaction- and Vomeronasal-Related Anatomical Structures in Aquatic and Semiaquatic Mammal Taxa

TAXON	MOB	CPE	ET	VNO
Monotremata				
Ornithorhynchidae	Yes	No	Reduced	Yes
Didelphimorphia				
Didelphidae	Yes	Yes	Yes	Yes
Afrosoricida				
Tenrecidae	Yes	Yes	Yes	Yes
Sirenia	Reduced	Reduced	Reduced	No
Rodentia				
Castoridae	Yes	Yes	Yes	Yes
Muridae	Yes	Yes	Yes	Yes
Thryonomyidae	Yes	Yes	Yes	Yes
Hydrochaeridae	Yes	Yes	Yes	Yes
Myocastoridae	Yes	Yes	Yes	Yes
Eulipotyphla				
Soricidae	Yes	Yes	Yes	Yes
Talpidae	Yes	Yes	Yes	Yes
Carnivora				
Viverridae	Yes	Yes	Yes	Yes
Herpestidae	Yes	Yes	Yes	Yes
Ursidae	Yes	Yes	Yes	Yes
Mustelidae	Yes	Yes	Yes	Yes
Pinnipedia	Yes	Yes	Reduced	Yes
Cetacea				
Archaeoceti	Yes	Yes	Yes	Unknown
Odontoceti	No	No	No	No
Mysticeti	Minimal	No	Reduced	No
Artiodactyla				
Hippopotamidae	Yes	Yes	Yes	Yes

NOTE: MOB, main olfactory bulb; CPE, cribriform plate of ethmoid; ET, ethmoturbinals; VNO, vomeronasal organ.

EVOLUTION OF THE CHEMICAL SENSES

Apart from highlighting the substantial lack of useful quantitative data on the chemosensory organs of mammals, this brief review shows that there are qualitative differences between different groups of aquatic and semiaquatic mammals with regards to their olfactory organs. In several members of these aquatic lineages, the relative and absolute size of the main olfactory organ has been reduced, and, by implication, some of their olfactory acuity has been lost. Table 7.1 summarizes the presence or absence, respectively, of various components of the main and the accessory olfactory organs in all the Recent and in one of the extinct taxa (i.e., the archaeocete whales) of aquatic and semiaquatic mammals.

Are the olfactory organs of aquatic and semiaquatic mammals significantly smaller than those of fully terrestrial mammals? Published data for the main olfactory bulb volumes, total

TABLE 7.2
Body Mass, Brain Volume, Main Olfactory Bulb Volume (MOB) and Accessory Olfactory Bulb Volume (AOB) of Semiaquatic Mammals

SPECIES	BODY MASS	TOTAL BRAIN VOLUME	MOB	AOB
Monotremata				
Ornithorhynchus anatinus	1040	8572	71.4	–
Afrosoricida				
Limnogale mergulus	92	1046	42.9	0.32
Potamogale velox	660	3878	86	2.31
Micropotamogale lamottei	64.2	743	34	0.06
Eulipotyphla				
Neomys fodiens	15.2	299	15.8	0.10
Desmana moschata	440	3620	141	0.22
Galemys pyrenaicus	57.5	1230	39	0.21

DATA SOURCES: *Ornithorhynchus*, Pirlot and Nelson (1978); all others, Stephan et al. (1981).

NOTE: Mass in grams; volume in cubic millimeters.

brain volumes, and body weights are available for only 41 Recent species of nonprimate mammals, representing monotremes, opossums, megabats, afrosoricids, eulipotyphlans, and tree shrews, the three latter groups being collectively referred to as insectivores in these publications (Stephan et al., 1974, 1981; Pirlot and Nelson, 1978). In addition, Stephan et al. (1981) also give accessory olfactory bulb volume data for the insectivores, but those data are not further analyzed here. The values for the seven semiaquatic species in the data set are summarized in Table 7.2. When olfactory bulb volume is related to total brain volume, the semiaquatic species do not differ significantly from terrestrial mammals (Fig. 7.4A). However, when bulb volume is plotted against body weight, a very clear difference between terrestrial and semiaquatic mammals becomes evident, with the semiaquatic species having relatively smaller olfactory bulb volumes (Fig. 7.4B). This difference between terrestrial and semiaquatic species is intriguing, but caution must be exercised when interpreting the results. The number of species in the sample is small and taxonomically limited, and truly large-bodied species are lacking (the largest included species being the short-beaked echidna, *Tachyglossus aculeatus*). To complicate

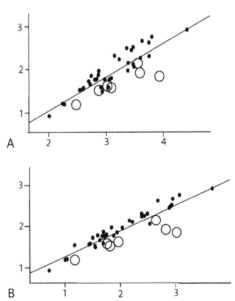

FIGURE 7.4. Relative size of the olfactory bulb in nonprimate mammals, as shown by plotting the olfactory bulb volume against total brain volume (A) and against body size (B). Terrestrial mammals are indicated as dots, semiaquatic mammals as open circles. (A) The least squares regression equation is log (bulb volume) = $-0.527 + 0.783$ log (brain volume), $n = 41$, $R^2 = 0.750$. ANCOVA results for semiaquatic versus terrestrial species as covariates: F value = 3.752, $P = 0.060$. (B) The least squares regression equation is log (bulb volume) = $0.639 + 0.622$ log (body mass), $n = 41$, $R^2 = 0.836$. ANCOVA results for semiaquatic versus terrestrial species as covariates: F value = 12.140, $P = 0.001$. Original data are from Stephan et al. (1974), Pirlot and Nelson (1978), and Stephan et al. (1981).

TABLE 7.3

Skull Areas (Skull Length × Basicranial Width) and Ethmoid Bone Areas
(Square Millimeters) of Aquatic and Semiaquatic Mammals

SPECIES	SKULL AREA	ETHMOID AREA
Afrosoricida		
Micropotamogale ruwenzorii	1125	58
Sirenia		
Dugong dugon	51570	158
Rodentia		
Castor fiber	7683	139
Arvicola terrestris	649	14
Ondatra zibethica	1727	25
Myocastor coypus	5082	77
Eulipotyphla		
Neomys fodiens	236	12
Carnivora		
Mustela vison	2287	70
Lutra lutra	6729	110
Enhydra lutris	13479	272
Callorhinus ursinus	28449	506
Zalophus californianus	25463	547
Cystophora cristata	39780	508
Phoca grypus	27086	402
Phoca hispida	16557	159
Artiodactyla		
Hippopotamus amphibius	187050	1940

DATA SOURCE: Pihlström et al., 2005.

matters further, it has been shown that among taxa with both semiaquatic and terrestrial species (e.g., shrews), the semiaquatic species tend to have larger body size than their terrestrial relatives, for reasons not related to sensory biology (Wolff and Guthrie, 1985). Based on current data, no robust conclusions regarding the effect of aquatic life on the size of the main olfactory organ can be drawn.

Pihlström et al. (2005) measured the area of the ethmoid bone in relation to skull area (i.e., skull length multiplied with basicranial width) in 134 nonprimate mammals, among which 16 are aquatic or semiaquatic. The values for these 16 species are summarized in Table 7.3. There is a small but significant difference in the average ethmoid bone:skull area ratios between terrestrial and nonterrestrial species, with the terrestrial species having larger ratios (Fig. 7.5). However, this difference is mainly due to the position of one single species, the dugong. By contrast, the five pinniped species in the sample do not markedly deviate from terrestrial mammals; this result is inconsistent with Sir Victor Negus's claim that the cribriform plate of seals is "insignificant" (Negus, 1958:55).

Olfaction, in general, seems to be of little importance to aquatic and semiaquatic mammals. Instead, other sensory systems have become more prominent, and in some lineages, entirely new mammalian sensory modalities have evolved. The most obvious changes have occurred in the toothed whales and the ornithorhynchid monotremes, respectively; the

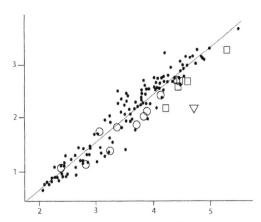

FIGURE 7.5. The relationship between ethmoid bone area and total skull area in nonprimate mammals. Total skull area is defined as skull length multiplied with basicranial width. Terrestrial mammals are indicated by dots, pinnipeds by open squares, dugong by an open triangle, and other semiaquatic mammals by open circles. Neither whales nor the platypus are included, as they lack a perforated cribriform plate. The least squares regression equation is log (ethmoid area) $= -1.116 + 0.890$ log (skull area), $n = 134$, $R^2 = 0.910$. ANCOVA results for semiaquatic versus terrestrial species as covariates: F value $= 6.642$, $P = 0.011$. Original data are from Pihlström et al. (2005).

loss of olfactory prowess is compensated by the evolution of echolocation in toothed whales, and by electroreception in the platypus (Scheich et al., 1986; Gregory et al., 1987, 1988; Pettigrew, 1999; Proske and Gregory, 2003; Wilkens and Hofmann, chapter 20 in this volume). In other aquatic and semiaquatic mammals the changes have been less dramatic; still, it has been proposed that the large whiskers of phocid seals may function not only as tactile identifiers of objects in the immediate vicinity, as they do in terrestrial mammals, but as a long-distance sense organ (Dehnhardt et al., 2001; Dehnhardt and Mauck, chapter 19 in this volume), and the presence of tactile body hair in the manatee has led to the suggestion that this animal, uniquely among mammals, effectively possesses a lateral line (Reep et al., 2002). Surprisingly, in baleen whales, no such compensating long-distance sensory system has yet been demonstrated. Less surprisingly, with one partial exception, "special" underwater senses are apparently also lacking in the remaining semiaquatic mammals, such as

otters, beavers, and water shrews. The partial exception is the star-nosed mole (*Condylura cristata*), which has several fleshy appendages at the tip of its nose; these appendages are used for identification of food items by touch (Catania and Kaas, 1996). However, it is questionable whether these appendages can be considered a primarily aquatic adaptation, as the star-nosed mole uses them both on land and underwater.

Generalizing, these evolutionarily younger semiaquatic taxa are presumably more similar to their terrestrial ancestors with regards to their sensory adaptations than are cetaceans, sirenians, pinnipeds, and the platypus. There may indeed exist, for each particular lineage, an inverse relationship between the relative size and importance of the main olfactory organ, and the length of evolutionary time at the group's disposal; however, we still have much to learn about the chemical senses in mammals, whether terrestrial or aquatic.

ACKNOWLEDGMENTS

I thank the editors for inviting me to contribute to this book. T. Reuter, M. Fortelius, and B. O'Hara provided valuable comments and suggestions on this manuscript. I also thank Drs. Sunil Bajpai and J. G. M. Thewissen for providing the photograph of the nasal cavity of a fossil whale specimen and for providing information relating to fossil whales. This research was supported by grants from Svenska Kulturfonden, Oskar Öflunds Stiftelse, and Ella och Georg Ehrnrooths Stiftelse.

LITERATURE CITED

Anthony, R. L. F., and G. M. Iliesco. 1926. Étude sur les cavités nasales des carnassiers. Proceedings of the Zoological Society of London 65: 989–1015.

Baron, G., H. D. Frahm, K. P. Bhatnagar, and H. Stephan. 1983. Comparison of brain structure volumes in Insectivora and primates. III. Main olfactory bulb (MOB). Journal für Hirnforschung 24:551–558.

Barton, R. A., A. Purvis, and P. H. Harvey. 1995. Evolutionary radiation of visual and olfactory brain systems in primates, bats and insectivores. Philosophical Transactions of the Royal Society of London B 348:381–392.

Bauchot, R., and H. Stephan. 1966. Donnees nouvelles sur l'encephalisation des insectivores et des prosimiens. Mammalia 30:160–196.

Bauchot, R., and H. Stephan. 1968. Etude des modifications encephaliques observees chez les insectivores adaptes a la recherché de nourriture en milieu aquatique. Mammalia 32:228–275.

Baxi, K. N., K. M. Dorries, and H. L. Eisthen. 2006. Is the vomeronasal system really specialized for detecting pheromones? Trends in Neurosciences 29:1–7.

Beer, G. R. de., and W. A. Fell. 1936. The development of the Monotremata. Part III. The development of the skull of *Ornithorhynchus*. Transactions of the Zoological Society of London 23:1–42.

Bhatnagar, K. P., and F. C. Kallen. 1974. Cribriform plate of ethmoid, olfactory bulb and olfactory acuity in forty species of bats. Journal of Morphology 142:71–90.

Breathnach, A. S. 1960. The cetacean central nervous system. Biological Reviews 35:187–230.

Brennan, P. A., and E. B. Keverne. 2004. Something in the air? New insights into mammalian pheromones. Current Biology 14:R81–R89.

Brodal, P. 1992. The Central Nervous System: Structure and Function. Oxford University Press, New York.

Buck, L. B. 2000. The molecular architecture of odor and pheromone sensing in mammals. Cell 100:611–618.

Buck, L., and R. Axel. 1991. A novel multigene family may encode odorant receptors: a molecular basis for odor recognition. Cell 65:175–187.

Burton, R. W., S. S. Anderson, and C. F. Summers. 1975. Perinatal activities in the grey seal (*Halichoerus grypus*). Journal of Zoology (London) 177:197–201.

Catania, K. C., and J. H. Kaas. 1996. The unusual nose and brain of the star-nosed mole. BioScience 46:578–586.

Catania, K. C., D. C. Lyon, O. B. Mock, and J. H. Kaas. 1999. Cortical organization in shrews: evidence from five species. Journal of Comparative Neurology 410:55–72.

Cave, A. J. E. 1988. Note on olfactory activity in mysticetes. Journal of Zoology (London) 214:307–311.

Crompton, A. W. 1958. The cranial morphology of a new genus and species of ictidosaurian. Proceedings of the Zoological Society of London 130:183–216.

Dart, R. A. 1923. The brain of the Zeuglodontidae (Cetacea). Proceedings of the Zoological Society of London 42:615–654.

Dehnhardt, G., B. Mauck, W. Hanke, and H. Bleckmann. 2001. Hydrodynamic trail-following in harbor seals (*Phoca vitulina*). Science 293:102–104.

Dobson, F. S., and P. Jouventin. 2003. How mothers find their pups in a colony of Antarctic fur seals. Behavioural Processes 61:77–85.

Doty, R. L. 2001. Olfaction. Annual Review of Psychology 52:423–452.

Døving, K. B., and D. Trotier. 1998. Structure and function of the vomeronasal organ. Journal of Experimental Biology 201:2913–2925.

Edinger, T. 1955. Hearing and smell in cetacean history. Monatsschrift für Psychiatrie und Neurologie 129:37–58.

Eisthen, H. L. 1992. Phylogeny of the vomeronasal system and of receptor cell types in the olfactory and vomeronasal epithelia of vertebrates. Microscopy Research and Technique 23:1–21.

Elliot Smith, G. 1895. Jacobson's organ and the olfactory bulb in *Ornithorhynchus*. Anatomischer Anzeiger 11:161–166.

Estes, R. D. 1972. The role of the vomeronasal organ in mammalian reproduction. Mammalia 36: 315–341.

Freitag, J., J. Krieger, J. Strotman, and H. Breer. 1995. Two classes of olfactory receptors in *Xenopus laevis*. Neuron 15:1383–1392.

Freitag, J., G. Ludwig, I. Andreini, P. Rössler, and H. Breer. 1998. Olfactory receptors in aquatic and terrestrial vertebrates. Journal of Comparative Physiology A 183:635–650.

Gaeth, A. P., R. V. Short, and M. B. Renfree. 1999. The developing renal, reproductive, and respiratory systems of the African elephant suggest an aquatic ancestry. Proceedings of the National Academy of Sciences USA 96:5555–5558.

Genschow, J. 1934. Über den Bau und die Entwicklung des Geruchsorganes der Sirenen. Zeitschrift für Morphologie und Ökologie der Tiere 28:402–444.

Gilad, Y., O. Man, S. Pääbo, and D. Lancet. 2003. Human specific loss of olfactory receptor genes. Proceedings of the National Academy of Sciences USA 100:3324–3327.

Gittleman, J. L. 1991. Carnivore olfactory bulb size: allometry, phylogeny and ecology. Journal of Zoology 225:253–272.

Gregory, J. E., A. Iggo, A. K. McIntyre, and U. Proske. 1987. Electroreceptors in the platypus. Nature 326:386–387.

Gregory, J. E., A. Iggo, A. K. McIntyre, and U. Proske. 1988. Receptors in the bill of the platypus. Journal of Physiology 400:349–366.

Gregory, W. K. 1947. The monotremes and the palimpsest theory. Bulletin of the American Museum of Natural History 88:1–52.

Gruhl, K. 1911. Beiträge zur Anatomie und Physiologie der Cetaceennase. Jenaische Zeitschrift für Naturwissenschaft 47:367–414.

Halpern, M., and A. Martínez-Marcos. 2003. Structure and function of the vomeronasal system: an update. Progress in Neurobiology 70:245–318.

Herness, M. S., and T. A. Gilbertson. 1999. Cellular mechanisms of taste transduction. Annual Review of Physiology 61:873–900.

Hillenius, W. J. 1992. The evolution of nasal turbinates and mammalian endothermy. Paleobiology 18:17–29.

Hillenius, W. J. 1994. Turbinates in therapsids: evidence for Late Permian origins of mammalian endothermy. Evolution 48:207–229.

Janis, C. M. 1988. New ideas in ungulate phylogeny and evolution. Trends in Ecology and Evolution 3:291–297.

Kellogg, R. 1928. The history of whales: their adaptation to life in the water (concluded). Quarterly Review of Biology 3:174–208.

Keverne, E. B. 1982. Chemical senses: smell; pp. 409–427 in H. B. Barlow and J. D. Mollon (eds.), The Senses. Cambridge University Press, Cambridge.

Keverne, E. B. 1999. The vomeronasal organ. Science 286:716–720.

Kowalewsky, S., M. Dambach, B. Mauck, and G. Dehnhardt. 2005. High olfactory sensitivity for dimethyl sulphide in harbour seals. Biology Letters 2: 106–109.

Kubota, K. 1968. Comparative anatomical and neurohistological observations on the tongue of the northern fur seal (Callorhinus ursinus). Anatomical Record 161:257–268.

Kuhn, H.-J. 1971. Die Entwicklung und Morphologie des Schädels von Tachyglossus aculeatus. Abhandlungen der Senckenbergischen Naturforschenden Gesellschaft 528:1–192.

Kvitek, R. G., A. R. DeGange, and M. K. Beitler. 1991. Paralytic shellfish poisoning toxins mediate feeding behavior of sea otters. Limnology and Oceanography 36:393–404.

Levin, M. J., and C. J. Pfeiffer. 2002. Gross and microscopic observations on the lingual structure of the Florida manatee Trichechus manatus latirostris. Anatomia, Histologia, Embryologia 31:278–285.

Leypold, B. G., C. R. Yu, T. Leinders-Zufall, M. M. Kim, F. Zufall, and R. Axel. 2002. Altered sexual and social behaviors in trp2 mutant mice. Proceedings of the National Academy of Sciences USA 99:6376–6381.

Lindemann, B. 1996. Taste reception. Physiological Reviews 76:719–766.

Lindemann, B. 2001. Receptors and transduction in taste. Nature 413:219–225.

Mackay-Sim, A., D. Duvall, and B. M. Graves. 1985. The West Indian manatee (Trichechus manatus) lacks a vomeronasal organ. Brain, Behavior, and Evolution 27:186–194.

Marino, L., T. L. Murphy, L. Gozal, and J. I. Johnson. 2001. Magnetic resonance imaging and three-dimensional reconstructions of the brain of a fetal common dolphin, Delphinus delphis. Anatomy and Embryology 203:393–402.

Meisami, E., and K. P. Bhatnagar. 1998. Structure and diversity in mammalian accessory olfactory bulb. Microscopy Research and Technique 43:476–499.

Mitchell, E., and R. H. Tedford. 1973. The Enaliarctinae, a new group of extinct aquatic Carnivora and a consideration of the origin of the Otariidae. Bulletin of the American Museum of Natural History 151:201–284.

Mombaerts, P. 2004. Genes and ligands for odorant, vomeronasal and taste receptors. Nature Reviews Neuroscience 5:263–278.

Morgane, P. J., and M. S. Jacobs. 1972. Comparative anatomy of the cetacean nervous system; pp. 117–244 in R. J. Harrison (ed.), Functional Anatomy of Marine Mammals, Vol. 1. Academic Press, London.

Mori, K., H. Nagao, and Y. Yoshihara. 1999. The olfactory bulb: coding and processing of odor molecule information. Science 286:711–715.

Muizon, C. de, and H. G. McDonald. 1995. An aquatic sloth from the Pliocene of Peru. Nature 375:224–227.

Musser, A. M. 2003. Review of the monotreme fossil record and comparison of palaeontological and molecular data. Comparative Biochemistry and Physiology A 136:927–942.

Musser, A. M., and M. Archer. 1998. New information about the skull and dentary of the Miocene platypus Obdurodon dicksoni, and a discussion of ornithorhynchid relationships. Philosophical Transactions of the Royal Society of London B 353:1063–1079.

Nachtigall, P. E., and R. W. Hall. 1984. Taste reception in the bottlenosed dolphin. Acta Zoologica Fennica 172:147–148.

Negus, V. 1958. The Comparative Anatomy of the Nose and Paranasal Sinuses. E. and S. Livingstone, Edinburgh.

Nummela, S., S. T. Hussain, and J. G. M. Thewissen. 2006. Cranial anatomy of Pakicetidae (Cetacea, Mammalia). Journal of Vertebrate Paleontology 26:746–759.

Oelschläger, H. A. 1989. Early development of the olfactory and terminalis systems in baleen whales. Brain, Behavior and, Evolution 34:171–184.

Oelschläger, H. A., and E. H. Buhl. 1985. Development and rudimentation of the peripheral olfactory system in the harbor porpoise Phocoena

phocoena (Mammalia: Cetacea). Journal of Morphology 184:351–360.

Oelschläger, H. A., E. H. Buhl, and J. F. Dann. 1987. Develoment of the Nervus terminalis in mammals including toothed whales and humans. Annals of the New York Academy of Sciences 519:447–464.

Paulli, S. 1900. Über die Pneumaticität des Schädels bei den Säugethieren. I. Über den Bau des Siebbeins. Über die Morphologie des Siebbeins und die der Pneumaticität bei den Monotremen und der Marsupialern. Gegenbaurs Morphologisches Jahrbuch 28:147–178.

Pettigrew, J. D. 1999. Electroreception in monotremes. Journal of Experimental Biology 202:1447–1454.

Pfeiffer, D. C., A. Wang, J. Nicolas, and C. J. Pfeiffer. 2001. Lingual ultrastructure of the long-finned pilot whale (*Globicephala melas*). Anatomia, Histologia, Embryologia 30:359–365.

Phillips, A. V. 2003. Behavioral cues used in reunions between mother and pup South American fur seals (*Arctocephalus australis*). Journal of Mammalogy 84:524–535.

Pihlström, H., M. Fortelius, S. Hemilä, R. Forsman, and T. Reuter. 2005. Scaling of mammalian ethmoid bones can predict olfactory organ size and performance. Proceedings of the Royal Society of London B 272:957–962.

Pilleri, G. 1959. Das Gehirn der Wassernager (*Castor canadensis, Ondatra zibethica, Myocastor coypus*). Acta Anatomica 39 (Suppl.): 96–123.

Pilleri, G. 1962. Zur Anatomie des Gehirnes von *Choeropsis liberiensis* Morton (Mammalia, Artiodactyla). Acta Zoologica 43:229–245.

Pilleri, G. 1983. Central nervous system, cranio-cerebral topography and cerebral hierarchy of the Canadian beaver (*Castor canadensis*); pp. 19–59 in G. Pilleri (ed.), Investigations on Beavers, Vol. I, Brain Anatomy Institute, Berne.

Pirlot, P., and T. Kamiya. 1985. Qualitative and quantitative brain morphology in the sirenian *Dugong dugong* Erxl. Zeitschrift für Zoologische Systematik und Evolutionsforschung 23:147–155.

Pirlot, P., and J. Nelson. 1978. Volumetric analyses of monotreme brains. Australian Zoologist 20: 171–179.

Proske, U., and E. Gregory. 2003. Electrolocation in the platypus: some speculations. Comparative Biochemistry and Physiology A 136:821–825.

Radinsky, L. B. 1968. Evolution of somatic sensory specialization in otter brains. Journal of Comparative Neurology 134:495–506.

Radinsky, L. B. 1975. Viverrid neuroanatomy: phylogenetic and behavioral implications. Journal of Mammalogy 56:130–150.

Radinsky, L. B. 1979. The fossil record of primate brain evolution. Forty-ninth James Arthur Lecture on the evolution of the human brain. American Museum of Natural History, New York 1–27.

Reep, R. L., C. D. Marshall, and M. L. Stoll. 2002. Tactile hairs on the postcranial body in Florida manatees: a mammalian lateral line? Brain, Behavior, and Evolution 350:1–14.

Repenning, C. A. 1976. Adaptive evolution of sea lions and walruses. Systematic Zoology 25: 375–390.

Rich, T. H., J. A. Hopson, A. M. Musser, T. F. Flannery, and P. Vickers-Rich. 2005. Independent origins of middle ear bones in monotremes and therians. Science 307:910–914.

Sam, M., S. Vora, B. Malnic, W. Ma, M. V. Novotny, and L. B. Buck. 2001. Odorants may arouse instinctive behaviours. Nature 412:142.

Sánchez-Villagra, M. R. 2001. Ontogenetic and phylogenetic transformations of the vomeronasal complex and nasal floor elements in marsupial mammals. Zoological Journal of the Linnean Society 131:459–479.

Sánchez-Villagra, M. R., and R. J. Asher. 2002. Cranio-sensory adaptations in small faunivorous semiaquatic mammals, with special reference to olfaction and the trigeminal system. Mammalia 66:93–109.

Scalia, F., and S. S. Winans. 1975. The differential projections of the olfactory bulb and accessory olfactory bulb in mammals. Journal of Comparative Neurology 161:31–56.

Scheich, H., G. Langner, C. Tidemann, R. B. Coles, and A. Guppy. 1986. Electroreception and electrolocation in platypus. Nature 319:401–402.

Slaughter, B. H. 1969. *Astroconodon*, the Cretaceous triconodont. Journal of Mammalogy 50:102–107.

Smith, C. U. M. 2000. Biology of Sensory Systems. John Wiley and Sons, Chichester.

Söllner, B., and R. Kraft. 1980. Anatomie und Histologie der Nasenhöhle der europäischen Wasserspitzmaus, *Neomys fodiens* (Pennant 1771), und anderer mitteleuropäischer Soriciden. Spixiana 3:251–272.

Sonntag, C. F. 1922. The comparative anatomy of the tongues of the Mammalia. VII. Cetacea, Sirenia, and Ungulata. Proceedings of the Zoological Society 44:639–657.

Sonntag, C. F. 1923a. The comparative anatomy of the tongues of the Mammalia. VIII. Carnivora. Proceedings of the Zoological Society 9:129–153.

Sonntag, C. F. 1923b. The comparative anatomy of the tongues of the Mammalia. IX. Edentata, Dermoptera, and Insectivora. Proceedings of the Zoological Society 34:515–529.

Sonntag, C. F. 1924a. The comparative anatomy of the tongues of the Mammalia. X. Rodentia. Proceedings of the Zoological Society 48:725–741.

Sonntag, C. F. 1924b. The comparative anatomy of the tongues of the Mammalia. XI. Marsupialia and Monotremata. Proceedings of the Zoological Society 49:743–755.

Spitzka, E. C. 1890. Remarks on the brain of the seals. American Naturalist 24:115–122.

Stephan, H. 1965. Der Bulbus olfactorius accessorius bei Insektivoren und Primaten. Acta Anatomica 62:215–253.

Stephan, H. 1967. Zur Entwicklungshöhe der Insektivoren nach Merkmalen des Gehirns und die Definition der "Basalen Insektivoren." Zoologischer Anzeiger 179:177–199.

Stephan, H., and O. J. Andy. 1964. Quantitative comparisons of brain structures from insectivores to primates. American Zoologist 4:59–74.

Stephan, H., and R. Bauchot. 1959. Le cerveau de Galemys pyrenaicus Geoffroy, 1811 (Insectivora Talpidae) et ses modifications dans l'adaptation a la vie aquatique. Mammalia 23:1–18.

Stephan, H., and H.-J. Kuhn. 1982. The brain of Micropotamogale lamottei Heim de Balsac, 1954. Zeitschrift für Säugetierkunde 47:129–142.

Stephan, H., P. Pirlot, and R. Schneider. 1974. Volumetric analysis of pteropid brains. Acta Anatomica 87:161–192.

Stephan, H., H. D. Frahm, and G. Baron. 1981. New and revised data on volumes of brain structures in insectivores and primates. Folia Primatologica 35:1–29.

Stephan, H., K. K. Mubalamata, and M. Stephan. 1986. The brain of Micropotamogale ruwenzorii (De Witte and Frechkop, 1955). Zeitschrift für Säugetierkunde 51:193–204.

Stowers, L., T. E. Holy, M. Meister, C. Dulac, and G. Koentges. 2002. Loss of sex discrimination and male-male aggression in mice deficient for TRP2. Science 295:1493–1500.

Sullivan, S. L. 2002. Mammalian chemosensory receptors. NeuroReport 13:A9–A17.

Switzer, R. C., J. I. Johnson, and J. A. W. Kirsch. 1980. Phylogeny through brain traits: relation of lateral olfactory tract fibers to the accessory olfactory formation as a palimpsest of mammalian descent. Brain, Behavior and, Evolution 17:339–363.

Thewissen, J. G. M., and E. M. Williams. 2002. The early radiations of Cetacea (Mammalia): evolutionary pattern and developmental correlations. Annual Review of Ecology and Systematics 33:73–90.

Trinh, K., and D. R. Storm. 2003. Vomeronasal organ detects odorants in absence of signaling through main olfactory epithelium. Nature Neuroscience 6:519–525.

Trinh, K., and D. R. Storm. 2004. Detection of odorants through the main olfactory epithelium and vomeronasal organ of mice. Nutrition Reviews 62:S189–S192.

Uhen, M. D. 2004. Form, function, and anatomy of Dorudon atrox (Mammalia, Cetacea): an archaeocete from the middle to late Eocene of Egypt. University of Michigan Papers on Paleontology 34:1–222.

Wakabayashi, Y., Y. Mori, M. Ischikawa, K. Yazaki, and K. Hagino-Yamagishi. 2002. A putative pheromone receptor gene is expressed in two distinct olfactory organs in goats. Chemical Senses 27:207–213.

Wall, W. P., and K. L. Heinbaugh. 1999. Locomotor adaptations in Metamynodon planifrons compared to other amynodontids (Perissodactyla, Rhinocerotoidea). National Park Service Paleontological Research 4:8–17.

Witmer, L. M. 2001. A nose for all reasons. Natural History 110 (June): 64–71.

Wolff, J. O., and R. D. Guthrie. 1985. Why are aquatic small mammals so large? Oikos 45:365–373.

Yamasaki, F., S. Komatsu, and T. Kamiya. 1978. Papillary projections at the lingual margin in the striped dolphin, Stenella coeruleoalba, with special reference to their development and regression. Journal of Morphology 157:33–47.

Zeller, U. 1988. The Lamina cribrosa of Ornithorhynchus (Monotremata, Mammalia). Anatomy and Embryology 178:513–519.

Zeller, U. 1989. Die Entwicklung und Morphologie des Schädels von Ornithorhynchus anatinus (Mammalia: Prototheria: Monotremata). Abhandlungen der Senckenbergischen Naturforschenden Gesellschaft 545:1–188.

Vision

8

The Physics of Light in Air and Water

Ronald H. H. Kröger

THE PHYSICS OF LIGHT

Light is electromagnetic radiation in a frequency range from about 3×10^{11} Hz to about 3.4×10^{16} Hz, including the infrared and ultraviolet regions (Hecht, 2002). The electromagnetic waves that make up light are not continuous, since light comes in tiny wave packages—quanta of light called photons. Light is thus not a continuous stream of waves that are high or low in amplitude (intensity). Instead, light is a rain of photons that may be a downpour (bright light) or a trickle (dim light). This characteristic of light is especially important under low-light conditions, when vision is dependent on the arrival of individual photons that are statistically distributed in both space and time (Rodieck, 1998).

Photons are characterized by two parameters: frequency and orientation of oscillation. In vision science, light is usually characterized by its wavelength, λ, in vacuum, and not by its frequency, f (the two are inversely related: $\lambda = c/f$, where c is the speed of light). The range of electromagnetic radiation visible to humans ranges approximately from 400 to 750 nm. Light of 400 nm is perceived as violet, 750 nm as red. Polarized light consists of photons that oscillate in the same direction. Sunlight is not polarized, in other words, there are about equal numbers of photons of all possible polarizations; but this light may be polarized in nature in several ways.

Photoreceptors (cells sensitive to light) function as photon counters, such that the quantity relevant to vision is the number of photons being absorbed by the photopigments (Rodieck, 1998). Many animal photoreceptors are able to respond to the absorption of a single photon. Enormous amplification is necessary

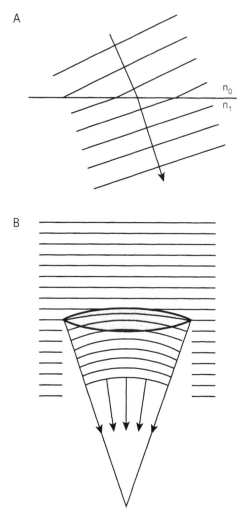

A

n_0
n_1

B

FIGURE 8.1. (A) Refraction. Light waves that travel between two transparent media of different refractive indices ($n_0 < n_1$) change their direction of propagation at the interface. This is the result of the lower speed of light in the medium of n_1. (B) Refraction in a lens. An appropriately shaped transparent body of high refractive index, a lens, will deform the wave fronts of light in such a way that they collapse on a single point, the focal point. This diagram does not illustrate diffraction of light at the edge of the aperture (free opening). Because of such diffraction, real optical systems focus light to a complex pattern instead of a point, and a smaller aperture leads to more diffraction. For instance, human visual resolution in air is limited by diffraction if pupil diameter is smaller than 2.4 mm.

energy as red photons (700 nm) (Hecht, 2002). A higher energy barrier can be overcome more easily by the absorption of an ultraviolet photon than by a red or infrared (>750 nm) photon. Ultraviolet-sensitive photopigments can therefore be more stable than photopigments sensitive to long wavelengths, and this makes ultraviolet pigments less sensitive to thermal noise. Photoreceptors always respond to stimuli in the same way because the biochemical amplification cascade generates the effect and the absorption of a photon only triggers that cascade.

In vacuum, light travels at the highest speed possible (c, which is about 300,000 km/s). When light passes through a transparent medium denser than vacuum, it is slowed down. The ratio of the speed of light in vacuum over its speed in a medium is called the refractive index n of the medium. The refractive index of air is 1.0003, which means that light is slowed only very slightly. Water is much denser than air, and the speed of light is reduced by about 25% ($n = 1.33$). Water with dissolved salt slows light down further; the refractive index of seawater is about 1.34 (Lide, 2004).

If light hits a transparent object at an angle other than 90°, any difference in refractive index causes the light to change its course: it is refracted (Fig. 8.1A). This effect can be used to focus light. If an appropriately shaped body, a lens, is placed in the path of light, the lens can deflect the incoming light in such a way that the wave fronts collapse to create a focal point (Fig. 8.1B). If there is a detector (e.g., a photoreceptor) at that position, it will receive light from an area as large as the aperture of the lens. Furthermore, light coming from different directions is focused at different locations, such that an image is created, which can be detected by an array of detectors such as the photoreceptor mosaic in the retina.

Vision depends on optical systems capable of creating images. Because of the large difference in refractive index between air ($n = 1.0003$) and water ($n = 1.33$ to 1.34), the designs of terrestrial and aquatic eyes are

to generate a useful signal from such a tiny amount of energy (Rodieck, 1998). The energy content of a photon is linearly related to its wavelength, such that photons in the near ultraviolet (350 nm) carry twice as much

TABLE 8.1
Physical Factors Relating to Vision in Air and Water and Their Effects

	AIR	WATER	EFFECTS ON EYE AND VISION
Refractive Index	Low	High	Differences in eye design, use of vision across media is difficult
Light levels	High at daytime	Decreasing with depth	Fast adaptive mechanisms in deep-diving animals
Available spectrum	Wide	Narrowing with depth	Color vision less useful in deep water
Static pressure	Low	Increasing with depth	No or little
Dynamic pressure	Low	High during fast swimming	Risk for vibrations and deformations of the eye ball in fast swimmers
Density	Low	High	Physical shocks during transitions between media
Heat capacity	Low	High	Cooling of the eye in water slows vision and increases light sensitivity
Capacity to suspend particles	Low	High	Risk of abrasion of cornea as well as scattering and absorption of light in turbid waters

different. In the terrestrial eye of a human, for example, the cornea accounts for about two-thirds of the total refractive power (Smith and Atchison, 1997) as a result of the large difference in refractive index between air and the eye (which mainly consists of fluids). The cornea of a human diver is optically ineffective in water, and vision becomes blurry because the refractive power of the human lens alone is not sufficient to focus light. In contrast, aquatic animals have powerful lenses able to focus light without any refractive power residing in the cornea. If an aquatic eye is exposed to air, the refractive power of the cornea is added, such that the total refractive power of the eye is too high, resulting in blurry vision (Land and Nilsson, 2002). It has been a particular challenge to natural evolution to find eye designs that are equally useful in both air and water.

LIGHT IN WATER

The physical conditions for vision in water are different from those in air (Table 8.1), and they are far more variable in aquatic habitats than on dry land. As a result, the visual systems of aquatic animals are adapted to a wide range of visual environments, and these environments need to be studied carefully before conclusions regarding the visual systems can be drawn. Moreover, many semiaquatic taxa and even the fully aquatic cetaceans use their eyes in air and water.

ABSORPTION AND SCATTERING

Clear ocean water is blue green because it most readily absorbs very short (ultraviolet to violet) and long (yellow to infrared) wavelengths (Jerlov, 1968; Lundgren and Höjerslev, 1971). This spectrally selective absorption leads to a narrowing of the available spectrum with

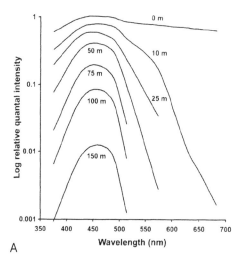

A

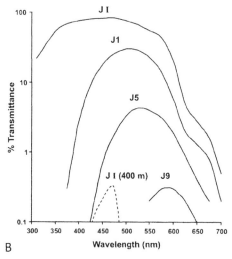

B

FIGURE 8.2. (A) Relative quantal light intensities in clear ocean water (Sargasso Sea) at different depths (after Lundgren and Höjerslev, 1971). Note that light intensity and spectral width rapidly decrease with depth. (B) Percent transmittance of 10 meters of different types of marine waters as functions of wavelength (after Jerlov, 1968). Water type JI is clear water of the open ocean; J1 to J9 are coastal waters with increasing loads of particles, dissolved substances, and microscopic organisms. Note that the amount of light transmitted, its spectral width, and the spectral maximum are different in different types of water. The water of the open ocean is clear and blue, while coastal waters are more turbid and green to brown in coloration. The *dashed line* shows the transmittance of 400 meters of clear blue ocean water.

increasing depth (Fig. 8.2A). Water also scatters light in a wavelength-dependent manner. Short wavelengths (violet to blue) are scattered more than middle to long wavelengths (green to red).

Absorption and scattering limit the distance sunlight can penetrate into water. While the light of the sun can easily reach the surface of Earth through tens of thousands of meters of air, there is virtually no sunlight below about 1000 meters of clear water (Warrant and Locket, 2004). This does not mean that the deep sea is completely dark. Many marine organisms emit light (bioluminescence), such that they provide light sources, and vision is useful even at depths not reached by sunlight (Warrant and Locket, 2004). In general, the underwater world is much darker and less colorful than most terrestrial habitats. Notable exceptions are the upper zones of tropical and subtropical coral reefs, which are environments of shallow and relatively clear water.

Scattering of light has additional profound effects. Light scattered toward an observer from behind an object back-illuminates the object. If the object is of light hue, back-illumination reduces contrast, whereas such illumination enhances contrast of dark objects. Light is also scattered between an object and the observer (Fig. 8.3). The resulting veiling effect reduces contrast, irrespective of the lightness of the object, and severely limits visual range. With sunlight as the source of illumination, maximum visual range in clear ocean water is about 40 meters (Lythgoe, 1979). Some of the light coming from the object itself is scattered toward the observer, blurring the image (Fig. 8.3). In water, contrasts and acuity of vision rapidly decrease with viewing distance even under optimal conditions.

In the pelagic (midwater) zone, predators and prey may come from any direction, and vision has to be optimized to such a situation. Viewed from above, an animal appears brighter than the background, because the animal's body reflects some of the downwelling sunlight upward. Viewed from below, the animal appears as a dark silhouette against the downwelling light. To minimize these effects, many aquatic animals use a strategy known as countershading. The dorsal body half is darkly pigmented, while the ventral half is of light hue. The

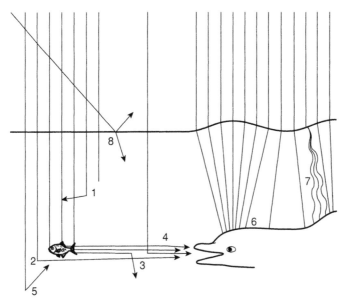

FIGURE 8.3. Optical factors that influence underwater vision. Incoming light may be absorbed or scattered out of the path to an object and be lost for vision (1). Light is scattered toward the eye of the observer from behind an object of interest (2), and this light generates background illumination. Some light from an object is scattered away on its way to the observer whereas downwelling light is scattered to the observer creating a veiling effect and reducing contrast (3). Other rays of light coming from the object reach the eye at a changed angle because of scattering and blur the image (4). Upwardly scattered light illuminates the underside of an object (5), and this is exploited in the countershading of certain aquatic animals. Wave refraction of light creates variations in intensity (6) and wavelength (7, *wavy lines* represent different colors). Both effects are most prominent close to the water surface. Partial polarization of light occurs at the water surface (8). The degree of polarization increases with the angle of incidence and decreases with water depth, the latter because of random scattering of light rays on their ways through the water column.

function of the dark dorsal body surface is obvious, because it minimizes the amount of light being reflected upward. The light ventral surface is somewhat more difficult to understand, because a shadow should be dark, regardless of the coloration of the animal. However, because water scatters light in all directions, some downwelling sunlight is scattered upward. Such light illuminates the lightly colored ventral body surface and is reflected downward, such that the shadow effect is reduced (Fig. 8.3).

These generalizations hold for the visual environment of the open ocean with clear blue water. In coastal regions, especially estuaries, the water may be heavily loaded with particles and colored dissolved substances, so-called *Gelbstoffe* (a German word meaning "yellow substances").

Particles mainly increase scattering, sometimes to a degree that the water appears milky. This wavelength-dependent scattering shifts the maximum energy of the available light toward long wavelengths (red). Gelbstoffe absorb short wavelengths and thus also lead to a red shift of the frequency spectrum. Coastal waters are therefore quite often not blue, but brownish in coloration (Fig. 8.2B; see Jerlov, 1968). Furthermore, visual range and depth of penetration by sunlight may be extremely short. Algae and photosynthetic bacteria may further change the color and reduce the transparency of water.

REFRACTION

The effects of the water surface are also significant because air and water are different in

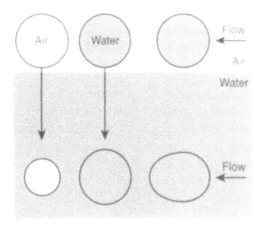

FIGURE 8.4. Effects of static and dynamic pressures. If an air-filled balloon is pulled underwater, the air is compressed by static pressure that rapidly increases with depth. The eye is essentially a fluid-filled structure and is not deformed by static pressure because water is largely incompressible. Flowing air has hardly any effect on an eye because of the low density and viscosity of air. In contrast, dynamic water pressure may deform the eyes of fast-swimming animals.

refractive index. This leads to some reflection of light at the water surface. More importantly, light penetrating into the water column is refracted at the surface. Waves may focus and spread light, leading to fluctuations in intensity (Fig 8.3). In addition, there are prismatic effects, leading to the separation of wavelengths (colors) (Fig. 8.3). Waves move, and lighting effects produced by waves are not static. The region close to the surface often abounds in wave-induced changes in intensity and color. Furthermore, light entering water at an angle other than 90° becomes partially polarized (Jerlov, 1976). With increasing depth, this polarization is reduced by random scattering. Light reflected from the body of a silvery fish may also be partially polarized (Warrant and Locket, 2004).

SPECIFIC WEIGHT

Water has a high specific weight, such that hydrostatic pressure rapidly increases with depth. At a depth of 10 meters, pressure is already twice as high as at the surface. Pressure as such has little effect on the eye because water is hardly compressible and the eye is water filled (Fig. 8.4); therefore, it can withstand enormous

static pressures without being deformed. There may, however, be more subtle effects of pressure on the molecular level in deep-diving species, because high pressure may affect proteins and other molecules (Douglas et al., 1998).

DENSITY AND VISCOSITY

The huge differences between air and water in density and viscosity have profound effects on eye design. An animal that moves rapidly through water may create a turbulent flow over its body surface, and this causes vibrations that interfere with steady vision. In addition, dynamic pressure during fast swimming may deform the cornea and eye. Hydrodynamic constraints have influenced many parameters of the visual systems of secondarily aquatic tetrapods, especially in fast swimmers.

SHOCKS AND ABRASION

All of the species discussed in this book have to break the water surface to breathe, and many aquatic birds forage on the wing and plunge-dive for food. Abrupt transitions from water to air, and even more so from air to water, lead to physical shocks. The eyes are exposed to enormous strains when a breaching humpback whale (*Megaptera novaeangliae*, Fig. 8.5) smashes back on the water surface, or a Northern Gannet *(Sula bassana)* plunge-dives for food. Protective anatomical or behavioral measures maintain the integrity of the delicate structures of the eye.

Large amounts of particles can, for long times, remain suspended in water. Such particles not only scatter light, they are also abrasive. Fast swimming in turbid water is similar to walking in a sandstorm, and eyes exposed to such conditions have specific adaptations against abrasion.

HEAT CAPACITY

Water has an unusually high heat capacity because of its unique molecular structure. In cold water, the body surface may be protected by fur, feathers, and/or fat. In contrast, the eyes are directly exposed to the environment and,

FIGURE 8.5. A breaching humpback whale smashes back onto the water surface. Such violent transitions between air and water generate shockwaves that endanger the delicate structures of the eyes. Photograph graciously provided by Thomas R. Kieckhefer, Pacific Cetacean Group.

in the open ocean. The corneas of aquatic species living in these areas are exposed to the salt in the environment. This is a particularly vexing problem if animals migrate between bodies of water differing in osmolarity.

ACKNOWLEDGMENTS

I thank Sentiel Rommel and Gadi Katzir for useful comments on earlier versions of this chapter.

consequently, drained of heat. The molecular processes of visual transduction are temperature dependent: low temperatures lead to slower transduction and reduced thermal noise, and thus increased sensitivity (Aho et al., 1988; Reuter and Peichl, chapter 10 in this volume). Some animals can regulate the temperature of their eyes to optimize between transduction speed and sensitivity. It is known that large swordfishes (Xiphias gladius) heat their retinas to increase the speed of vision (Fritsches et al., 2005). Adaptations for temperature control seem also to be present in the eyes of some secondarily aquatic tetrapods.

OSMOLARITY

The osmolarity of water is a function of the amount of dissolved salts and other substances. The osmolarity of seawater in the open ocean is about 1000 mOsm/L. Osmolarity is much lower in estuaries and inshore areas, such as the Baltic Sea, which receives high inputs of freshwater from the continents. In areas where evaporation is high and freshwater influx low, such as in the Red Sea, salt content is considerably higher than

LITERATURE CITED

Aho, A.-C., K. Donner, C. Hyden, L. O. Larsen, and T. Reuter. 1988. Low retinal noise in animals with low body temperature allows high visual sensitivity. Nature 334:348–350.

Douglas, R. H., J. C. Partridge, and N. J. Marshall. 1998. The eyes of deep-sea fish. I. Lens pigmentation, tapeta and visual pigments. Progress in Retinal and Eye Research 17:597–636.

Fritsches, K. A., R. W. Brill, and E. J. Warrant. 2005. Warm eyes provide superior vision in swordfishes. Current Biology 15:55–58.

Hecht, E. 2002. Optics. Addison Wesley, San Francisco.

Jerlov, N. G. 1968. Optical Oceanography. Elsevier, New York.

Jerlov, N. G. 1976. Marine Optics. Elsevier, Amsterdam.

Land, M. F., and D.-E. Nilsson. 2002. Animal Eyes. Oxford University Press, Oxford.

Lide, D. R. 2004. CRC Handbook of Chemistry and Physics. CRC Press, Boca Raton, FL.

Lundgren, B., and N. Höjerslev. 1971. Daylight measurements in the Sargasso Sea. Report No. 14. Köbenhavn Universitet, Institut for fysisk Oceanografi, Copenhagen.

Lythgoe, J. N. 1979. The Ecology of Vision. Clarendon Press, Oxford.

Rodieck, R. W. 1998. The First Steps in Seeing. Sinauer, Sunderland, MA.

Smith, G., and D. A. Atchison. 1997. The Eye and Visual Optical Instruments. Cambridge University Press, Cambridge.

Warrant, E J., and N. A. Locket. 2004. Vision in the deep sea. Biological Reviews 79:671–712.

9

Comparative Anatomy and Physiology of Vision in Aquatic Tetrapods

Ronald H. H. Kröger and Gadi Katzir

Vertebrate eyes are of the simple or camera type, unlike the complex or faceted eyes of most arthropods (Land and Nilsson, 2002). In camera eyes, the optical system creates an inverted image of the environment. That image is encoded and preprocessed by the retina before visual information is transmitted to the brain in highly condensed form. Here we present the optical elements of a typical tetrapod eye and discuss the functional differences of vision in air and water, with examples from secondarily aquatic reptiles and birds. For further reading the reader is referred to standard textbooks (e.g., Davson, 1990; Kaufman and Alm, 2003), and details on the various eye

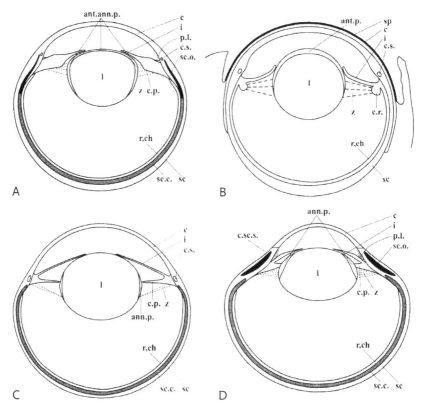

FIGURE 9.1. Diagrams of the eyes of a turtle (A), snake (B), crocodilian (C), and bird (D). In turtles (A), the sauropsid annular pad (ann.p.) extends to the anterior surface of the lens (l) and is therefore called an anterioannular pad (ant.ann.p.). In snakes (B), the eyes rotate under a rigid spectacle (sp), which protects the cornea (c) and is replaced when the skin is shed. The lens has an anterior pad (ant.p.) instead of an annular pad. Even terrestrial snakes have almost spherical lenses that otherwise are typical for aquatic eyes. The general design of crocodilian eyes (C) is characteristic for nocturnal terrestrial vertebrates. The eye design shown in (D) is typical for many sauropsids such as lizards and birds. Other abbreviations: c.p., ciliary process; c.r., ciliary roll; c.s., canal of Schlemm; c.sc.s., corneoscleral sulcus; i, iris; r, retina; ch, choroid; p.l., pectinate ligament; sc, sclera; sc.c., scleral cartilage; sc.o., scleral ossicles, and z, zonules of Zinn. Eyes redrawn after Beer (1898), Walls (1942), Rochon-Duvigneaud (1943), and Duke-Elder (1958).

types of vertebrates are given by Walls (1942), Rochon-Duvigneaud (1943), and Duke-Elder (1958).

COMPARATIVE ANATOMY AND FUNCTION OF THE EYE

CORNEA

The cornea (Fig. 9.1) is the interface between the eye and the environment. It is a thin but strong sheet of tissue consisting of several layers of transparent cells and extracellular structures. The cornea bulges outward because of the intraocular pressure that stabilizes the cornea and the entire eye. In air, a curved cornea has considerable positive (i.e., focusing) refractive power. In contrast, there is little refraction of light at the cornea in water because the refractive index of water is similar to that of the cornea.

IRIS

The iris is a thin sheet of pigmented tissue that acts like a diaphragm of a camera (a "'stop"). It is usually heavily vascularized and equipped with several groups of muscles that can change the size and shape of its free aperture, the pupil. There are two important optical effects of pupil size and shape. Firstly, the amount of light entering the eye is regulated. Secondly, light is diffracted, an effect that may limit visual resolution if pupil size is small.

LENS AND CILIARY MUSCLE

The crystalline lens is the other refractive element of the eye in air, and usually the only one in water. The lens is suspended by thin fibers, the zonules of Zinn that stretch from the lens to the ciliary body at the boundary between retina and iris. Muscles in the ciliary body can adjust the tension of the zonules. Changing zonule tension changes the shape of the lens and, consequently, its refractive power. This mechanism of adjusting the eye's refractive power to different viewing distances is termed accommodation. It can, in principle, also overcome the change in refractive power of the cornea between air and water. Powerful accommodative mechanisms are present in birds and some reptiles. In mammals, the ciliary muscle consists of smooth muscle tissue, whereas in birds this muscle is striated, while still being innervated by the autonomic nervous system. Three muscle fiber groups are recognized in birds. The anterior group (Crampton's muscle) increases the overall corneal power by drawing the periphery of the cornea posteriorly and increasing the steepness of the central cornea while flattening the peripheral cornea. The posterior group (Brücke's muscle) reduces the lens radius of curvature by pulling the ciliary body toward the cornea. The internal fiber group (Müller's muscle) assists the anterior and posterior fiber groups in both corneal and lenticular accommodation by pulling the peripheral cornea posteriorly and the base plate of the ciliary body anteriorly (Glasser et al., 1994; Pardue and Sivak, 1997).

SCLEROTIC RING

Reptile and bird eyes usually have a ring of bones around the pupil (e.g., Walls, 1942), the sclerotic ring (Fig. 9.1A and D), which is located anterior to the ciliary region. These bones maintain that shape of the often prominent corneoscleral sulcus in birds. This ring fossilizes well and provides information on the approximate size of the eyes in extinct reptiles, including the secondarily aquatic ichthyosaurs (Fernández et al., 2005).

ANTERIOR AND POSTERIOR CHAMBERS

The space between cornea and lens, the anterior chamber of the eye, is filled with a fluid called aqueous humor. In the healthy eye, constant intraocular pressure is maintained by various mechanisms that stringently regulate the influx and efflux of aqueous humor. The posterior, or vitreous, chamber is the space between lens and retina, and it is filled with the vitreous humor, a gelatinous substance. The vitreous humor usually has a very low concentration of proteins and about the same refractive index as the aqueous humor.

RETINA

Distal to the vitreous humor there is the retina, which encodes and preprocesses the optical image (discussed in detail by Reuter and Peichl [chapter 10 in this volume]), and the retinal epithelium (RE), which has important functions in the regeneration of visual pigments after bleaching by light and the maintenance of the retina (e.g., Rodieck, 1998). In animals, such as humans, that use their eyes in bright light, the cells of the RE are heavily pigmented by melanin granules and are thus referred to as the retinal pigment epithelium. Animals active in dim light often have a reflective layer just beyond the retina. Such a layer is called a tapetum lucidum. Tapeta lucida are responsible for the bright eye shine that can be observed if the eyes of nocturnal or crepuscular animals are illuminated with strong light. There are several types of tapeta lucida. In a retinal tapetum, reflective material is located within the cells of the RE. In a choroidal tapetum cellulosum there are reflective crystals in a layer of cells distal to the RE cells. A choroidal tapetum fibrosum consists of extracellular fibers in a regular matrix (Nicol, 1981; Ollivier et al., 2004). All tapeta have the same effect: light is reflected back into the photoreceptors after its first passage through them. This essentially doubles the exposure of the photosensitive cells to the light and thus increases the chance of a photon to be absorbed by the photopigment.

CHOROID AND SCLERA

Beyond the RE there is a plexus of blood vessels, the choroid, which supplies nutrients to the retina. Finally, the sclera is the outer shell of the eye in all regions except for the cornea. The sclera consists of tough connective tissue and gives the globe its shape as long as the eye is inflated by intraocular pressure. In some vertebrates it is reinforced with cartilage (Fig. 9.1A, C, and D). A critical point in the sclera is where the optic nerve leaves the eye. The nerve passes through the lamina cribrosa, a network of fibers spanning the perforation in the sclera. If intraocular pressure is too high (glaucoma), the optic nerve may be damaged at its passage through the lamina cribrosa.

ACCESSORY STRUCTURES

In addition to the eye, the peripheral visual system consists of a number of accessory structures that have important roles in visual function. A set of six extraocular muscles attaches to the sclera and can rotate the eye in all three axes of space. Additionally, there is a retractor bulbi muscle that can retract the eye into the orbit for protection. This muscle is particularly well developed in cetaceans (Kröger and Kirschfeld, 1989) but is absent or has changed function in snakes, birds, and some higher primates (Duke-Elder, 1958).

Movable lids are present in mammals and many, but not all, reptiles. The presence of a movable upper eyelid is highly variable in birds. Many tetrapods (and most sharks) have a nictitating membrane, also called the third eye lid (Walls, 1942). Such a membrane is usually thin and can be rapidly extended to cover the cornea. It has special functions in some aquatic tetrapods. Glands around the eye produce tears that help to keep the eye clean and supply the cornea with nutrients. Tears also have special functions in some aquatic tetrapods.

OPTICAL PARAMETERS

An optical system can be described using the simplifications for paraxial optics. By doing so, one assumes that the whole optical system behaves in the same way as a small central area around the axis of rotational symmetry, the optical axis of the system. This type of optical analysis has limited power to predict the performance of animal eyes but is nevertheless useful to understand some basic and important principles, which are best explained graphically (Fig. 9.2).

ANATOMY AND FUNCTION IN AQUATIC TETRAPODS

REPTILES

TURTLES

Chelonian eyes (Fig. 9.1A) are rather typical sauropsid eyes. The sclera is reinforced by both cartilage (posterior segment) and ossicles (around the ciliary region). In the marine leatherback turtle (*Dermochelys coriacea*, Chelidae) (see Table 1.2 in this volume) the scleral cartilage is of enormous thickness, such that the eye in axial section is similar to the eye of a large cetacean (Rochon-Duvigneaud, 1943). The cornea's curvature is almost in continuation of the curvature of the sclera, such that the corneoscleral sulcus is insignificant (Duke-Elder, 1958), which is in contrast to the design of a typical lizard or bird eye. The lack of a sulcus may be an adaptation to streamlining demands. The retina is rather primitively organized (Duke-Elder, 1958). Walls (1942) states that chelonians do not have tapeta lucida, while such a reflective layer distal to the retina is described by Wyneken (2001) in marine turtles.

The lower lid can be closed over the eye, and a nictitating membrane is present. A few turtles have a transparent window in the lower lid (Duke-Elder, 1958) that may be considered an intermediate toward the evolution of a spectacle,

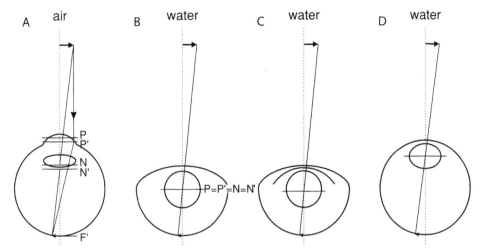

FIGURE 9.2. Diagrams of the eyes of a diurnal land-mammal (A), typical fish (B), cetacean (C), and sirenian (D), illustrating key optical parameters. Terrestrial eyes (A) have two refractive elements, the cornea and lens (figure modified after Smith and Atchinson, 1997). This leads to a separation of the anterior and posterior nodal points (N and N', respectively). The distance P'–F' (P', posterior principal plane; F', posterior focal plane) is the posterior focal length and the distance N'–F' is the posterior nodal distance (PND). A long PND leads to high magnification and low light-gathering ability. In a typical fish eye (B), the lens is the only refractive element. Since the lens typically is symmetrical and borders to media of the same refractive index on both sides, P (anterior principal plane), P', N, and N' coincide. PND is considerably shorter than in (A) leading to less magnification and higher light-gathering ability. In cetaceans (C), the cornea has negative refractive power (diverging lens), and the PND is even shorter than in a fish eye of equal size. The sirenian eye (D) is long relative to its total size, and the lens has less refractive power than in fishes and cetaceans.

as is present in snakes (see below) and some lizards. The retractor bulbi muscle that allows for retraction of the bulbus into the orbit for protection is well developed. When the eye is retracted, it also rotates, which brings the sensitive cornea out of harm's way (Walls, 1942). In marine and estuarine species, the lachrymal (tear) gland is very well developed. Its main function is to maintain the osmotic balance of the body in hypertonic environments by expelling salts (Cowan, 1974; Robinson and Dunson, 1976).

The lens is almost spherical in marine species and somewhat flattened in semiaquatic turtles (Walls, 1942). It is very pliable, which facilitates accommodation by changes in the shape of the lens. The lens, being equipped with a thin annular pad that extends to the anterior surface of the lens, is approached by the well-developed ciliary processes that touch the lens in the fully accommodated state (Duke-Elder, 1958). Marine turtles are emmetropic (normal sighted) in water and severely myopic

(nearsighted) in air. In contrast, freshwater turtles are emmetropic in air and have a large enough accommodative range to fully compensate for the lack of refraction at the cornea in water (Beer, 1898; Walls, 1942). A muscular ring at the margin of the iris may (Fritzberg, 1912) or may not (Beer, 1898) be involved in deforming the lens during accommodation.

SQUAMATES

Little is known about eye design and visual abilities in secondarily aquatic lizards, except that most lizard eyes are very similar to those of birds in general design. The eyes of snakes are well known (Fig. 9.1B), and they differ considerably from the eyes of all other reptiles. Several features are reminiscent of primarily aquatic eyes. The lens is almost spherical and usually rigid, as in fishes. Accommodation is achieved by moving the lens anteriorly, as in elasmobranchs and amphibians (Walls, 1942). These and other features led Caprette et al. (2004) to the conclusion that snakes may have originated from an aquatic ancestor. This is, however, contradicted by Vidal and Hedges (2004), who from molecular evidence conclude that snakes must have originated from a terrestrial lineage of reptiles. The aberrant eye design is in that case explained as the result of a burrowing evolutionary phase (Walls, 1940, 1942; Bellairs and Underwood, 1951; Underwood, 1970; Davis et al., 2001; Vidal and Hedges, 2004). In any case, all extant aquatic and semiaquatic snakes are found in modern families. This indicates that they are at least secondarily aquatic, or—if snakes indeed have an aquatic ancestry—tertiarily aquatic.

The snake eye is characterized by an almost spherical crystalline lens and the spherical shape of the globe. The typical reptilian scleral cartilage and ossicles are lacking. The lids are fused to a transparent spectacle that closely follows the shape of the curved cornea. The spectacles, just as the rest of the integument, are shed during ecdysis (molting). Since the cornea and spectacle are curved, some refractive power resides in the air-tissue interface. This power is lost when the eye is submerged, or—if the eye is primarily adapted to vision in water—added when the animal surfaces. Either way, amphibious vision is a challenge.

Vision in the truly aquatic species, namely, the wart snakes (Acrochordidae), water snakes (Colubridae, Homalopsinae), sea snakes (Elapidae, Hydrophiinae), and sea kraits (Elapidae, Laticaudinae), has received little scientific attention (Mattison, 2003) (for a brief summary of reptilian diversity, see Table 1.2 in this volume). From behavioral experiments it is known that *Enhydris polylepis* (Homalopsinae) mainly uses visual cues to locate prey (Shine et al., 2004a). In *Emydocephalus annulatus* (Hydrophiidae), males rely on vision to find females (Shine, 2005), while prey (fish eggs) is mainly found by chemical cues (Shine et al., 2004b). An interesting sexual dimorphism is present in *Acrochordus arafurae* (Acrochordidae). Females grow much larger than males and are ambush predators on large fish in deep water, using visual and/or vibrational cues. The smaller males

mainly use olfaction to find small fish in shallow water (Vincent et al., 2005).

A few studies have been performed on the accommodative capabilities of semiaquatic colubrid snakes in the subfamily of Natricinae. The common trend is that accommodative mechanisms capable of overcoming the difference in refractive power of the eye between air and water are present only in species that are specialized for capturing aquatic prey (specialists). In species only moderately dependent on aquatic prey (generalists), pupil constriction is used to increase depth of field and thus improve aquatic vision to some degree (Schaeffel and de Queiroz, 1990; Schaeffel and Mathis, 1991). The notion that aquatic specialists have better underwater visual capabilities is supported by behavioral differences between specialists and generalists in the genus *Thamnophis* (Colubridae, Natricinae) (Alfaro, 2002).

Of particular interest is *Natrix tesselata* (Colubridae, Natricinae), a European semiaquatic snake with high affinities to water and aquatic prey (Mattison, 2003). Already Beer (1898) noted that the lens of this species changes shape upon electrical stimulation of the eye, and the author hypothesized that this trait would enable the animals to achieve large amounts of accommodation. Schaeffel and Mathis (1991) confirmed that some semiaquatic snakes, among them *N. tesselata*, can change the refractive power of the eye by more than 100 diopters. The authors also conclude that the usual ophidian way of accommodation—by moving the lens—is not sufficient to explain this capability. To meet the challenges of amphibious vision, *N. tesselata* has evolutionarily reinvented the usual reptilian way of accommodation by deformation of the lens. It is, however, not clear yet how snakes achieve movements—and in some species, deformation—of the lens (reviewed in Ott, 2006).

Very little is known about other visual adaptations to the aquatic lifestyle in snakes. The keratinized cover of the eye is exchanged during moulting, which may provide protection against abrasion. It also makes the eye insensitive to different osmolarities and prevents dehydration when the eye is exposed to air.

CROCODILIANS

Crocodilians use their eyes (Fig. 9.1C) in air to capture prey at the water surface or on land. Although many species also take fish underwater, they appear to use other senses than vision to detect aquatic prey (e.g., Fleishman and Rand, 1989; Olmos and Sazima, 1990; Dehnhardt and Mauck, chapter 18 of this volume). Even extremely aquatic crocodilians such as gharials *(Gavialis gangeticus)* cannot focus their eyes in water (Fleishman et al., 1988).

The eyes are located high in the skull, such that a floating crocodilian can scan the environment with just the eyes and nostrils exposed to air. Since many species live in highly turbid waters, they are well concealed and hard to detect by potential prey. This facilitates their main hunting strategies, namely, motionless ambush and stealthy approach until striking distance is reached (e.g., Thorbjarnarson, 1993).

Since eye design is basically terrestrial, it is not discussed in detail here. However, one adaptation is worth mentioning, namely, the crocodile tear. It is a feature of crocodilians that has made its way into a popular phrase ("crying crocodile tears") and given name to a human pathological condition (crocodile tear syndrome, Bogorad's syndrome [Bogorad, 1979]). The large Harderian gland of a crocodilian produces copious amounts of a viscous fluid when the eye is exposed to air for a prolonged time. This may happen if a crocodilian feeds on terrestrial prey that for any reason cannot be dragged into water. Under normal circumstances, the secretion lubricates the cornea and may help to suppress bacterial growth if the animal lives in muddy water (Britton, 2006).

ICHTHYOSAURS

Vertebrate eyes do not usually fossilize well. There is, however, good evidence for vision in ichthyosaurs, of which some species had really enormous eyes. The record holder is *Temnodon-*

tosaurus (Lower Jurassic), of which a sclerotic ring 253 mm in diameter has been found (Motani et al., 1999). The animals were probably diving deep and would have benefited from highly sensitive eyes. Motani et al. (1999) state that *Temnodontosaurus* had an *f*-number (diameter of the pupil/focal length) of as low as 0.76. Such a small *f*-number should make the eye very sensitive to extended light sources. If one compares ichthyosaur and fish eyes, it is doubtful, however, that this estimate is correct. Refractive index at the center of a spherical fish lens exceeds the index of pure dried protein (Kröger et al., 1994). These lenses have relative focal lengths (focal length/radius of the lens) of about 2.3, such that the lowest possible *f*-number of the eye is 1.15. It is hardly conceivable that ichthyosaurs could achieve even higher refractive index at the center of the lens, which would be necessary to reduce the *f*-number below 1.15.

Small fish eyes of just a few millimeters in diameter and huge ichthyosaur eyes with the same *f*-number have the same image brightness (photons per mm^2). However, if both eyes also have the same angular resolution, each "pixel" in the large ichthyosaur eye covers a much larger surface and thus receives more light than in the small fish eye. This means that the ichthyosaur eye is considerably more sensitive despite the fact that image brightness is the same.

Humphries and Ruxton (2002) hypothesize that ichthyosaurs may have needed huge eyes in order to achieve high acuity in combination with high sensitivity. However, the large eyes of some ichthyosaurs could collect tremendous amounts of visual information. It is highly unlikely that the ichthyosaur brain was capable of handling that much information. The human eye is about 21 mm in diameter (Davson, 1990). Despite being large brained, humans use the maximum resolution an eye of that size can achieve only in a small area of the retina—the fovea—which has an angular subtense of about 2° (Smith and Atchison, 1997). Since the potential information capacity is proportional to the square of the angular subtense, even the human brain would be by far overwhelmed by visual information if we had the highest possible resolution in the entire visual field of 180°. Ichthyosaurs had, because of the lateral positions of the eyes, total visual fields of both eyes close to 360°. If ichthyosaurs had their huge eyes because of a need for high resolution, they could have used the maximum resolution of their eyes only in a tiny fraction of the visual field.

The visual task that is best solved with a large eye is when an animal has to detect small light sources, such as bioluminescence emitted by other animals (e.g., Beebe, 1934; Herring, 1990, 2002) against a dark background. In such a situation, what is relevant is only how much light enters the eye (assuming that the optical quality of the eye is high), and an ichthyosaur eye with its large pupil outperforms a small fish eye by orders of magnitude. Spatial resolution has to be high only in the outer retina. As soon as a reliable visual signal is secured in the outer retina, spatial resolution can be dramatically reduced in the inner retina such that only a manageable amount of information has to be transmitted to the brain. It is therefore likely that *Temnodontosaurus* and its almost equally large-eyed relative *Ophthalmosaurus* (Motani et al., 1999) from the late Middle and Upper Jurassic (Motani, 1999) hunted bioluminescent prey at great depths.

BIRDS

Birds are primarily diurnal and rely on vision for their locomotion. Consequently, the evolution of their morphological features (feathers, specialized skeleton and respiratory systems) was paralleled by that of exceptionally large eyes and large brain areas devoted to the analysis of visual information (Davies and Green, 1994; Frost et al., 1994; Lee, 1994). The view that bird ancestors were terrestrial animals was recently drawn into question by the discovery of a rich fauna of aquatic birds (You et al., 2006).

Our discussion of aquatic birds is focused on species that operate visually in water or through the water surface, as these behaviors require different eye adaptations. This discussion largely excludes waders (Charadriiformes), ducks (e.g., *Anas* spp., *Spatula* spp.), ibises, and spoonbills *(Plegadis* spp. and *Platalea* spp. respectively), and flamingoes *(Phoenicopterus* spp.) because these birds find food in water by using mechano- and chemoreception.

Plunge-diving, pursuit-diving, and striking (see Table 1.3 of this volume) are three modes of hunting used by birds. Plunge-divers and strikers spend prolonged periods in visual search of food that is on or just below the water surface. Underwater prey often is detected and the capturing movements are initiated from the air, such as in herons (Ardeidae), gannets (Sulidae), brown pelicans *(Pelecanus occidentalis)*, kingfishers (Alcedinidae), sea eagles *(Haliaeetus* spp.), and ospreys *(Pandion haliaetus)* (for a brief summary of aquatic bird diversity see Table 1.3 in this volume). In many cases, prey is detected while flying or hovering above the water surface, sometimes at heights exceeding 10 meters, and plunge-diving ensues (Cramp and Simmons, 1977). Such dives may be followed by an immediate return to the surface, as in pied kingfishers (Labinger et al., 1991), or by an underwater pursuit, as in gannets (Garthe et al., 2000). Gulls *(Larus* spp.) as well as shearwaters and petrels (e.g., *Oceanodroma* spp., Hydrobatidae) procure their prey on, or just beneath, the surface while they float or dive directly from a short distance above it (Hart, 2004).

Refraction (see Kröger, chapter 8 in this volume) is a major problem that birds hunting across the air-water interface have to deal with. Light rays traveling across the air-water interface are refracted due to the difference in respective index of refraction. To a bird in the air, the apparent position of a fish will be closer to the surface than the real position (Katzir and Intrator, 1987). At angles of incidence equal to or greater than about 49°, rays from an underwater object will undergo complete internal reflection, and thus aerial viewing of underwater objects is restricted to angles of approximately 98°. This is the aerial counterpart of "Snell's window," the restricted view of the aerial world to an underwater observer just beneath the surface. Chromatic dispersion (Fig. 8.3 in this volume) leads to differential illumination of underwater objects and inherent chromatic biases of underwater targets, when viewed from the air. The optical effects of waves, in addition to the effects on underwater illumination (Fig. 8.3 in this volume), produce continuous changes in apparent features of underwater objects encompassing changes of size, contour, contrast, texture, and apparent position and thus of apparent motion.

HERONS

Foraging herons (Ardeidae) typically wade in shallow water and, once they detect submerged prey, aim at it from the air and strike rapidly with their bill. This behavior implies that the rapid ballistic final stage of prey capture requires correction for refraction of light at the water-air interface. Little egrets *(Egretta garzetta)*, reef herons *(E. gularis)*, and squacco herons *(Ardeola ralloides)* demonstrate a clear capacity to correct for image displacement for all reachable prey distances and depths. Night herons *(Nycticorax nycticorax)* are less able to do so, while cattle egrets *(Bubulcus ibis)* are literally unable to capture submerged prey. These results are closely correlated with foraging patterns and motor capacities. The former three species are agile, strike at fish from a wide range of angles and surface conditions, and frequently do so while running. Night herons are relatively slow and more restricted in their strike situations. Cattle egrets rarely, if ever, hunt in water and may have lost the capacity to correct for refraction during their unique evolution from aquatic to terrestrial foraging, along with the evolution of African grasslands (Katzir et al., 1999) and the birds' close association with hoofed mammals.

KINGFISHERS

Of the kingfishers (Alcedinidae, Coraciiformes), several species specialize on aquatic prey, and the strictly piscivorous Pied kingfisher *(Ceryle rudis)* is probably the most pelagic of these. It will detect a fish from the air, hover several meters above the surface keeping its head extremely stable, and then plunge-dive. It is capable of correcting for refraction-induced image displacement and determines most accurately the size, depth, distance from the substrate, and motion path of its prey (Labinger et al., 1991; Katzir and Camhi, 1993; Katzir et al., 2002).

In kingfishers, which as most other birds have two foveae in each eye, angular separations between these foveae is exceptionally large, up to 40° and 50° in blue-winged kookaburras *(Dacelo leachii)* and azure kingfisher *(Ceyx azureus)*, respectively, compared with approximately 15° in raptors such as eagles. The kingfishers' binocular foveae are well aligned with the bill: "much like laser sighting a gun" (Moroney and Pettigrew, 1987; Schwab and Hart, 2004). Ganglion cell density of the temporal (binocular) fovea is relatively low, suggesting moderate acuity, while the central (monocular) fovea has a high ganglion cell density, suggesting high acuity levels. It may well be that aquatic kingfishers, such as the pied kingfisher, sights its prey with the monocular fovea while hovering above it. As it plunge-dives, the ability to determine three-dimensional prey motion may become more important than mere acuity (Moroney and Pettigrew, 1987; Schwab and Hart, 2004).

PENGUINS AND ALBATROSSES

Information on eye structure and related aspects in pursuit-divers mostly pertains to penguins (Sphenisciformes) (Sivak and Milodot, 1977; Sivak and Vrablic, 1979; Howland and Sivak, 1984; Martin and Young, 1984; Sivak et al., 1987; Martin, 1999). Penguins are pursuit-divers at great depth (up to hundreds of meters) and have to cope with the loss of refractive power of the cornea when the eye is submerged. Penguins have relatively flat corneas providing refractive powers of only 11 to 30 diopters (Sivak and Milodot, 1977; Howland and Sivak, 1984; Martin, 1999). Similarly, the optical design in the gray-headed albatross *(Diomedea melanophris)* and the black-browed albatross (*D. chrysostoma*; Procellariiformes,

Diomedeidae) also points to amphibious capacities, and their relatively flat corneas provide only 23 diopters of refractive power. These values are considerably lower than those of most other avian species having eyes of comparable sizes. These low values reduce the amount of accommodation necessary to achieve emmetropia underwater. Furthermore, the penguin lens is relatively spherical in its resting state (Sivak, 1980; Martin, 1998, 1999), attesting to the predominance of aquatic rather than aerial demands on the optical system.

A schematic eye model of Humboldt's penguin *(Spheniscus humboldti)* suggested an aquatic design, being emmetropic in water and myopic (28 diopters) in air, yet individuals of three penguin species measured by photorefraction were found to be emmetropic in air and slightly hyperopic in water to a degree well within the power of the lens to compensate for (Sivak, 1976; Howland and Sivak, 1984; Martin and Young, 1984; Sivak et al., 1987).

In the Humboldt penguin the monocular visual fields in air are approximately 160° in width, and the binocular field is approximately 120° long and 30° wide with the bill placed close to its center. In the king penguin, *Aptenodytes patagonicus*, the respective values are 180°, 180°, and 30°. Upon immersion the visual fields undergo a marked decrease in size, and while retained just above the bill, the binocular overlap is abolished at the plane of the bill itself. This implies that in their underwater pursuit of prey, the penguins may use binocular information for tasks such as distance estimation, but not for guidance of the bill tip. Interestingly, the anterior eye structure and visual field topography in albatrosses in both air and water show marked similarity with those of the Humboldt penguin. At present, information on visual fields and binocularity in other pursuit-divers such as loons, grebes, and cormorants is not available. Retinal adaptations in penguins are discussed by Reuter and Peichl (chapter 10 in this volume).

Cormorants and penguins, and maybe other marine birds, have Harderian glands that pro-

duce viscous secretions that protect the corneas from the osmotic effects of seawater (Walls, 1942).

OTHER AQUATIC BIRDS

The corneas of pursuit-diving birds that have been investigated are rather steeply curved, except for those of penguins. This implies that lenticular accommodation is necessary, and indeed such capacities have been demonstrated in Anseriformes, Pelecaniformes, Alcidae, and Passeriformes.

The eyes of the hooded merganser (*Mergus cucullatus*, Anatidae), red-headed duck (*Aythya americana*, Anatidae), double-crested cormorant (*Phalacrocorax auritus*, Phalacrocoracidae), black guillemot (*Cepphus grille*, Alcidae), and the dipper (*Cinclus mexicanus*, Passeriformes), when stimulated chemically or electrically, exhibit changes in refractive state of 40 to 80 diopters by deformation of the malleable lens (Fig. 9.3). In some species, accommodation is achieved by the pressing of the malleable lens against the iris sphincter muscles, which forms a rigid ring. This results in the bulging of the lens through the pupil, a lenticonus (Hess, 1909, 1913, cited in Glasser and Howland, 1996; Sivak et al., 1977; Levy and Sivak, 1980; Sivak and Vrablic, 1982). In contrast, the range of accommodation among terrestrial birds (e.g., pigeon *[Columba livia]*, chicken *[Gallus gallus]*) is only 10 to 20 diopters. A unique evolutionary case is that of the dippers (Cinclidae); these birds are songbirds (Passeriformes) that have evolved into underwater foragers and visually pursue their invertebrate prey by running on the bottom of fast moving streams (Goodge, 1960).

In the pursuit-diving hooded merganser most of the ciliary muscle fibers are in the posterior and internal fiber groups, with a relatively small anterior fiber group. In contrast, in three terrestrial species, the pigeon, kestrel *(Falco sparverius)*, and chicken, most of the muscle fibers are in the anterior muscle group. The pursuit-diver thus shows a predominance of lenticular accommodation and a minimum of

FIGURE 9.3. Eyes of two cormorants (Phalacrocoracidae) viewed in water from dorsal with slightly different angles in (A) and (B). The lens protrudes through pupil in water but not in air and compensates for the lack of refraction of light at the cornea. Photographs by Gadi Katzir.

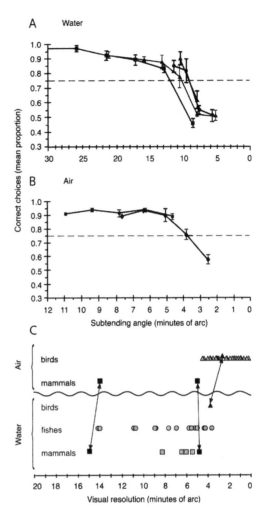

FIGURE 9.4. Visual resolution in great cormorants in water (A) and air (B). Resolution was based using high-contrast, square-wave gratings in a forced choice situation and is determined by the intercept of the line with the binomial probability level (hatched horizontal line). (C) Comparison of visual resolution in great cormorants, other birds (approximately 60 species), 16 species of fish, and aquatic mammals (Air: *Amblonyx cinerea, Zalophus californianus.* Water: *Amblonyx cinerea, Zalophus californianus, Eumetopias jubatus, Orcinus orca, Phoca vitulina, Lagenorhynchus obliquidens*). Reprinted from Strod et al. (2004), with permission from Elsevier.

corneal accommodation, while in terrestrial birds the reverse is true. Moreover, in the hooded merganser an especially large number of muscle fibers resides in the pupillary rim of the iris, an area responsible for lens squeezing, and the base plate of the iris is the most densely packed with collagen among the abovementioned four species (Glasser et al., 1995; Pardue and Sivak, 1997).

Great cormorants *(Phalacrocorax carbo)* use visual cues to detect and capture underwater prey (Strod, 2002). In clear water, cormorants are able to detect a fish if its image on the retina subtends approximately 10 minutes of arc (a 3 cm long fish at about 3 meters). Visual resolution of cormorants in clear water was approximately 7 minutes of arc, while in air it was approximately 4 minutes of arc (Fig. 9.4). Compared with other birds, the cormorants' resolu-

tion in air is in the midrange (equal to a chicken), while their underwater resolution is equal to the higher values reported for fishes (Strod et al., 2004; Fig. 9.4 in this chapter). Cormorants thus "make the best of both worlds," and while they can cope with the demands of rapid, visually guided actions when airborne, they are also most capable of detecting or even out-competing fish visually underwater.

Voluntarily diving great cormorants *(P. carbo sinensis)* were fully accommodated, providing a lenticular compensation for approximately 60 diopters of corneal power lost upon submergence, and double-crested cormorants *(P. auritus)* were also found to be emmetropic in air and in water (Sivak et al., 1977; Katzir and Howland, 2003). The rate of accommodative change in great cormorants (more than 50 diopters in 40 to 80 ms upon eye submergence, i.e., more than 1000 diopters/s) is exceptionally high compared with other vertebrates.

Two aspects of change in pupil size, related to actual submergence and to changes in light levels, may be considered. A marked (three- to fivefold) pupil constriction was described for eyes of cormorants, the hooded merganser, and the red-headed duck (e.g., Levy and Sivak, 1980) when stimulated chemically or electrically. This was attributed to the process of forming a rigid iris plate against which the lens is pushed to form the required lenticonus. However, in free-diving great cormorants the pupils undergo a marked increase in diameter at the moment of submergence (Fig. 9.5) (Katzir and Howland, 2003). Furthermore, in air and bright light, penguins, cormorants (Fig. 9.5), loons, and other birds show small pupils. While a small pupil produces a sharper image of greater depth of field (Land and Nilsson, 2002), it is also possible that the stopped-down pupil keeps the retina under low-light conditions and thereby better adapted to the rapidly decreasing light level during a dive (Martin, 1999). In a rapid descent to depths of between 100 and 300 meters, a penguin will experience a decrease of light levels of more than 5 $\log_{10}$ units. This is similar to moving between broad sunlight to moonlight levels in about 1 minute (Martin, 1999). Known retinal processes of dark adaptation, except for seals (see below), are much slower, and as the duration of dives are only several minutes, the preadaptation may be significant. Pupil dilation will increase retinal image brightness (approximately 2 to 3 $\log_{10}$ units)

FIGURE 9.5 Head of great cormorant in air *(upper)*, and underwater *(middle and lower)*. Note the small pupil in air and the large pupil underwater (as visualized by the reflecting tapetum lucidum). Photographs courtesy of Sari Giladi and Kevin van Doorn, reproduced with permission of the Company of Biologists.

and thus maximize sensitivity and resolution at the rapidly decreasing light levels encountered. These same arguments may apply to dramatic changes in ambient light levels encountered in waters rich in sediment and organic matter, in which light attenuation is dramatic.

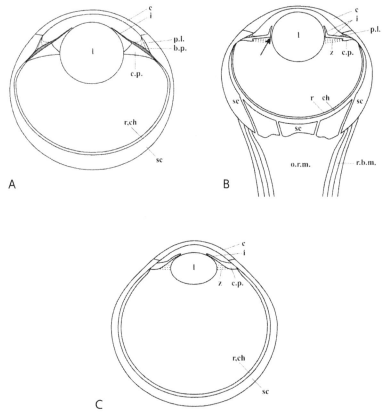

FIGURE 9.6. Diagrams of the eyes of a pinniped (A), cetacean (B), and sirenian (C). The cornea (c) of some pinnipeds is more flat than indicated in this diagram. Pinnipeds usually have a base plate (b.p.) of connective tissue in the ciliary body and base of the iris. The lens (l) is almost spherical as is typical of aquatic mammals. In cetaceans (B), the cornea has the shape and function of a diverging lens, which increases light-gathering ability. The iris (i) is attached to the equator of the lens (arrow). The strong retractor bulbi muscle (r.b.m.) may function in accommodation and temperature control in the eye. The lens in sirenians (C) is flat and there is a large posterior chamber resulting in high magnifications and low light-gathering ability, consistent with the diurnal lifestyle of sirenians. Other abbreviations: c.p., ciliary process; ch, choroid; o.r.m., ocular rete mirabile; p.l., pectinate ligament; r, retina; sc, sclera; and z, zonules of Zinn. Figure modified after Kröger (1989), Kröger and Kirschfeld (1989, 1993), Pütter (1903), and Mass and Supin (2003, 2005).

SKIMMERS: A SPECIAL CASE

The skimmers (Rhynchopidae, Charadriiformes) do not operate visually through or under the water surface, but show a visual adaptation that most likely is related to their aquatic lifestyle. They procure aquatic prey while at low height flying over the water surface with the lower, elongated mandible partially submerged (Tomkins, 1951) and are the only birds known to have slit pupils (Zusi and Bridge, 1981). In other vertebrate groups, slit pupils are often correlated with nocturnal or crepuscular lifestyles (Walls, 1942; Malmström

and Kröger, 2006). Although also skimmers are active mainly at low light levels (Fasola and Canova, 1993; McNeil et al., 1992), other nocturnal birds such as owls do not have slit pupils. The slit pupils of skimmers seem to be an adaptation to their peculiar way of catching prey just below the water surface.

MAMMALS

PINNIPEDS

A large comparative study of the anatomy of the eyes of marine mammals (Fig. 9.6), including adult eyes of four species of earless seal, the walrus, and one species of sea lion, was performed by Pütter (1903). He found in most pinniped species 100,000 to 200,000 nerve fibers in the optic nerve. The domestic cat, which relies rather heavily on vision, also falls into this range (Stone, 1978; Stone and Campion, 1978). A notable exception among pinnipeds is the southern elephant seal (Mirounga leonina). It has about 750,000 fibers in the optic nerve, which is in the same range as in humans (e.g., Repka and Quigley, 1989). Vision seems to be important for pinnipeds, and it may be of high importance in some species.

As in terrestrial carnivores, pinniped eyes point to frontal directions. In some species the periphery of the cornea is considerably thicker than the central region. This may mean that in such species the cornea has negative refractive power in water, which widens the visual field (see section on cetacean eyes). The anterior surface of the cornea is flattened to different degrees, depending on species (Pütter, 1903; Dawson et al., 1987a; Mass and Supin, 2003, 2005). A flat anterior surface of the cornea works like a diving goggle, because it has no refractive power in either water or air. However, such a mechanism cannot explain the capability of amphibious vision in all species. At least in harbor seals (Phoca vitulina) (Johnson, 1893, 1901; Hanke et al., 2006) and harp seals (Phoca groenlandica) (Piggins, 1970) the cornea is flattened only in the dorsoventral meridian. Jamieson (1971) explains this peculiar corneal shape that leads to strong astigmatism in air as an adaptation to hydrodynamic requirements.

Strong corneal curvature in the nasotemporal meridian is correlated with a slit-shaped pupil that is narrow in the nasotemporal direction. These observations suggest that the animals use a pinhole effect that gives long depth of focus in the nasotemporal direction and a goggle effect in the dorsoventral direction to achieve sufficiently sharp images in air (Walls, 1942; Jamieson and Fisher, 1977; Hanke et al., 2006). Piggins (1970) and Hanke et al. (2006) observed accommodative changes in refractive

state of the eye in water in P. vitulina and P. groenlandica, respectively. To what extent accommodation contributes to amphibious vision in pinnipeds is unknown at present.

An extreme case of corneal flattening is present in the California sea lion (Zalophus californianus) (Dawson et al., 1987b). The flat area is sufficiently large to allow for equivalent aerial and aquatic vision in high to moderate light intensities. At low intensities, however, aerial spatial resolution is considerably lower than aquatic resolution (Schusterman and Balliet, 1971), probably because the pupil opens to such an extent that the steeply curved periphery of the cornea comes into play.

The anterior surfaces of seal corneas consist of flattened epithelial cells (Dawson, 1980; Welsch et al., 2001) that may be under constant renewal. This may be an adaptation to the problem of abrasion. Some seal populations live in waters that are extremely turbid because of particles kept in suspension by the tides, and abrasion is a serious problem in such habitats. Seals also have nictitating membranes (Johnson, 1893, 1901; Pütter, 1903) but do not seem to use them to clean the eyes (Johnson, 1893).

Compared to the eyes of most terrestrial mammals, seal eyes have thick scleras. The sclera is thinned to different degrees, depending on species, at the equator of the eye (Fig. 9.6A). The functional significance of this feature is not understood. The iris is equipped with strong muscles, of which the sphincter muscle is the strongest (Pütter, 1903). All seals studied have highly efficient light-dark adaptation based on prominent changes in pupil size and fast retinal mechanisms (Levenson and Schusterman, 1997, 1999). The northern elephant seal (Mirounga angustirostris) is the record holder in the speed of adaptation. It dark adapts almost completely within 1 minute (Levenson and Schusterman, 1999). Elephant seals are known to perform deep dives (exceeding 1500 meters) and regularly forage at depths between 300 and 600 meters (e.g., Bennett et al., 2001). Rapid dark adaptation and high retinal resolution may enable them to resolve small bioluminescent sources at considerable distances. Once the heading toward a bioluminescent cephalopod or fish is determined by vision, the seal may swim into its hydrodynamic trail and find the prey with the help of its whiskers system (Dehnhardt et al., 2001; Dehnhardt and Mauck, chapter 18 in this volume), even if the victim detects the danger and abstains from generating further bioluminescent flashes. Such a strategy can, of course, be successful only at night or at depths not reached by daylight. This nicely agrees with observations that elephant seal dives are deeper during daytime and shallower at night (Campagna et al., 1999; Bennett et al., 2001). In the seal eye, there is a seamless transition between the iris and the ciliary body. The ciliary processes reach the lens and are connected to it (Pütter, 1903). The pinniped ciliary body is characterized by an angled base plate (Pütter, 1903) consisting of tough connective tissue (Fig. 9.6A). On the posterior side of the base plate there is a weak muscle of circular fibers that seems to be absent in some species (Pütter, 1903; Sivak et al., 1989; West et al., 1991). On the anterior side of the base plate there is a well-developed muscle consisting of meridional fibers. Ciliary body and iris form a dome-shaped structure, with the rim of the iris touching the anterior surface of the crystalline lens (Fig. 9.6A). The lens is almost or perfectly spherical, as in most eyes that are primarily used in water. The meridional muscle of the ciliary body (called musculus tensor chorioideae by Pütter [1903], and ciliary muscle by Sivak et al. [1989] and West et al. [1991]) may have a role in accommodation. When its fibers contract, the angle of the base plate may change, and this may result in an axial movement of the lens. There is, however, no experimental evidence for this hypothesis at the present time.

All species studied have a reflecting tapetum lucidum of the cellular type that is of greenish or bluish color. The tapetum extends from the center of the retina to about the equator of the eye (Pütter, 1903; Nagy and Ronald, 1970; Jamieson and Fisher, 1971). Behind the eye and enclosing the optic nerve there is a small plexus of blood vessels in earless seals. In sea lions, the position of this plexus is occupied by a pad of fatty tissue (Pütter, 1903). Both structures may have a similar function in dampening shocks that occur when a seal jumps into the water or porpoises swim at high speed.

Visual acuity in seals—in the species where it is known—varies between about 5 and 8 minutes of arc (Dehnhardt, 2002; most values from measurements in air), which is as good as or better than in the house cat (Berkley, 1973). Seals are also capable of quickly solving rather complex tasks of mental rotation of visually inspected objects (Mauck and Dehnhardt, 1997; Stich et al., 2003) and conceptualize complex visual information (Mauck and Dehnhardt, 2005). Vision doubtlessly is an integral and important component of the sensory world of pinnipeds.

CETACEANS, GENERAL

Most cetacean species have well-developed eyes (Fig. 9.6B), and the largest extant vertebrate eyes are present in cetaceans (Walls, 1942). Reduced eyes are found only in the freshwater dolphins of the Ganges, Indus, and Yangtze river systems (Platanistidae and Lipotidae respectively) (Leatherwood et al., 1983). These rivers are characterized by extremely high loads of suspended particles that make vision basically useless (Evans, 1987).

A typical cetacean eye has a posteriorly thickened sclera, especially in large species (Pütter, 1903). In small species, such as the harbor porpoise (Phocoenidae), this characteristic is less prominent. Posterior to the eye and protruding into a thinner region of the sclera there is a rete mirabile, a plexus of blood vessels. The supply of blood to the unusually thick choroid within the eye originates exclusively from the ocular rete (Kröger, 1989; Kröger and Kirschfeld, 1989). The rete is enclosed by the retractor bulbi muscle, which in cetaceans is well developed and consists of four symmetrical parts that enclose the ocular rete (Pütter, 1903).

The eyes can be covered by thick lids; a nictitating membrane is absent (Johnson, 1901). Ocular glands produce copious amounts of a viscous secretion, the so-called whale tear (e.g., Young and Dawson, 1992). With the eye in air, the secretion rapidly accumulates to an irregular layer on the cornea that should impair acute vision and has so far spoiled all attempts to exactly determine corneal shape in living cetaceans. Dawson et al. (1987a) worked on dolphins *(Tursiops truncatus)* in air and tried—unsuccessfully—to rinse away the whale tear. Kröger (1989) and Kröger and Kirschfeld (1989) could get a few incomplete measurements on dolphins *(T. truncatus)* that used their eyes in air for only a few seconds at a time. The whale tear may not be an optical problem under natural conditions, because cetaceans rarely expose their eyes to air for more than a few seconds, most often for much shorter periods. In water, the whale tear is optically ineffective and does not interfere with vision. Instead, it protects the cornea from abrasive particles and may have a role in maintaining the osmotic balance of the corneal cells in various environments, as well as reducing drag forces on the eye during fast swimming (Tarpley and Ridgway, 1991).

Optical models have been generated for the eyes of harbor porpoises *(Phocoena phocoena)* by Matthiessen (1886) and Kröger (1989). Both authors found that the thick, almost spherical crystalline lens has more than sufficient refractive power to focus light on the retina. With the lens as the only refractive element, harbor porpoises in water would be myopic by about 5 diopters. The enigma was resolved when Kröger and Kirschfeld (1992, 1993, 1994) found that the harbor porpoise cornea is a dispersive lens having negative refractive power of about −5 diopters. Because most cetacean corneas are similar in shape to the harbor porpoise cornea (Pütter, 1903; Waller, 1980; Zhu et al., 2001), it is likely that negative refractive power residing in the cornea is a feature typical for cetacean eyes. A negative lens (cornea) combined with a positive lens (crystalline lens) increases the visual field, a feature that may be of importance for cetaceans because the eyes are positioned laterally and deeply inserted into the body (Fig. 9.7).

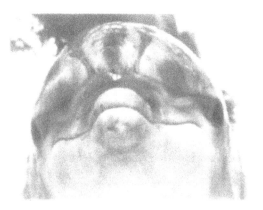

FIGURE 9.7. Frontal view of bottlenose dolphin *(Tursiops truncatus)*. Note the lateral position of the eyes, which are lodged deeply in the head. Photograph taken at the Duisburg Zoo in Germany by RK.

Another adaptation that increases the angle of vision with high resolution is the presence of two areas of high ganglion cell densities in the retina that are displaced from the optical axis roughly rostrally and caudally (Dral, 1975, 1977, 1983; Mass et al., 1986; Mass and Supin, 1990, 1995, 1997; Murayama et al., 1992, 1995; Supin et al., 2001). With two directions of gaze in each eye, little rotation of the eye is necessary to scan a large visual field with high resolution (Kröger and Kirschfeld, 1993). Even a narrow binocular field in front of the animal seems to be possible.

Cetaceans have reflective tapeta of the fibrous type (Pütter, 1903; Walls, 1942; Duke-Elder, 1958) that are different in coloration between species (Dawson et al., 1987b; Zhu et al., 2001). The pupils show extensive changes in size and shape. In dim light, the pupil is large and round. With increasing light intensity, it constricts to a shallow U-shape. There have been speculations on the functional significance of this state of the pupil with regard to amphibious vision (Rivamonte, 1976; Dral, 1985), but the most likely explanation is that the eye in water needs more protection from downwelling than upwelling light. In very bright light, two small rostrocaudally displaced apertures remain (Fig. 9.8). Their functional significance is understood as an adaptation to the two areas of high ganglion cell densities in the retina. The path of light through the eye has

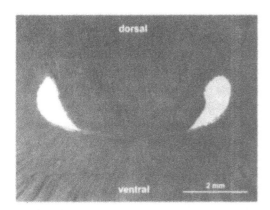

FIGURE 9.8. Posterior view of the iris of a Risso's dolphin *(Grampus griseus)* with the lens removed. The pupil is almost completely closed. Note that two separate lens openings are formed. Photograph courtesy of Leo Peichl.

to be optimized for these retinal areas of high spatial resolution, and that is achieved by the equally rostrocaudally displaced pupillary openings. These openings let light that reaches the areas of high retinal cell densities pass through the central region of the lens, which is least affected by optical aberrations (Kröger and Kirschfeld, 1993).

Despite their fully aquatic lifestyles, both mysticetes and odontocetes use their eyes also in air for various reasons. Dolphins may be guided to food sources by looking for circling sea birds. Baleen whales may use spy-hopping (i.e., lifting of the head over the water surface in a vertical posture) for orientation (Taber, 1984). Dolphins in captivity use aerial vision to receive signals from their trainers and in various performances (observed by RK). The bottlenose dolphin *(Tursiops truncatus)* has a spatial resolution of 8.2 minutes of arc in water and 12.5 minutes of arc in air (Herman et al., 1975). Most of this difference in resolution can be explained by the fact that the size of the retinal image is smaller in air than in water because of the difference in refractive index between these media.

The ability of amphibious vision in cetaceans is surprising, since no accommodative mechanism is known. Most authors state that the ciliary muscle is absent or rudimentary (Pütter, 1903; Rochon-Duvigneaud, 1943; Pil-

leri, 1964; Dral, 1972; Gao and Zhou, 1986; West et al., 1991), while Waller described ciliary muscles in *T. truncatus* and *Globicephala melaena* (Waller, 1980, 1992). Many suggestions have been made to resolve the enigma, such as the constricted pupil acting as a single (Dawson et al., 1972) or double pinhole (Herman et al., 1975), multiple focal lengths of the lens (Rivamonte, 1976; Dral, 1985), a ramp retina (Dawson, 1980), and changes in corneal shape by changes in intraocular pressure (Kröger and Kirschfeld, 1989). However, none of the suggested mechanisms is supported by convincing experimental results.

Many cetaceans engage in spectacular aerial displays. Most notable are the spinner dolphins *(Stenella longirostris* and *S. clymene)*, which often jump from the water and repeatedly spin around the longitudinal body axis, and the humpback whales *(Megaptera novaeangliae)*, which lift their enormous bodies almost entirely from the water and create gigantic splashes on reentry (Fig. 8.5 in this volume). During fast pursuit, small odontocetes often jump clear from the water for breathing. When a cetacean hits the water surface on reentry, shock waves are created that are a danger to the delicate structures of the eye. Furthermore, male cetaceans, just like many of their terrestrial artiodactyl cousins, engage in agonistic head-butting, which also puts the eyes at risk (Lusseau, 2003). Protection can be achieved by retraction of the eye into the orbit and thick lids being closed over the eye. It is also conceivable that the thick choroid and the ocular rete mirabile are involved in dampening shock waves inside the eye.

To counteract the loss of heat in cold water, it may be that the unusually large and symmetrical retractor bulbi muscle can generate heat that is relocated into the eye via the ocular rete, which is enclosed by this muscle and communicates with the thick choroid within the eye. A similar hypothesis was formulated by Pütter (1903). It has been shown that control over intraocular temperature for optimum speed and sensitivity of the retina is of advantage to large fishes facing similar visual challenges as cetaceans (Fritsches et al., 2005).

ODONTOCETES

Many odontocetes are fast swimmers, and top speeds of 40 km/h have been measured (Rohr et al., 2002). At high speeds, vibrations caused by turbulences and dynamic pressure on the eye can influence vision. The role of vibrations is difficult to assess, because it is not known how turbulent the flow of water is over the heads of various odontocetes. More is known on the adaptations to dynamic pressure. Its effect is reduced by the lateral positions and deep insertion of the eyes into the body (Fig. 9.7). Furthermore, intraocular pressure was unusually high, about 70 mm Hg, in a terminally ill bottlenose dolphin *(Tursiops truncatus)* (Dawson, 1980). In healthy animals of the species *T. truncatus* and *Grampus griseus* (Risso's dolphin) intraocular pressure was lower (about 30 mm Hg), but still considerably higher than in other mammals (Dawson et al., 1992). High intraocular pressure stabilizes the eye. However, an intraocular pressure of 70 mm Hg in a human eye would immediately damage the optic nerve where it exits the bulbus through the lamina cribrosa and thus lead to blindness. The upper limit for normal intraocular pressure in humans is considered to be 21 mm Hg (Shiose and Kawase, 1986). The ocular rete mirabile of cetaceans that feeds the rete-like choroid within the bulbus may be an adaptation that reduces the pressure gradient across the lamina cribrosa (Kröger and Kirschfeld, 1989). Even in the terminally ill animal with exceptionally high intraocular pressure, subsequent histological analysis revealed no damage to the optic nerve at the lamina cribrosa (Dawson, 1980). In the healthy animals, intraocular pressure showed conspicuous periodic fluctuations. The functional significance of this feature is as yet unclear (Dawson et al., 1992).

Vision is certainly of importance to odontocetes, but probably not the most important sense. The animals are color blind (Peichl et al., 2001, 2002; Levenson and Dizon, 2003; Reuter

and Peichl, chapter 10 in this volume) and equipped with sophisticated echolocation systems that provide detailed information on the environment. The pygmy sperm whale *(Kogia breviceps)* seems to be a special case. The species has more than 1 million axons in the optic nerve, compared to about 100,000 to 200,000 in other odontocete species (Dawson, 1980). Physeterids (sperm whales) are deep divers (Leatherwood et al., 1983; Evans, 1987) and therefore may use vision in similar contexts as elephant seals seem to do and ichthyosaurs probably did, that is, to detect bioluminescent prey against a dark background at great depths.

Deviations from the typical odontocete eye design are present mainly in river dolphins. The eyes are reduced to light-sensitive pits in the South Asian river dolphins (*Platanista* spp., Platanistidae) (Pilleri, 1979). In contrast, the eyes are fully functional in the Amazon river dolphin (*Inia geoffrensis*, Iniidae) and show a number of interesting adaptations. The lens is yellow, the tapetum lucidum is almost nonexistent, and there is only one area of high ganglion cell density in the retina (Mass and Supin, 1989). Furthermore, the eyes are more frontally directed than in any other cetacean species. The whale tear seems to be absent in *Inia* (Kröger, 1984), which is surprising because the animals are exposed to heavy loads of suspended particles in their riverine habitats. The Chinese river dolphin (*Lipotes vexillifer*, Lipotidae) has small and degenerate eyes that, however, seem to be functional (Zhou et al., 1980).

MYSTICETES

Little is known on visual function and the role of vision in mysticetes. Many features of eye anatomy and retinal morphology are similar to those of odontocete eyes (Pütter, 1903; Zhu et al., 2001), such that visual function also may be similar. Large species have enormously thickened scleras. The functional role of this feature has still to be revealed. The thickened sclera certainly stabilizes the eye because of its high mass. It may also have an insulating effect that would enable the animals to regulate the temperature of the retina independently of body temperature.

SIRENIANS

Eye design and visual function in seacows (sirenians) is poorly studied. The eyes (Fig. 9.6C) are small, but functional, and vision is exclusively used in water. Thick lids can be closed over the eye, which can be retracted into the orbit (Walls, 1942). The nictitating membrane is present (Pütter, 1903). The tear gland is absent, but the Harderian gland is well developed. It secretes a viscous fluid that seems to protect the cornea from damage by seawater (Dexler and Freund, 1906). The cornea is unusual in that it is permanently vascularized, a condition that is considered to be pathological in other animals (Harper et al., 2005). The relative nodal length of the eye is long because the lens is surprisingly flat (Walls, 1942; Piggins et al., 1983; Mass et al., 1997). This indicates relatively high image magnification and narrow field of view. It also indicates that the eye is not primarily designed for use under low-light conditions, which agrees with the absence of a reflective tapetum (Walls, 1942).

The ciliary muscle is rudimentary such that accommodative changes in focal length of the eye seem not to be possible (West et al., 1991). This is in agreement with observations of slow-speed collisions with nearby objects by manatees (Hartman, 1979; Bauer et al., 2003). The eye seems to be set for distance vision, which is consistent with the relatively high magnification of the retinal image. Vision in sirenians seems to be of secondary importance, although it is of relevance in some behavioral contexts (Hartman, 1979).

EVOLUTIONARY IMPLICATIONS

As described above, there are different uses of vision in secondarily aquatic tetrapods, and different clades have found different evolutionary solutions to the problems associated with taking vision back into the water. Two factors inter-

play to shape eye design: shared ancestry and shared ecology.

SHARED ANCESTRY

The terrestrial ancestor from which the aquatic forms were derived may have had structures that could easily be evolutionarily modified to facilitate vision in water and/or amphibious vision. A good example is the powerful accommodative mechanism of birds (Walls, 1942; Duke-Elder, 1958). This has probably made it relatively easy for aquatic birds to evolve good underwater visual capabilities. Similarly, cetaceans and pinnipeds have further refined sensory systems that were already highly developed in their ancestors (i.e., hearing and the whiskers system, respectively).

SHARED ECOLOGY

If a successful group of animals (e.g., odontocetes) had developed a certain sensory modality (e.g., echolocation) to efficiently find and capture prey in water, selection might drive competing clades to find other ways of life leading to evolutionary divergence (Ricklefs, 1979), or it may drive these other clades into improving their methods to pursue the same prey, leading to an evolutionary arms race. It is possible that such new ways of life, for example, huge size combined with filter-feeding in mysticetes, hydrodynamic trail-following in pinnipeds (see Dehnhardt and Mauck, chapter 18 in this volume), and high visual acuity with the use of weak color cues in penguins, are the results of such exclusionary selection.

SHARED ANCESTRY AND ECOLOGY

As can be understood from the examples given above, ancestry and ecology usually interact. A good example of this is color vision. Terrestrial mammals had lost two of four cone visual pigments, which reduced their color vision capabilities to a mere shadow of the advanced tetrachromatic color vision systems of their piscine and reptilian ancestors (e.g., Bowmaker, 1998; Reuter and Peichl, chapter 10 in this volume). When some mammals returned to the watery realm, the predators among them, cetaceans and pinnipeds, did not invest in color vision at all. Instead, they used the brain's limited capacity for other advanced sensory systems, echolocation and whiskers, respectively. For aquatic grazers such as sirenians, dichromatic color vision was equally adaptive as it was for their terrestrial relatives and was retained. The only penguin species studied so far still has most of the superior color vision capacity of its terrestrial ancestors, especially in the blue-green region of the spectrum (Bowmaker and Martin, 1985). Penguins can find food by color that escapes their mammalian competitors and use the ability to discriminate between wavelengths in other behavioral contexts.

Secondarily aquatic tetrapods that are highly (or even obligatorily) aquatic are usually more specialized than more amphibious taxa. For instance, hippopotamuses and dolphins are very different morphologically, and only molecular evidence is able to discern the close relationship between these animals (Luo, 2000). This makes it often difficult to distinguish the roles of ancestry and ecology as the reasons for homoplastic or homologous adaptations in extant species. In most extant artiodactyls and cetaceans the eyes are positioned laterally. This would suggest that this trait was conserved from the ancestors of the aquatic forms. However, the now-identified closest terrestrial (semiaquatic, in fact) relative of cetaceans, the hippopotamus, has the eyes positioned high in the skull, similar to eye positions in crocodiles. Furthermore, the fossil record shows that the earliest cetaceans (pakicetids [Nummela et al., 2006], ambulocetids [Thewissen et al., 1996]) had eyes in similar positions. This indicates that although the comparison between extant species (artiodactyls and cetaceans) may suggest ancestry as the reason for lateral eye positions in cetaceans, there also seems to have been a strong ecological factor that has led to a relocation of the eyes from high in the skull back to lateral positions.

The frontally directed eyes of pinnipeds are often interpreted as the result of the group's ancestry from terrestrial carnivores with eyes facing rostrally. Keeping the history of eye positions in cetaceans in mind, it is doubtful, however, that this is the only reason for rostrally facing eyes in extant pinnipeds. The large sizes of the eyes in all pinnipeds except for the walrus imply that these animals invest considerably in vision and that it is important for survival. It may be, however, that frontally directed eyes in the terrestrial ancestors of pinnipeds facilitated the transition from land to water toward the currently present division of labor between vision and the whiskers system.

There can be several reasons for a group of animals to go through a radical change in habitat and lifestyle as the change from terrestrial to aquatic existence. One such reason is the availability of food. Other reasons must have been present in the group of animals that went through the transition, since the animals cannot sacrifice a current selection advantage in order to be able to in the future colonize a new habitat. All intermediate forms have to survive competition in their transitional niche. Sometimes there are, however, traits in a group of animals that make it particularly easy to realize a dramatic transition in lifestyle. Such traits are called preadaptations in the modern meaning of the word (Futuyma, 1997). The eyes of snakes seem to be an example of this. Being very reminiscent of aquatic eyes, it apparently was easy to achieve the transition from vision in air to vision in water. Snake eyes, however, also raise a warning sign for such reasoning. The semiaquatic species in the subfamily of Natricinae reevolved powerful accommodation by deformation of the lens to be successful (see above), with all the consequences for the biology and mechanics of lens fiber cells, intraocular muscles, and control by the central nervous system. The "preadaptation" in eye design seems to have been of limited value, because the eyes nevertheless are severely defocused in water. Unfortunately, very little is known about eye structure and visual function in fully aquatic snakes, such that we do not know whether they may have followed similar evolutionary paths as their semiaquatic natricine relatives.

INGENIOUS EVOLUTION

Depending on ancestry, ecology, or combinations thereof, there rather frequently are different solutions to similar problems in secondarily aquatic tetrapods. An example from vision concerns having well-focused images in both water and air. Pinnipeds have solved the problem by having flat corneas that have little or no refractive power in both media. Light is focused only by the powerful lens. Freshwater turtles and the pursuit-divers among aquatic birds have curved corneas and compensate for the lack of corneal refraction in water by powerful accommodative mechanisms that encompass dramatic changes in lens shape. The same is true for natricine snakes, which is surprising since their ancestors had lost the powerful accommodative mechanism that is typical for sauropsids. In some birds, such as penguins and albatrosses, there is a compromise solution with corneas that are only slightly curved such that accommodative demand upon submergence is low. Even the completely aquatic cetaceans have solved the problems associated with amphibious vision. Their solution is different from those described above and at present still incompletely understood.

Air-breathing animals can stay submerged only for limited periods of time. If they forage at great depths they have to quickly move from the sunlit surface to depths that are hardly, if at all, reached by sunlight. Deep-diving pinnipeds, cetaceans, and penguins show dramatic changes in pupil size. At the surface the pupils are tiny to protect the retinas from too much light and to keep them in an at least a partially dark-adapted state. In addition, elephant seals have extremely fast mechanisms to dark-adapt the retina. Similar mechanisms may be present in deep-diving cetaceans and penguins. To maximize photon capture at great depths, cetaceans and pinnipeds have reflective tapeta lucida. However, the tapeta are different in these groups, being of the fibrous type in cetaceans and of the cellular type in pinnipeds. Functionally, however, they are very similar. In birds, tapeta lucida have so far been described only for the nocturnal goatsuckers (Caprimulgiformes) (Arnott and Nicol, 1975; Rojas et al., 2004). It may be, however, that even diving birds have layers of reflective material distal to the retina, possibly of yet another type than is present in cetaceans and pinnipeds.

Very few bodies of water are osmotically balanced to the body fluids of secondarily aquatic tetrapods. The eyes of aquatic snakes are covered by keratinized spectacles, such that the osmolarity of the external medium is of little relevance. The spectacles are furthermore replaced when a snake molts, such that even the problem of abrasion by particles suspended in water is solved. The eyes of crocodilians, some birds, cetaceans, and sirenians are protected by viscous secretions of ocular glands. In pinnipeds, of which some species inhabit extremely turbid waters, the cells of the corneal epithelium may be under constant renewal.

In many cases, it is not known how specific problems are solved. A good example for this are physical shocks that occur upon hitting the water surface from the air. The eyes of pinnipeds and cetaceans can be retracted into the orbits and covered by thick lids for protection. Pinnipeds and cetaceans furthermore have ocular retia mirabilia or fatty pads that may have functions in dampening shock waves. In birds, however, these structures are absent. It is not clear how penguins protect their eyes when they porpoise at high swimming speed or cast themselves into the water from an ice shelf. The world of secondarily aquatic tetrapods is doubtlessly very different from human experiences, and many interesting questions still await answers.

ACKNOWLEDGMENTS

RK thanks, in alphabetical order, Alexander Costidis, Dan-E. Nilsson, Leo Peichl, Tom Reuter, Sentiel Rommel, and Eric Warrant for useful comments and contributions. RK extends special thanks to the team of the Biology Library at Lund University, which brought to him countless articles and books at lightning speed. RK is supported by the Swedish Research Council, the Wallenberg Foundation, and the Swedish Foundation for Strategic Research.

GK wishes to thank Tamir Strod, Ruth Almon, Ido Izhaki, Howard (Howie) Howland, Kevin van Doorn, and Sari Giladi for their contribution to the research, which was supported by grants from the Israel Science Foundation and by the Israel Ministry of Science.

LITERATURE CITED

Alfaro, M. E. 2002. Forward attack modes of aquatic feeding garter snakes. Functional Ecology 16: 204–215.

Arnott, H. J., and J. A. C. Nicol. 1975. An electron microscopic study of the tapetum lucidum in the eyes of goatsuckers. Transactions of the American Microscopical Society 94:433.

Bauer, G. B., D. E. Colbert, J. C. I. Gaspard, B. Littlefield, and W. Fellner. 2003. Underwater visual acuity of Florida manatees (Trichechus manatus latirostris). International Journal of Comparative Psychology 16:130–142.

Beebe, W. 1934. Half Mile Down. Harcourt, Brace and Company, New York.

Beer, T. 1898. Die Akkommodation des Auges bei den Reptilien. Pflügers Archiv für die gesamte Physiologie des Menschen und der Tiere 69:507–568.

Bellairs, A. D., and G. Underwood. 1951. The origin of snakes. Biological Reviews of the Cambridge Philosophical Society 26:193–237.

Bennett, K. A., B. J. McConnell, and M. A. Fedak. 2001. Diurnal and seasonal variations in the duration and depth of the longest dives in southern elephant seals (Mirounga leonina): possible physiological and behavioural constraints. Journal of Experimental Biology 204:649–662.

Berkley, M. A. 1973. Grating resolution and refraction in cat estimated from evoked cerebral potentials. Vision Research 13:403–415.

Bogorad, F. A. 1979. The symptom of crocodile tears. Introduction and translation by Austin Seckersen. Journal of Historical Medicine and Allied Sciences 34:74–79.

Bowmaker, J. K. 1998. Evolution of colour vision in vertebrates. Eye 12:541–547.

Bowmaker, J. K., and G. R. Martin. 1985. Visual pigments and oil droplets in the penguin, Spheniscus humboldti. Journal of Comparative Physiology A 156:71–78.

Britton, A. 2006. Crocodilian Biology Database. Available at www.flmnh.ufl.edu/cnhc/cbd.html.

Campagna, C., M. A. Fedak, and B. J. McConnel. 1999. Post-breeding distribution and diving behavior of adult male southern elephant seals from Patagonia. Journal of Mammalogy 80: 1341–1352.

Caprette, C. L., M. S. Y. Lee, R. Shine, A. Mokany, and J. F. Downhower. 2004. The origin of snakes (Serpentes) as seen through eye anatomy. Biological Journal of the Linnean Society 81:469–482.

Cowan, F. B. M.. 1974. Observations of extrarenal excretion by orbital glands and osmoregulation in Malaclemys terrapin. Comparative Biochemistry and Physiology A 48:489–500.

Cramp, S., and K. E. L. Simmons. 1977. The Birds of the Western Palearctic, Handbook of the Birds of Europe, the Middle East and North Africa, Vol. I. Oxford University Press, Oxford.

Davies, M. N. O., and P. R. Green (eds.). 1994. Perception and Motion Control in Birds. Springer-Verlag, Berlin.

Davis, R. W., L. A. Fuiman, T. M. Williams, and B. J. Le Boeuf. 2001. Three-dimensional movements and swimming activity of a northern elephant seal. Comparative Biochemistry and Physiology A 129:759–770.

Davson, H. 1990. Physiology of the Eye; pp. 53–100. Macmillan, London.

Dawson, W. W. 1980. The cetacean eye; in L. M. Herman (ed.), Cetacean Behavior: Mechanisms and Functions. Wiley, New York.

Dawson, W. W., L. A. Birndorf, and J. M. Perez. 1972. Gross anatomy and optics of the dolphin eye. Cetology 10:1–12.

Dawson, W. W., J. P. Schroeder, and S. N. Sharpe. 1987a. Corneal surface properties of two marine mammal species. Marine Mammal Science 3:186–197.

Dawson, W. W., J. P. Schroeder, and J. F. Dawson. 1987b. The ocular fundus of two cetaceans. Marine Mammal Science 3:1–13.

Dawson, W. W., J. P. Schroeder, J. C. Dawson, and P. E. Nachtigall. 1992. Cyclic ocular hypertension in cetaceans. Marine Mammal Science 8:135–142.

Dehnhardt, G. 2002. Sensory systems; pp. 116–141 in A. R. Hoelzel (ed.), Marine Mammals Biology. Blackwell Science, Oxford.

Dehnhardt, G., B. Mauck, W. Hanke, and H. Bleckmann. 2001. Hydrodynamic trail-following in harbor seals (Phoca vitulina). Science 293:102–104.

Dexler, H., and L. Freund. 1906. Contributions to the physiology and biology of the dugong. American Naturalist 40:49–72.

Dral, A. D. G. 1972. Aquatic and aerial vision in the bottle-nosed dolphin. Netherlands Journal of Sea Research 5:510–513.

Dral, A. D. G. 1975. Some quantitative aspects of the retina of *Tursiops truncatus*. Aquatic Mammals 2:28–31.

Dral, A. D. G. 1977. On the retinal anatomy of Cetacea (mainly *Tursiops truncatus*); pp. 81–134 in R. J. Harrison (ed.), Functional Anatomy of Marine Mammals. Academic Press, London.

Dral, A. D. G. 1983. The retinal ganglion cells of *Delphinus delphis* and their distribution. Aquatic Mammals 10:57–68.

Dral, A. D. G., H. R. Duncker, and G. Fleischer 1985. Amphibious vision in dolphins; pp. 707–709 in Duncker and Fleischer (eds.), Fortschritte der Zoologie. Gustav Fischer, Stuttgart.

Duke-Elder, S. 1958. The eye in evolution; in S. Duke-Elder (ed.), System of Ophthalmology, Vol. 1. Klimpton, London.

Evans, P. G. H. 1987. The Natural History of Whales and Dolphins. Christopher Helm, London.

Fasola, M., and L. Canova, 1993. Diet activity of resident and immigrant waterbirds at Lake Turkana, Kenya. Ibis 135:442–450.

Fernández, M. S., F. Archuby, M. Talevi, and R. Ebner. 2005. Ichthysaurian eyes: paleobiological information content in the sclerotic ring of *Caypullisaurus* (Ichthyosauria, Ophthalmosauria). Journal of Vertebrate Paleontology 25:330–337.

Fleishman, L. J., and A. S. Rand. 1989. *Caiman crocodilus* does not require vision for underwater prey capture. Journal of Herpetology 23:296.

Fleishman, L. J., H. C. Howland, M. J. Howland, A. S. Rand, and M. L. Davenport. 1988. Crocodiles don't focus underwater. Journal of Comparative Physiology A 163:441–443.

Fritsches, K. A., R. W. Brill, and E. J. Warrant. 2005. Warm eyes provide superior vision in swordfishes. Current Biology 15:55–58.

Fritzberg, W. 1912. Beiträge zur Kenntnis des Akkommodationsapparates bei Reptilien. Archiv für vergleichende Ophthalmologie 3:292–322.

Frost, B. J., D. R. Wylie, and Y. C. Wang. 1994. The analysis of motion in the visual systems of birds; pp. 248–269, 270–291 in M. N. O. Davies and P. R. Green (eds.), Perception and Motion Control in Birds. Springer-Verlag, Berlin.

Futuyma, D. J. 1997. Evolutionary Biology. Sinauer Associates, Sunderland, MA.

Gao, A., and K. Zhou. 1986. Anatomical and histological studies of the eyes of the finless porpoise *Neophocaena phocaenoides*. Acta Zoologica Sinica 32:248–254.

Garthe, S., S. Benvenuti, and W. A. Montevecchi. 2000. Pursuit plunging by northern gannets *(Sula bassana)* feeding on capelin *(Mallotus villosus)*. Proceedings of the Royal Society B 267:1717–1722.

Glasser, A., and H. C. Howland. 1996. A history of studies of visual accommodation in birds. Quarterly Review of Biology 71:475–509.

Glasser, A., D. Troilo, and H. C. Howland. 1994. The mechanism of corneal accommodation in chicks. Vision Research 34:1549–1566.

Glasser, A., C. J. Murphy, D. Troilo, and H. C. Howland. 1995. The mechanism of lenticular accommodation in the chick eye. Vision Research 35:1525–1540.

Goodge, W. R. 1960. Adaptations for amphibious vision in the dipper *(Cinclus mexicanus)*. Journal of Morphology 107:79–91.

Hanke, F. D., G. Dehnhardt, F. Schaeffel, and W. Hanke. 2006. Corneal topography, refractive state, and accommodation in harbor seals *(Phoca vitulina)*. Vision Research 46:837–847.

Harper, J. Y., D. A. Samuelson, and R. L. Reep. 2005. Corneal vascularization in the Florida manatee *(Trichechus manatus latirostris)* and three-dimensional reconstruction of vessels. Veterinary Ophthalmology 8:89–99.

Hart, N. S. 2004. Microspectrophotometry of visual pigments and oil droplets in a marine bird, the wedge-tailed shearwater *Puffinus pacificus*: topographic variations in photoreceptor spectral characteristics. Journal of Experimental Biology 207: 1229–1240.

Hartman, D. S. 1979. Ecology and Behavior of the Manatee *(Trichechus manatus)* in Florida. Special Publication No. 5. American Society of Mammalogists, Lawrence, KS.

Herman, L. M., M. F. Peacock, M. P. Yunker, and C. J. Madsen. 1975. Bottlenosed dolphin: double slit pupil yields equivalent aerial and underwater diurnal acuity. Science 189:650–652.

Herring, P. 2002. The Biology of the Deep Ocean. Oxford University Press, Oxford.

Herring, P. J. 1990. Bioluminescent communication in the sea; pp. 245–264 in P. J. Herring, A. K. Campbell, M. Whitfield, and L. Maddock (eds.), Light and Life in the Sea. Cambridge University Press, Cambridge.

Howland, H. C., and J. G. Sivak. 1984. Penguin vision in air and water. Vision Research 24:1905–1909.

Humphries, S., and G. D. Ruxton. 2002. Why did some ichthyosaurs have such large eyes? Journal of Experimental Biology 205:439–441.

Jamieson, G. 1971. The functional significance of corneal distortion in marine mammals. Canadian Journal of Zoology 49:421–423.

Jamieson, G. S., and H. D. Fisher. 1971. The retina of the harbour seal, *Phoca vitulina*. Canadian Journal of Zoology 49:19–23.

Jamieson, G. S., and H. D. Fisher. 1977. The pinniped eye: a review; pp. 245–261 in R. J. Harrison (ed.), Functional Anatomy of Marine Mammals. Academic Press, London.

Johnson, G. L. 1893. Observations on the refraction and vision of the seal's eye. Proceedings of the Zoological Society of London 719–723.

Johnson, G. L. 1901. Contributions to the comparative anatomy of the mammalian eye based on ophthalmoscopic examination. Philosophical Transactions of the Royal Society of London B 194:1–30.

Katzir, G., and J. Camhi. 1993. Escape response of black mollies (Poecilia sphenops) from predatory dives of a pied kingfisher, Ceryle rudis. Copeia 2:549–553.

Katzir, G., and H. C. Howland. 2003. Corneal power and underwater accommodation in great cormorants. Journal of Experimental Biology 206: 833–841.

Katzir, G., and N. Intrator. 1987. The striking of underwater prey by a reef heron Egretta gularis schistacea. Journal Comparative Physiology A 160:517–523.

Katzir, G., T. Strod, R. Schechtman, S. Hareli, and Z. Arad. 1999. Cattle egrets are less able to cope with light refraction than are other herons. Animal Behaviour 57:687–694.

Katzir, G., Z. Labinger, and Y. Benjamini. 2002. Prey size selectivity and prey depth in the pied kingfisher (Ceryle rudis L.). Bird Behaviour 2:43–52.

Kaufman, P. L., and A. Alm. 2003. Adler's Physiology of the Eye. Mosby, St. Louis.

Kröger, R. H. H. 1984. Untersuchungen zum Spielverhalten von Delphinen unter Gefangenschaftsbedingungen, Thesis, Universität Hamburg.

Kröger, R. H. H. 1989. Dioptrik, Funktion der Pupille und Akkommodation bei Zahnwalen, Dissertation, Eberhard-Karls-Universität Tübingen.

Kröger, R. H. H., and K. Kirschfeld. 1989. Visual accommodation in cetaceans; pp. 38–40 in P. G. H. Evans and C. Smenk (eds.), European Research on Cetaceans. European Cetacean Society, Leiden.

Kröger, R. H. H., and K. Kirschfeld. 1992. The cornea as an optical element in the cetacean eye; pp. 97–106 in J. A. Thomas, R. A. Kastelein, and A. Ya. Supin (eds.), Marine Mammal Sensory Systems. Plenum Press, New York.

Kröger, R. H. H., and K. Kirschfeld. 1993. Optics of the harbor porpoise eye in water. Journal of the Optical Society of America A 10:1481–1489.

Kröger, R. H. H., and K. Kirschfeld. 1994. Refractive index in the cornea of a harbor porpoise (Phocoena phocoena) measured by two-wavelengths laser-interferometry. Aquatic Mammals 20:99–107.

Kröger, R. H. H., M. C. W. Campbell, R. Munger, and R. D. Fernald. 1994. Refractive index distribution and spherical aberration in the crystalline lens of the African cichlid fish Haplochromis burtoni. Vision Research 34:1815–1822.

Labinger, Z., Y. Benjamini, and G. Katzir. 1991. Prey size choice in captive pied kingfishers (Ceryle rudis). Animal Behaviour 42:969–975.

Land, M. F., and D.-E. Nilsson. 2002. Animal Eyes. Oxford University Press, Oxford.

Leatherwood, S., R. R. Reeves, and L. Foster. 1983. The Sierra Club Handbook of Whales and Dolphins. Sierra Club Books, San Francisco.

Lee, D. N. 1994. An eye or ear for flying; pp. 270–291 in M. N. O. Davies and P. R. Green (eds.), Perception and Motion Control in Birds. Springer-Verlag, Berlin.

Levenson, D. H., and A. Dizon. 2003. Genetic evidence for the ancestral loss of short-wavelength-sensitive cone pigments in mysticete and odontocete cetaceans. Proceedings of the Royal Society of London B 270:673–679.

Levenson, D. H., and R. J. Schusterman. 1997. Pupillometry in seals and sea lions: ecologcal implications. Canadian Journal of Zoology 75:2050–2057.

Levenson, D. H., and R. J. Schusterman. 1999. Dark adaptation and visual sensitivity in shallow and deep diving pinnipeds. Marine Mammal Science 15:1303–1313.

Levy, B., and J. G. Sivak. 1980. Mechanisms of accommodation in the bird eye. Journal of Comparative Physiology A 137:267–272.

Luo, Z. 2000. In search of the whales' sisters. Nature 404:235–237.

Lusseau, D. 2003. The emergence of cetaceans: phylogenetic analysis of male social behaviour supports the Cetartiodactyla clade. Journal of Evolutionary Biology 16:531–535.

Malmström, T., and R. H. H. Kröger. 2006. Pupil shapes and lens optics in the eyes of terrestrial vertebrates. Journal of Experimental Biology 209:18–25

Martin, G. R. 1998. Eye structure and amphibious foraging in albatrosses. Proceedings of the Royal Society of London 265:665–671.

Martin, G. R. 1999. Eye structure and foraging in King Penguin Aptenodytes patagonicus. Ibis 141:444–450.

Martin, G. R., and S. R. Young. 1984. The eye of the Humboldt penguin, Spheniscus humboldti: visual fields and schematic optics. Proceedings of the Royal Society B 223:197–222.

Mass, A. M., and A. Ya. Supin. 1989. Distribution of ganglion cells in the retina of the Amazon river dolphin Inia geoffrensis. Aquatic Mammals 15:49–56.

Mass, A., and A. Ya. Supin. 1990. Best vision zones in the retinae of some cetaceans; pp. 505–517 in J. A. Thomas and R. A. Kastelein (eds.), Sensory

Abilities of Cetaceans, Nato ASI Series A. Plenum Press, New York.

Mass, A. M., and A. Y.a. Supin. 1995. Ganglion cell topography of the retina in the bottlenosed dolphin, *Tursiops truncatus*. Brain, Behavior, and Evolution 45:257–265.

Mass, A. M., and A. Ya. Supin. 1997. Ocular anatomy, retinal ganglion cell distribution, and visual resolution in the gray whale, *Eschrichtus gibbosus*. Aquatic Mammals 23:17–28.

Mass, A. M., and A. Ya. Supin. 2003. Retinal topography of the harp seal *Pagophilus groenlandicus*. Brain, Behavior, and Evolution 62:212–222.

Mass, A. M., and A. Ya. Supin. 2005. Ganglion cell topography and retinal resolution of the Steller sea lion *(Eumetopias jubatus)*. Aquatic Mammals 31:393–402.

Mass, A. M., A. Ya. Supin, and A. N. Severtsov. 1986. Topographic distribution of sizes and density of ganglion cells in the retina of a porpoise, *Phocoena phocoena*. Aquatic Mammals 12:95–102.

Mass, A. M., D. K. Odell, D. R. Ketten, and A. Ya. Supin. 1997. Retinal topography and visual acuity in the Florida manatee, *Trichechus manatus latirostris*. Doklady Akademii Nauk 355:427–430.

Matthiessen, L. 1886. Ueber den physikalisch-optischen Bau des Auges der Cetaceen und der Fische. Pflüger's Archiv 38:521–528.

Mattison, C. 2003. Snakes of the World. Facts on File, New York.

Mauck, B., and G. Dehnhardt. 1997. Mental rotation in a California sea lion *(Zalophus californianus)*. Journal of Experimental Biology 200:1309–1316.

Mauck, B., and G. Dehnhardt. 2005. Identity concept formation during visual multiple-choice matching in a harbor seal *(Phoca vitulina)*. Learning and Behavior 33:428–436.

McNeil, R., and P. G.-C. J. D. Drapeau. 1992. The occurrence and adaptive significance of nocturnal habits in waterfowl. Biological Reviews of the Cambridge Philosophical Society 67:381–419.

Moroney, M. K., and J. D. Pettigrew. 1987. Some observations on the visual optics of kingfishers (Aves, Coraciformes, Alcedinidae). Journal of Comparative Physiology 160:137–149.

Motani, R. 1999. Phylogeny of the Ichthyopterygia. Journal of Vertebrate Paleontology 19:473–496.

Motani, R., B. M. Rothschild, and W. Wahl Jr. 1999. Large eyeballs in diving ichthyosaurs. Nature 402:747.

Murayama, T., H. Somiya, I. Aoki, and T. Ishii. 1992. The distribution of ganlion cells in the retina and visual acuity of Minke whale. Nippon Suissan Gakkaishi 58:1057–1061.

Murayama, T., H. Somiya, I. Aoki, and T. Ishii. 1995. Retinal ganglion cell size and distribution predict visual capabilities of Dall's porpoise. Marine Mammal Science 11:136–149.

Nagy, A. R., and K. Ronald. 1970. The harp seal, *Pagophilus groenlandicus* (Erxleben, 1777). VI. Structure of retina. Canadian Journal of Zoology 48:367–370.

Nicol, J. A. C., 1981. Tapeta lucida of vertebrates; pp. 401–431 in J. M. Enoch and F. L. Tobey Jr. (eds), Vertebrate photoreceptor optics, Springer-Verlag, Berlin.

Nummela, S., S. T. Hussain, and J. G. M. Thewissen. 2006. Cranial anatomy of Pakicetidae (Cetacea, Mammalia). Journal of Vertebrate Paleontology 26:746–759.

Ollivier, F. J., D. A. Samuelson, D. E. Brooks, P. A. Lewis, M. E. Kallberg, and A. M. Komaromy, 2004. Comparative morphology of the tapetum lucidum (among selected species). Veterinary Ophthalmology 7:11–22.

Olmos, F., and I. Sazima. 1990. A fishing tactic in floating Paraguayan caiman: the cross-posture. Copeia 1990:875–877.

Ott, M. 2006. Visual accommodation in vertebrates: mechanisms, physiological response and stimuli. Journal of Comparative Physiology A 192:97–111 .

Pardue, M. T., and J. G. Sivak. 1997. The functional anatomy of the ciliary muscle in four avian species. Brain, Behavior, and Evolution 49:295–311.

Peichl, L., G. Behrmann, and Association for Research in Vision and Ophthalmology (R. H. H.) Kröger. 2001. For whales and seals the ocean is not blue: a visual pigment loss in marine mammals. European Journal of Neuroscience 13:1520–1528.

Peichl, L., K. Kovacs, and C. Lydersen. 2002. Absence of S-cones in the retinae of further marine mammals (whales and seals). Association for Research in Vision and Ophthalmology (ARVO) Annual Meeting Abstract Search and Program Planner 2002, Abstract No. 3762, Rockville, MD.

Piggins, D. J. 1970. Refraction of the harp seal, *Pagophilus groenlandicus* (Erxleben 1777). Nature 227:78–79.

Piggins, D., W. R. A. Muntz, and R. C. Best. 1983. Physical and morphological aspects of the eye of the manatee *Trichechus inunguis* Natterer 1883: (Sirenia: Mammalia). Marine Behaviour and Physiology 9:111–130.

Pilleri, G. 1964. Zur Morphologie des Auges vom Weißwal, *Delphinapterus leucas* (Pallas). Hvalrådets skrifter 47:3–16.

Pilleri, G. 1979. The blind Indus dolphin, *Platanista indi*. Endeavour 3:48–55.

Pütter, A. 1903. Die Augen der Wassersäugethiere. Zoologische Jahrbücher, Anatomie und Ontogenie 17:99–402.

Repka, M. X., and H. A. Quigley. 1989. The effect of age on normal human optic nerve fiber number and diameter. Ophthalmology 96:26–32.

Ricklefs, R. E. 1979. Ecology. Thomas Nelson and Sons, London.

Rivamonte, L. A. 1976. Eye model to account for comparable aerial and underwater acuities of the bottlenose dolphin. Netherlands Journal of Sea Research 10:491–498.

Robinson, G. D., and W. A. Dunson. 1976. Water and sodium balance in the estuarine diamondback terrapin *(Malaclemys)*. Journal of Comparative Physiology B 105:129–152.

Rochon-Duvigneaud, A. 1943. Les Yeux et la Vision des Vertébrés. Masson, Paris.

Rodieck, R. W. 1998. The First Steps in Seeing Sinauer, Sunderland, MA.

Rohr, J. J., F. E. Fish, and J. W. Gilpatrick. 2002. Maximum swim speeds of captive and free-ranging delphinids: critical analysis of extraordinary performance. Marine Mammal Science 18:1–19.

Rojas, L. M., Y. M. Ramirez, G. Marin, and R. McNeil. 2004. Visual capability in Caprimulgiformes. Ornitologia Neotropical 15 (Suppl. S): 251–260.

Schaeffel, F., and A. de Queiroz. 1990. Alternative mechanisms of enhanced underwater vision in the garter snakes *Thamnophis melanogaster* and *T. couchii*. Copeia 1:50–58.

Schaeffel, F., and U. Mathis. 1991. Underwater vision in semi-aquatic European snakes. Naturwissenschaften 78:373–375.

Schusterman, R. J., and R. F. Balliet. 1971. Aerial and underwater visual acuity in the California sea lion *(Zalophus californianus)* as a function of luminance. Annals of the New York Academy of Sciences 188:37–46.

Schwab, I. R., and Hart N. S. 2004. Halcyon days. British Journal of Ophthalmology. 88:613.

Shine, R. 2005. All at sea: aquatic life modifies materecognition modalities in sea snakes *(Emydocephalus annulatus*, Hydrophiidae). Behavioral Ecology and Sociobiology 57:591–598.

Shine, R., G. P. Brown, and M. J. Elphick. 2004a. Field experiments on foraging in free-ranging water snakes *Enhydris polyopis* (Homalopsinae). Animal Behaviour 68:1313–1324.

Shine, R., X. Bonnet, M. J. Elphick, and E. G. Barrott. 2004b. A novel foraging mode in snakes: browsing by the sea snake *Emydocephalus annulatus* (Serpentes, Hydrophiidae). Functional Ecology 18:16–24.

Shiose, Y., and Y. Kawase. 1986. A new approach to stratified normal intraocular pressure in a general population. American Journal of Ophthalmology 101:714–722.

Sivak, J. G. 1976. The role of the flat cornea in the amphibious behaviour of the blackfoot penguin *(Spheniscus demersus)*. Canadian Journal of Zoology 54:1341–1345.

Sivak, J. G. 1980. Avian mechanisms for vision in air and water. Trends in Neurosciences 12:314–317.

Sivak, J. G., and M. Milodot. 1977. Optical performance of the penguin eye in air and in water. Journal of Comparative Physiology A 119:241–247.

Sivak, J. G., and O. E. Vrablic. 1979. The anatomy of the eye of the Adelie penguin with special reference to optical structure and intraocular musculature. Canadian Journal of Zoology 57:346–352.

Sivak, J. G., O. E. Vrablic. 1982. Ultrasonic of intraocular muscles of diving and nondiving ducks. Canadian Journal of Zoology 60:1588–1666.

Sivak, J. G., J. L. Lincer, and W. Bobier. 1977. Amphibious visual optics of the eyes of the double crested cormorant *(Phalacrocorax auritus)* and the brown pelican *(Pelecanus occidentalis)*. Canadian Journal of Zoology 55:782–788.

Sivak, J. G., H. C. Howland, and P. McGill-Harelstad. 1987. Vision of the Humboldt penguin *(Spheniscus humboldti)* in air and water. Proceedings of the Royal Society of London B 229:647–472.

Sivak, J. G., H. C. Howland, J. West, and J. Weerheim. 1989. The eye of the hooded seal, *Cystophora cristata*, in air and water. Journal of Comparative Physiology A 165:771–777 .

Smith, G., and D. A. Atchison. 1997. The Eye and Visual Optical Instruments. Cambridge University Press, Cambridge.

Stich, K. P., G. Dehnhardt, and B. Mauck. 2003. Mental rotation of perspective stimuli in a California sea lion *(Zalophus californianus)*. Brain, Behavior, and Evolution 61:102–112.

Stone, J. 1978. The number and distribution of ganglion cells in the cat's retina. Journal of Comparative Neurology 180:753–771.

Stone, J., and J. E. Campion. 1978. Estimate of the number of myelinated axons in the cat's optic nerve. Journal of Comparative Neurology 180: 799–806.

Strod, T. 2002. Visual capacity and prey preference in the great cormorant, *Phalacrocorax carbo sinensis*. Ph.D. Thesis. The Technion, Israel Institute of Technology, Haifa.

Strod, T., Z. Arad, I. Izhaki, and G. Katzir. 2004. Cormorants keep their power visual resolution in a pursuit-diving bird under amphibious and turbid conditions. Current Biology 14:376–377.

Supin, A. Ya., V. V. Popov, and A. M. Mass. 2001. The Sensory Physiology of Aquatic Mammals. Kluwer Acadamic Publishers, Boston.

Taber, A. B. 1984. Grey whale; pp. 56–61 in D. W. MacDonald (ed.), World of Animals: Sea Mammals. Equinox, Oxford.

Tarpley, R. J., and S. H. Ridgway. 1991. Orbital gland structure and secretions in the Atlantic bottlenose dolphin (Tursiops truncatus). Journal of Morphology 207:173–184.

Thewissen, J. G. M., S. I. Madar, and S. T. Hussain. 1996. Ambulocetus natans, an Eocene cetacean (Mammalia) from Pakistan. Courier Forschungsinstitut Senckenberg 191:1–86.

Thorbjarnarson, J. B. 1993. Fishing behavior of spectacled caiman in the Venezuelan llanos. Copeia 1993:1166–1171.

Tomkins, I. T. 1951. Method of feeding of the black skimmer, Rynchops nigra. Auk 68:236–239.

Underwood, G. 1970. The eye; pp. 1–97 in C. Gans and T. S. Parsons (eds.), Morphology B. Academic Press, London.

Vidal, N., and S. B. Hedges. 2004. Molecular evidence for a terrestrial origin of snakes. Proceedings of the Royal Society of London B 271:S226–S229.

Vincent, S. E., R. Shine, and G. P. Brown. 2005. Does foraging mode influence sensory modalities for prey detection in male and female filesnakes, Acrochordus arafurae? Animal Behaviour 70:715–721.

Waller, G. N. H. 1980. The Visual System of Toothed Whales (Mammalia, Cetacea, Odontoceti). Ph.D. Thesis, Corpus Christi College, Cambridge.

Waller, G. N. H. 1992. Ciliary muscles in the eye of the long-finned pilot whale Globicephala melaena (Traill, 1809). Aquatic Mammals 18:36–39.

Walls, G. L. 1940. Ophthalmological implications for the early history of the snakes. Copeia 1940:1–8.

Walls, G. L. 1942. The Vertebrate Eye and its Adaptive Radiation. McGraw-Hill, New York.

Welsch, U., S. Ramdohr, B. Riedelsheimer, R. Hebel, R. Eisert, and J. Plotz. 2001. Microscopic anatomy of the eye of the deep-diving Antarctic Weddell seal (Leptonychotes weddellii). Journal of Morphology 248:165–174.

West, J. A., J. G. Sivak, C. J. Murphy, and K. M. Kovacs. 1991. A comparative study of the anatomy of the iris and ciliary body in aquatic mammals. Canadian Journal of Zoology 69:2594–2607.

Wyneken, J. 2001. The Anatomy of Sea Turtles. NOAA Technical Memorandum NMFS-SEFSC-470. U.S. Department of Commerce, Washington, DC.

You, H. L., M. C. Lamanna, J. D. Harris, L. M. Chiappe, J. O'Connor, S. Ji, J.-C. Lü, C.-X. Yuan, D-G. Li, X. Zhang, K. J. Lacovara, P. Dodson, and Q. Ji. 2006. A nearly modern amphibious bird from the Early Cretaceous of northwestern China. Science 312:1640–1643.

Young, N. M., and W. W. Dawson. 1992. The ocular secretions of the bottlenose dolphin Tursiops truncatus. Marine Mammal Science 8:57–68.

Zhou, K. Y., G. Pilleri, and Y. M. Li. 1980. Observations on Baiji (Lipotes vexillifer) and finless porpoise (Neophocaena asiaeorientalis) in the lower reaches of the Chang Jiang. Scientia Sinica 23:785–794.

Zhu, Q., D. J. Hillman, and W. G. Henk. 2001. Morphology of the eye and surrounding structures of the bowhead whale, Balaena mysticetus. Marine Mammal Science 17:729–750.

Zusi, R. L., and D. Bridge. 1981. On the slit pupil of the black skimmer. Journal of Field Ornithology 52:338–340.

10

Structure and Function of the Retina in Aquatic Tetrapods

Tom Reuter and Leo Peichl

Kröger and Katzir (chapter 9 in this volume) have described the optical apparatus of the eye and its adaptations to function in water. Unlike these peripheral structures, no structural changes in the anatomy or histology of the retina are necessary for underwater vision at shallow depths, where the illumination usually is relatively bright and spectrally broad. Thus the retinae of ducks, and freshwater and sea turtles are comparable to those of terrestrial birds and tortoises (Granda and Dvorak, 1977; Zueva, 1982; Jane and Bowmaker, 1988; Bowmaker et al., 1997; Bartol and Musick, 2001; Loew and Govardovskii, 2001). The vertebrate retina is an evolutionarily conservative structure, and in fact shallow-water fish and many terrestrial tetrapods have strikingly similar retinae (Archer et al., 1999).

On the other hand, seals, whales, and penguins, which forage hundreds of meters below the ocean surface, show interesting molecular and cellular adaptations related to the low light levels and to the limited spectral range of the light available at greater depths. Generally, retinal adaptations in deep-diving tetrapods are similar to the adaptations of the retinae in

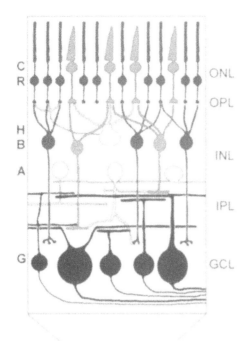

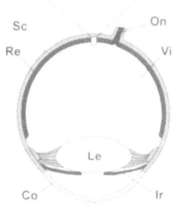

FIGURE 10.1. The mammalian retina and eye. *Below*, a cross section through the mammalian eye, showing its basic anatomy. Abbreviations: Co, cornea; Le, lens; Ir, iris and pupil; Vi, vitreous; Sc, sclera, choroid and pigment epithelium; Re, retina; and On, optic nerve. *Above*, a section through the mammalian retina showing basic neuron types and their connections. Light enters the retina from the ganglion cell side (from bottom in drawing) and is detected by the rod (R, dark gray) and cone (C, light gray) photoreceptors located in the outer nuclear layer (ONL). In the mammalian retina, these photoreceptors connect to separate bipolar cell types (B), the rod bipolars and cone bipolars located in the inner nuclear layer (INL, shown in different shading), establishing a rod pathway and a cone pathway. Separate processes of horizontal cells (H, pale) modulate the signal transfer at these synapses in the outer plexiform layer (OPL). Cone bipolars synapse with dendrites of various ganglion cell types (G) in the

terrestrial tetrapods with crepuscular or nocturnal lifestyles.

GENERALIZED ANATOMY AND FUNCTION OF THE RETINA

The vertebrate retina is a well-layered structure with the photoreceptors (rods and cones) lying close to the pigment epithelium, choroid, and sclera, and the retinal ganglion cells lying close to the vitreous (Fig. 10.1). As a result of this design, light passes through the transparent neural layers of the retina before reaching the photoreceptors. Rods are for low-light vision and cone photoreceptors for daylight and color vision. Following receptor photoactivation, the rod and cone signals are transmitted to bipolar and horizontal cells, which in turn pass them to a complex network of amacrine and ganglion cells. The neural network performs a preliminary analysis reducing redundancy and emphasizing biologically relevant features such as brightness contrast, color contrast, and movement within the mosaic of light patterns projected on the retina (Rodieck, 1998). The axons of the retinal ganglion cells form the optic nerve, and thus all visual information reaches the brain via the ganglion cells.

In this section, we discuss the anatomy and physiology of the retina, providing the basics for understanding of the next section. In that section, we discuss the particular specializations of aquatic tetrapods. The focus is on the input and output stages of retinal processing, the photoreceptors, and the ganglion cells. The order of the topics is

ganglion cell layer (GCL) and modulatory amacrine cell types (A) in the inner plexiform layer (IPL). Rod bipolars do not make direct contact with the ganglion cell, instead signals from rod bipolars reach the ganglion cells via specific amacrine cells. The outer nuclear layer, inner nuclear layer, and ganglion cell layer contain the neuronal somata, the outer and inner plexiform layers the neuronal processes and synapses. Each ganglion cell sends an axon into the optic nerve to reach the visual brain centers. (For details of the retinal wiring, see Wässle, 2004.)

from smaller to larger: from molecules and cells to the organization of the histological mosaic across the retina.

PHOTORECEPTORS

ROD AND CONE VISUAL PIGMENTS

The outer segments of rods and cones are filled by invaginated cell membranes (discs) that are densely packed with light-sensitive visual pigment molecules. Each pigment molecule consists of a protein (opsin) enclosing a chromophore molecule called retinal (vitamin A aldehyde). Vision occurs when these molecules are activated by the energy provided by absorbed photons. Preceding photoactivation, retinal is in its 11-*cis* configuration and fits snugly into an intramolecular pocket of the opsin, but on absorption of a photon it turns to the all-*trans* configuration, and this isomerization results in a stepwise change of the opsin structure and a phototransduction cascade of chemical reactions electrically activating the whole receptor cell (Rodieck, 1998).

Opsin molecules, like other proteins, are under continuous evolutionary change, and thus there are thousands of tetrapod opsins. In contrast, in vertebrate eyes there are only two chromophores: 11-*cis*-retinal (retinal$_1$) and 11-*cis*-3-dehydroretinal (retinal$_2$). Visual pigments using retinal$_1$ are called rhodopsins, and those using retinal$_2$ porphyropsins. All avian and mammalian rod and cone pigments are rhodopsins, while fish, amphibian, or reptile pigments may be either rhodopsin or porphyropsin, or a mixture of both (Bridges, 1972).

The light absorbance of a visual pigment is determined by an interaction between opsin and chromophore and is tuned to a particular wavelength of maximum absorbance termed λ_{max}. An opsin molecule consists of roughly 350 amino acids, but only a few of these, facing the chromophore pocket, have been shown to be implicated in the spectral tuning of the pigment (Fasick and Robinson, 1998;

Bowmaker and Hunt, 1999; Hunt et al., 2001). Thus one or a few point mutations may dramatically change the absorption spectrum of a pigment.

The shapes of the absorbance curves are similar in all visual pigments across taxa and can thus be characterized by just two parameters: the wavelength of the absorption peak, or λ_{max}, and information about chromophore type, retinal$_1$ or retinal$_2$ (Govardovskii et al., 2000). Retinal$_2$ pigments have a roughly 25% lower absorbance at λ_{max} but have instead a broader curve than retinal$_1$ pigments (Dartnall, 1968). Across species, the absorbance spectra of cone pigments are widely distributed over the visible spectrum, with peaks from 360 nm in the UV to 630 nm in the red, to meet species-specific requirements for chromatic vision. In contrast, the rod pigment spectra are clustered within a narrow range from 470 to 540 nm (Fig. 10.2) (Lythgoe, 1991; Hart, 2001a).

COLOR VISION

Color vision is the capacity of a visual system to respond differently to light of different wavelengths. It is based on the presence of two or more photoreceptor types, and networks of nerve cells in the retina and the brain comparing the strength of the signals stemming from different cone types. Vertebrates usually have a single rod type and two to four cone types. The cone types are characterized by different visual pigments absorbing at different spectral positions (Fig. 10.2) (Neumeyer, 1991). The basic pattern of vertebrate cone pigments is a system with four opsins, each characterizing a different spectral cone type (Table 10.1) (Yokoyama, 2000; Ödeen and Håstad, 2003; Collin and Trezise, 2004). Here we distinguish between the following cone pigment/opsin classes: S1 (λ_{max} at very short wavelengths, UV to violet), S2 (λ_{max} at short wavelengths, violet to blue), M (λ_{max} at middle wavelengths, green), and L (λ_{max} at long wavelengths, green to yellow and red) (Table 10.1).

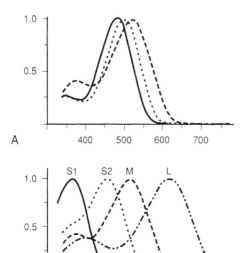

FIGURE 10.2. Absorbance spectra of visual pigments in aquatic species. (A) Rod visual pigment absorbance spectra in Sowerby's beaked whale *Mesoplodon bidens* (rhodopsin, λ_{max} 484 nm, *solid line*), the frog *Rana* (rhodopsin, λ_{max} 502 nm, *dotted line*) and *Rana*'s tadpole (porphyropsin, λ_{max} 523 nm, *dashed line*). (B) Absorbance spectra of the four cone visual pigment types in the red-eared slider turtle *Trachemys scripta elegans* (Emyidae). S1, ultraviolet-sensitive pigment, λ_{max} 372 nm; S2, blue-sensitive pigment, λ_{max} 458 nm; M, green-sensitive pigment, λ_{max} 515 nm; L, red-sensitive pigment, λ_{max} 617 nm. All curves are normalized to their respective maximum absorbance (for details, see Loew and Govardovskii, 2001). Wavelength in nanometers is shown on the *x* axis, and normalized absorbance on the *y* axis.

A theoretical analysis indicates that under the condition of satisfactory daylight, an arrangement with four cone types with sensitivity peaks evenly distributed over the spectrum, from UV to red, provides the animal with very good chromatic discrimination (Govardovskii and Vorobyev, 1989; Vorobyev, 1997; Hart and Vorobyev, 2005). This type of excellent color vision occurs in many fishes, reptiles, and birds (Fig. 10.2) (Govardovskii and Zueva, 1977; Hárosi and Hashimoto, 1983; Neumeyer, 1998; Hart, 2001a), but not in mammals (Jacobs, 1993). The S2, M, and L cones of many reptiles and birds are further provided with small optical filters, intensely colored "oil droplets," in their inner segments. The oil droplets act as cutoff filters absorbing blue light, in some cases also green and yellow light, suppressing the short-wave branch of the spectral sensitivity. These cones thus possess individual colored "sunglasses." This helps to produce narrowband, spectrally tuned color receptors (Donner, 1953; Baylor and Hodgkin, 1973; Arnold and Neumeyer, 1987; Loew and Govardovskii, 2001). The oil droplets have a cost, however, as they slightly reduce the sensitivity of cone vision by decreasing the number of photons reaching the visual pigments in the outer segments.

Depending on species, not all four cone types may be present. In most mammals, only two cone types are found: S1 (UV, violet, or blue) and L (green or yellow); see reviews by Jacobs (1993), Jacobs et al. (1998), and Yokoyama (2000) (Fig. 10.3A and B). The Old World primates, including humans, have "reinvented" three-cone color vision from its two-cone ancestry by having two variants of L cones (green- and red-sensitive) in addition to blue-sensitive S1 cones (Nathans et al., 1986). Amphibians lack M opsins but have S1, S2, and L cones (Table 10.1) (Hisatomi and Tokunaga, 2002). The number of different cone types is often used to characterize color vision as mono-, di-, tri-, or tetrachromatic. However, the effective mode of color vision must always be demonstrated in behavioral experiments (Neumeyer, 1986, 1991; Goldsmith, 1991).

PHOTORECEPTOR SENSITIVITY AND TEMPORAL SUMMATION

In this section we stress the importance of temporal summation of photoreceptor signals, and we point out that summation increases sensitivity but may impair spatiotemporal resolution, that is, the capacity to follow rapid movements (Aho et al., 1993b). We also stress that detection of a signal is based on signal versus noise discrimination, and that visual sensitivity is affected by intrinsic photoreceptor noise (Barlow, 1957; Baylor et al., 1980; Aho et al., 1988).

The summation time of a photoreceptor is the length of the period during which different

TABLE 10.1

Photoreceptor and Visual Pigment Data of Adult Terrestrial, Amphibious, and Aquatic Vertebrates

HABITAT, BIOLOGY	A1/A2 PIGMENT	ROD PIGMENT	CONE PIGMENTS				COLORED OIL DROPLETS	TESTED COLOR VISION
			S1	S2	M	L		
Teleost Fish								
Tetrapturus audax striped marlin	Marine	A1	436*	488*	–	531*	a	ND
Carassius auratus goldfish	Aquatic freshwater	A2	360	450	535	620	a	Tetrachromatic
Amphibia								
Rana sp. frog	Amphib. freshwater	A1	431	433	502*	562	a	Dichromatic?
Ambystoma tigrinum axolotl salamander	Amphib. freshwater	A2	~360 –370	440	–	613	a	ND
Reptilia								
Anolis carolinensis green anole lizard	Terrest. diurn.	A2	365	462	503	625	p	p
Gekko gecko tokay gecko	Terrest. noct.	A1	364	467*	–	521*	a	p
Trachemys scripta eleg. slider turtle	Amphib. freshwater	A2	372	458	515	617	p	Tetrachromatic
Alligator mississipp. alligator	Amphib. crepusc.	A1	444*	503*	535*	566	a	ND
Aves								
Columba livia pigeon	Terrest. diurn.	A1	410	453	507	567	p	Tetrachromatic?
Gallus gallus domest. chicken	Terrest. diurn.	A1	415	455	508	571	p	Tetrachromatic?
Anas platyrhynchos mallard duck	Amphib. diurn.	A1	415	452	506	567	p	ND

TABLE 10.1 (continued)

Photoreceptor and Visual Pigment Data of Adult Terrestrial, Amphibious, and Aquatic Vertebrates

| | HABITAT, BIOLOGY | A1/A2 PIGMENT | ROD PIGMENT | CONE PIGMENTS | | | | COLORED OIL DROPLETS | TESTED COLOR VISION |
				S1	S2	M	L		
Puffinus pacificus wedge-tailed shearwater	Amphib. diurn.	A1	505	405	450	505	565	p	ND
Spheniscus humboldti Humboldt penguin	Amphib. diurn.	A1	504	403	450	?	543	p	ND
Mammalia									
Loxodonta africana African elephant	Terrest.	A1	496	419	—	—	552	a	ND
Canis lupus wolf and dog	Terrest.	A1	~500	~430	—	—	~555	a	Dichromatic
Capra hircus goat	Terrest.	A1	~500	~443	—	—	~553	a	ND
Bos taurus cattle	Terrest.	A1	499	~450	—	—	~555	a	ND
Trichechus manatus Florida manatee	Aquatic, coastal	A1	502	~410	—	—	556	a	Dichromatic?
Ursus maritimus polar bear	Amphib.	A1	~500	~440	—	—	~553	a	ND
Enhydra lutris sea otter	Amphib.	A1	~500	~440	—	—	~545	a	ND
Zalophus californianus California sea lion	Amphib.	A1	~500	—	—	—	~560	a	Dichromatic?
Odobenus rosmarus walrus	Amphib.	A1	~500	—	—	—	~560	a	ND

Phoca vitulina harbor seal	Amphib.	A1	~501	–	–	–	548	a	ND
Pagophilus groenlandicus harp seal	Amphib.	A1	498	–	–	–	548	a	ND
Mirounga angustirostris northern elephant seal	Amphib.	A1	483	–	–	–	~552	a	ND
Tursiops truncatus bottlenose dolphin	Marine	A1	488	–	–	–	524	a	Dichromatic?
Globicephala melas long-finned pilot whale	Marine	A1	488	–	–	–	~530	a	ND
Phocoena phocoena harbor porpoise	Marine	A1	+	–	–	–	522	a	ND
Mesoplodon bidens Sowerby's beaked whale	Marine	A1	484	–	–	–	+	a	ND
Eschrichtius robustus gray whale	Marine	A1	497	–	–	–	?	a	ND

DATA SOURCES: Kelber et al. (2003); *Alligator* (Sillman et al., 1991); *Ambystoma* (Makino et al., 1991; Ma et al., 2001); *Anas* (Jane and Bowmaker, 1988); *Anolis* (Kawamura and Yokoyama, 1998); *Bos* (Jacobs et al., 1998); *Canis* (Jacobs, 1993); *Capra* (Jacobs et al., 1998); *Carassius* (Neumeyer, 1998; Palacios et al., 1998); *Columba* (Neumeyer, 1991; Hart, 2001a); *Enhydra* (Levenson et al., 2006); *Eschrichtius* (McFarland, 1971); *Gallus* (Hart, 2001a); *Gekko* (Loew, 1994; Roth and Kelber, 2004); *Globicephala* (Fasick and Robinson, 2000; Newman and Robinson, 2005); *Loxodonta* (Yokoyama et al., 2005); *Mesoplodon* (Fasick and Robinson, 2000; Newman and Robinson, 2005); *Mirounga* (Southall et al., 2002; Levenson et al., 2006); *Odobenus* (Levenson et al., 2006); *Pagophilus* (Newman and Robinson, 2005; Levenson et al., 2006); *Phoca* (Fasick and Robinson, 2000; Newman and Robinson, 2005); *Phocoena* (Newman and Robinson, 2005); *Puffinus* (Hart, 2004); *Rana* (Liebman and Entine, 1968; Koskelainen et al., 1994; Hisatomi et al., 1999); *Spheniscus* (Bowmaker and Martin, 1985); *Tetrapturus* (Fritsches et al., 2003b); *Trachemys* (Neumeyer, 1998; Loew and Govardovskii, 2001); *Trichechus* (Fasick and Robinson, 2000; Griebel and Peichl, 2003; Newman and Robinson, 2005); *Tursiops* (Fasick and Robinson, 1998; Fasick et al., 1998; Griebel and Peichl, 2003; Levenson et al., 2006); *Ursus* (Levenson et al., 2006); *Zalophus* (Griebel and Peichl, 2003; Levenson et al., 2006).

NOTE: Table lists chromophore type (vitamin A1 or vitamin A2 aldehyde); wavelengths of peak absorption (in nanometers) of rod and cone visual pigments; cone opsin type (S1, S2, M, and L); presence (p) or absence (a) of colored oil droplets; and behaviorally tested occurrence of color vision. *, opsin type is not known; +, an opsin type is present, but its peak absorption has not been determined; –, absence of an opsin type. In behavioral color vision tests:?, suggestive but not firm evidence; ND, no proper behavioral test has been carried out. The 433 nm cone pigment of *Rana* frogs is situated in rodlike cells, and their 502 nm "cones" may be ontogenetically young rods mistaken for cones. The rodlike pigment of *Anolis* is situated in a cone type. *Anolis* and *Gekko* possess no rods.

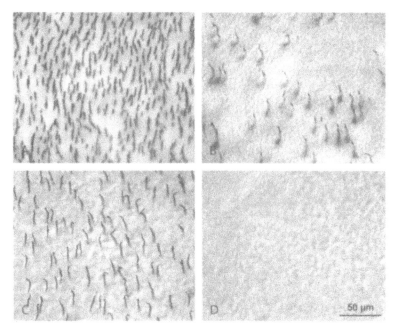

FIGURE 10.3. Spectral cone types in the ferret, *Mustela vison* (Mustelidae, A, B) and the northern fur seal, *Callorhinus ursinus* (Otariidae, C, D) made visible by immunocytochemical labeling for the L-cone opsin (A, C) and for the S-cone opsin (B, D). Micrographs were taken from flat-mounted retinae, focused at the photoreceptor layer; the opsin label reveals the cone outer segments. The ferret has a substantial population of L cones (A) and a sparser population of S cones (B). The northern fur seal has L cones (C) but completely lacks S cones (D). The spaces between the cones are occupied by the much more numerous rods; the rod-to-cone ratio in the ferret is about 14:1, in the northern fur seal it is about 100:1. All fields are from peripheral retina and shown at the same magnification.

absorbed photons contribute to a single-cell response, similar to exposure times in photography. Longer exposure times allow production of an image under low-light conditions but do not provide sharp images of moving objects.

Both rods and cones adapt to ambient light levels by adjusting sensitivity over time, but generally rods are more sensitive and rod systems can sum photons over at least hundreds of milliseconds, and at low temperatures even over 3 to 4 seconds (Aho et al., 1993a). Cones are less sensitive, have shorter summation times, and are about five times faster (Rodieck, 1998). The cone-dominated retinae of diurnal reptiles and birds optimize speed and spatiotemporal resolution.

ROD NOISE AND BODY TEMPERATURE

The sensitivity of a retina is affected by photoreceptor noise, potentially lowering the signal-to-noise ratio of a given visual stimulus, and thus decreasing sensitivity. In rod vision, relevant noise stems from thermal (dark) isomerizations of rhodopsin molecules (Baylor et al., 1980). A molecular mechanism for these dark isomerizations has recently been proposed (Ala-Laurila et al., 2004). The visual effect of this noise can be described as a haze of diffuse light, reducing sensitivity by decreasing contrast (Copenhagen et al., 1987; Donner, 1989b; Donner et al., 1990). This type of visual noise decreases with decreasing temperature. Indeed, in poikilotherm tetrapods visual sensitivity increases dramatically when temperatures drop. The increase may be 10-fold with a temperature drop of 10°C, the reason being a combination of lower noise levels and longer summation times (Aho et al., 1993a).

GANGLION CELLS

TEMPORAL AND SPATIAL SUMMATION

Ganglion cells receive photoreceptor inputs via a chain of interneurons: the bipolar cells, horizontal cells, and amacrine cells (Fig. 10.1). Despite these interneurons, the temporal characteristics (the integration time) of a ganglion cell seems to be determined by the kinetics of the photoreceptors that for the moment are driving the cell, namely, the cones at daylight, or the rods at low light (for rods, see Donner, 1989a).

Usually, a ganglion cell collects input from many photoreceptors. The anatomical basis of this spatial summation, in other words, the patch of retina from which the cell collects receptor signals, is called the (ganglion cell) receptive field. It is determined mainly by the extent of the dendritic tree of the cell (T-shaped structures extending from the ganglion cell somata, shown in Fig. 10.1). Large ganglion cells have wide dendritic trees and thus large receptive fields. They sum over many photoreceptors (efficient spatial summation), leading to high signal-to-noise ratios and high sensitivity, but reduced spatial resolution (Copenhagen et al., 1987, 1990; Donner, 1989b). Retinae dominated by large ganglion cells correspond to electronic images with large pixel size, or "grainy" photographs of sensitive film.

In contrast, high spatial resolution (high visual acuity) can be achieved by ganglion cells with small receptive fields. Because of the smaller summation area, such cells achieve a good signal-to-noise ratio only at higher light intensities. Not surprisingly, the smallest receptive fields and best visual acuities occur in diurnal species (Rodieck, 1998; Wässle, 2004).

GANGLION CELL TYPES AND TOPOGRAPHIES

In any retina there are many thousands, and in the retinae of some birds even up to a few millions, of ganglion cells (Table 10.2), often encompassing more than a dozen functional types. A specialized part of the central retina is

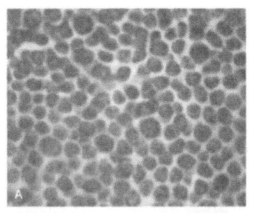
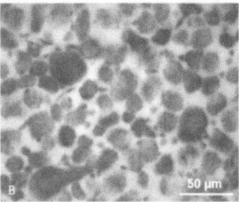

FIGURE 10.4. Peak retinal ganglion cell density in the area centralis of the dog, *Canis lupus* (A) and the bearded seal, *Erignatus barbatus* (B). Micrographs were taken from flat-mounted, cresyl-violet-stained whole retinae, focused on the ganglion cell layer. In the seal, cell density is markedly lower, and soma sizes are larger. Both pictures are at the same magnification.

the fovea, which lacks blood vessels and has a high concentration of photoreceptors. Ganglion cells in the fovea may be so small that their input is dominated by a single cone (midget system, reviewed by Wässle [2004]). At the other end of the range, the largest tetrapod ganglion cells are often found in the peripheral retina, and these have receptive fields covering several square millimeters, and thousands of photoreceptors (Wässle, 2004). The retinal distributions of different ganglion cell types are indicative of how the visual space is analyzed, and the areas with a high density of small ganglion cells give us useful estimates of the visual acuity of each particular animal species (Fig. 10.4; Table 10.2).

TABLE 10.2
Eye Diameter, Posterior Nodal Distance (PND), Peak Retinal Ganglion Cell (RGC) Density, Total Number of RGCs, and
Visual Acuity in Amphibious and Aquatic Vertebrates and Some Related Terrestrial Tetrapods

	AXIAL LENGTH OF EYE (MM)	PND (MM)	PEAK RGC DENSITY $(1/MM^2)$	TOTAL NO. OF RGCS PER EYE $\times 10^3$	CALC. VIS. ACUITY (C/DEG)	OBS. VIS. ACUITY (C/DEG)
Teleost Fishes						
Makaira nigricans blue marlin	33.4	24.2	1,600	ND	8.3	ND
Carassius auratus goldfish	5.0	3.4	5,800	~200	2.2	~2
Amphibia						
Rana pipiens leopard frog	6.5	3.4	11,000	~450	3.1	2.8
Bufo marinus cane toad	11.8	6.1	16,000	950	6.8	ND
Reptilia						
Ctenophorus ornatus dragon lizard	3.7	2.5	55,000	ND	5.1	ND
Chameleo chameleon chameleon	7.0	5.9	13,200	320	5.9	ND
Trachemys scripta elegans slider turtle	7.0	4.2	20,000	370	5.2	ND
Aves						
Columba livia pigeon	11.6	7.9	41,000	2.400	14.0	~15
Gallus gallus domest. chicken	13.3	9.0	25,000	>2,000	12.5	7.2
Puffinus puffinus Manx shearwater	11.8	8.0	21,500	770	10.3	ND
Mammalia						
Loxodonta africana African elephant (juvenile)	33.5	19.1	600	400	4.1	2.9
Canis lupus wolf and dog	23.5	13.4	14,000	200	13.8	12
Felis catus domestic cat	22.0	12.5	10,000 (7,500 β-cells)	200	11.0 (9.4)	6–9
Mustela putoris f. furo ferret	8.5	4.8	4,350	80	2.8	1.9
Capra hircus goat	26.0	14.8	3,600	1,250	7.8	3
Bos taurus cattle	30.0	17.1	6,000	ND	11.6	ND
Sus scrofa domestic pig	21.5	14.5	6,600	ND	10.0	ND
Equus caballus horse	40.6	27.0	6,500	615	19.0	15–25
Trichechus manatus Florida manatee	16.0	10.5	270	66	1.5	~1.5
Enhydra lutris sea otter	10.7	7.0	4,225	125	4.2	ND

TABLE 10.2 (*continued*)

	AXIAL LENGTH OF EYE (MM)	PND (MM)	PEAK RGC DENSITY (1/MM²)	TOTAL NO. OF RGCS PER EYE × 10³	CALC. VIS. ACUITY (C/DEG)	OBS. VIS. ACUITY (C/DEG)
Eumetopias jubatus Steller sea lion	32.5	19.0	2,000	177	7.2	4.2
Callorhinus ursinus northern fur seal	41.4	22.5	1,330	265	7.2	ND
Odobenus rosmarus walrus	30.0	12.5	1,200	102	3.6	ND
Pagophilus groenlandicus harp seal	42.0	25.5	2,000	440	10.0	ND
Tursiops truncatus bottlenose dolphin	25.0	14.5	700	220	3.4	3.7
Pseudorca crassidens false killer whale	29.8	17.0	475	ND	3.3	ND
Lagenorhynchus obliquidens Pacific white-sided dolphin	22.8	13.0	552	ND	2.7	ND
Phocoena phocoena harbor porpoise	19.0	11.5	700	130	2.7	ND
Delphinapterus leucas beluga	23.7	13.5	464	ND	2.6	ND
Eschrichtius robustus gray whale	55–60	23.0	200	174	2.9	ND

DATA SOURCES: Acuities observed in behavioral experiments are taken from Rahmann (1967), some eye axial lengths are taken from Howland et al. (2004) and Walls (1942). PNDs and ganglion cell densities are taken from the following papers and from studies cited therein: *Bos* (Hebel, 1976); *Bufo* (Nguyen and Straznicky, 1989); *Callorhinus* (Mass and Supin, 1992); *Canis* (Peichl, 1992); *Capra* (González-Soriano et al., 1997); *Carassius* (Neumeyer, 2003); *Chameleo* (El Hassni et al., 1997); *Columba* (Hughes, 1977; Hayes, 1982; Hodos et al., 1991); *Ctenophorus* (Barbour et al., 2002); *Delphinapterus* (Murayama and Somiya, 1998); *Enhydra* (Mass and Supin, 2000); *Equus* (Timney and Keil, 1992); *Eschrichtius* (Mass and Supin, 1997); *Eumetopias* (Mass and Supin, 2005); *Felis* (Peichl and Wässle, 1979); *Gallus* (Ehrlich, 1981); *Lagenorhynchus* (Murayama and Somiya, 1998); *Loxodonta* (Stone and Halasz, 1989; behavioral acuity is that of an adult *Elephas*); *Makaira* (Fritsches et al., 2003a, 2005); *Mustela* (Henderson, 1985); *Odobenus* (Supin et al., 2001); *Pagophilus* (Mass and Supin, 2003); *Phocoena* (Supin et al., 2001); *Pseudorca* (Murayama and Somiya, 1998); *Puffinus* (Hayes et al., 1991; Martin and Brooke, 1991); *Rana* (Aho, 1997); *Sus* (Hebel, 1976); *Trachemys* (Granda and Dvorak, 1977; Peterson and Ulinski, 1979); *Trichechus* (Supin et al., 2001; Bauer et al., 2003); *Tursiops* (Supin et al., 2001).

NOTE: Eye diameter is the axial length from cornea to sclera. The PND is the distance between the retina and the nodal point of the optical system (composed of lens and cornea), it determines the image size on the retina. PND and the peak ganglion cell density can be used to calculate an "anatomical" visual acuity (Calc. vis. acuity) (Pettigrew et al., 1988; Supin et al., 2001). Visual acuity is here defined as the finest grid of black and white stripes resolved by the animal. The width of one black plus one white stripe represents one cycle, and the unit of acuity is the number of resolved cycles per visual degree (Land and Nilsson, 2002). Some acuities were observed in behavioral experiments (Obs. vis. acuity). The acuities of the teleosts, the cetaceans, the pinnipeds, the Florida manatee, and the water turtle refer to acuity in water, whereas all other acuities are in air (cf. Kröger, chapter 8 in this volume). ND, no data.

Thus the local neuron density and hence processing capacity is not homogeneous across the retina. True foveae are found only in some reptiles, birds, and primates, but in most mammalian (and vertebrate) retinae, there is a region of maximal cone and ganglion cell density termed the area centralis or macula (Figs. 10.4 and 10.5). There the visual scenery is sampled with a maximal "pixel" density, providing maximal visual acuity (Peichl and Wässle, 1979; Wässle, 2004). In addition to the more-or-less circular area centralis, many mammals also

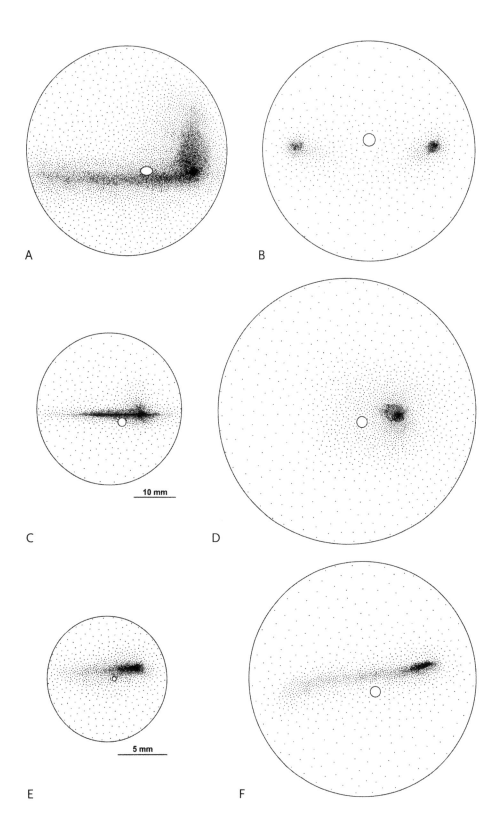

A

B

C

10 mm

D

E

5 mm

F

FIGURE 10.5. (Opposite page.) Comparison of retinal ganglion cell distributions in some marine mammals *(right column)* and related terrestrial species *(left column):* (A) goat *(Capra hircus);* (B) bottlenose dolphin *(Tursiops truncatus);* (C) wolf *(Canis lupus);* (D) harp seal *(Pagophilus groenlandicus);* (E) ferret *(Mustela vison)* (F) sea otter *(Enhydra lutris).* In these schematic maps, dot density qualitatively represents ganglion cell density. All are left retinae seen from the side of the vitreous, temporal (lateral) is to the right and dorsal is at the top. Each species has an area centralis of peak ganglion cell density in temporal retina, and some have two regions of relatively high ganglion cell density or a horizontally extended "visual streak." *Open circles* represent the optic nerve head ('blind spot'). Retinae A through D are shown at the same magnification (scale bar in C), whereas retinae E and F are more enlarged (scale bar in E). Data for goat from González-Soriano et al. (1997); for bottlenose dolphin from Supin et al. (2001); for wolf from Peichl (1992); for harp seal from Mass and Supin (2003); for ferret from Henderson (1985); and for sea otter from Mass and Supin (2000).

---

possess a horizontally extended band of high ganglion cell density across the retina, termed the visual streak (Fig. 10.5A, C, and E). Visual streaks are common in larger species living in open habitats, such as artiodactyls, perissodactyls, rabbits, and some carnivores.

In his terrain theory, Hughes (1977) has argued that in an open habitat new objects (predators or prey) commonly first appear at the horizon, and that enhanced visual acuity along the horizon is an adaptive advantage for both prey and predators—if they are large enough to see the horizon above the vegetation. Artiodactyls have an additional dorsally oriented zone of high ganglion cell density above the area centralis, which samples the ground in front of the animal and may be useful for detecting obstacles during running (Fig. 10.5A). In contrast, forest-dwellers such as primates or cats do not possess a pronounced visual streak. Typical visual streaks are also found in shallow-water fish. Species inhabiting open water and having an uninterrupted view of the sand-water horizon possess a prominent visual streak, while those living in enclosed environments do not (Collin and Pettigrew, 1988a, 1988b).

VISUAL ACUITY

The maximum ganglion cell density at the area centralis, the size of the eye, and the position of the lens (posterior nodal distance [PND]) (see Kröger and Katzir, chapter 9 in this volume) can be used to calculate an anatomical visual acuity (Supin et al., 2001; Land and Nilsson, 2002). Table 10.2 presents the calculated acuities for a range of vertebrates. In the mammalian group, acuities are given for the Florida manatee, four

pinnipeds, the sea otter, and six cetaceans, together with seven terrestrial relatives of these aquatic or amphibious species, and the horse as a particularly large-eyed terrestrial species (the listed terrestrial species are more or less crepusclar, at least not distinctly diurnal). Figure 10.6 plots the calculated visual acuities of these 20 mammals as a function of eye size (PND). Assuming that the peak ganglion cell density is the same in all these retinae, the expected calculated acuity would be directly proportional to PND, and the data in Figure 10.6 would lie on a straight line from the lower left to the upper right corner of the graph, from the ferret to the horse. For seven of the eight land mammals included (i.e., all relatives of the manatee, seals, and whales, for which data are available), the points in fact cluster in a relatively coherent region around this hypothetical line. The aquatic species, however, deviate by having a relatively constant calculated acuity independent of eye size (see below).

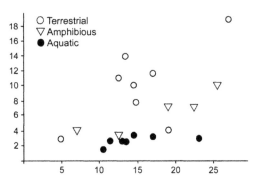

FIGURE 10.6. Relationship between calculated anatomical visual acuity (*y* axis, c/deg) and posterior nodal distance (PND) (*x* axis, mm) in mammals (data taken from Table 10.2). The graph compares seven aquatic *(black circles)* and five amphibious *(open triangles)* species with eight closely related terrestrial *(open circles)* species.

Some caution must be exercised in interpreting the data in Figure 10.6. Not all ganglion cells contribute to visual acuity. In well-studied species, particularly primates and cat, it has been established that only the ganglion cell type with the smallest receptive fields determines visual acuity (Wässle, 2004). Thus only their peak density should be used for an anatomical estimate of visual acuity. However, for most terrestrial and all aquatic and amphibious mammals (and most of the other vertebrates) no ganglion cell classification exists, and we do not know which type and proportion of the ganglion cells are responsible for visual acuity. So, in accordance with other authors we have based our calculations on total ganglion cell densities, and the derived acuities have to be considered overestimates. Indeed, behaviorally determined acuities, which are available for a few species, tend to be lower (Table 10.2).

THE RETINA IN AQUATIC TETRAPODS

PHOTORECEPTORS

ROD VISUAL PIGMENTS

Dissolved organic substances give the water in many small lakes and streams a color of very weak tea, so, the spectrum of light passing through a body of this type of freshwater is successively shifted toward the red (long-wave) end. This is because the humus colloids producing the red shift are absorbing wavelengths below 500 nm much more than they are absorbing wavelengths above 550 nm (Lythgoe, 1979). Freshwater tetrapods, such as frog tadpoles (*Rana temporaria*, Ranidae) and aquatic slider turtles (*Trachemys scripta elegans*, Emyidae), have retinal$_2$ visual pigments (porphyropsins), which for the rods leads to a red shift of 20 nm in λ_{max} (Fig. 10.2; Table 10.1), and a distinct improvement of sensitivity in typical lake water.

Such adaptation is also apparent in the pigment and spectral sensitivity shifts from phorphyropsin to rhodopsin during the metamorphosis of *R. temporaria* aquatic tadpoles to land-living frogs (Reuter, 1969). In bullfrogs

(*R. catesbeiana*) a similar pigment change is observed during metamorphosis. In addition, the relatively aquatic adult bullfrogs show an opposite shift by developing a porphyropsin field in the dorsal part of their retinae during a few winter months (Reuter et al., 1971). The dorsal retina of a bullfrog may look down into the water. Birds and mammals do not have porphyropsins, and none of the available data support the notion that ducks, wading birds, otters, or beavers would have red-shifted rhodopsins (Archer et al., 1999; Hart, 2001a).

There is a dramatic difference between light transmission in lakes, where humus colloids absorb and scatter almost all light within 10 meters, and in clear ocean water, where at some wavelengths, 10% of the photons penetrate down to depths of about 100 meters. Distilled water and clear ocean water have their best transmission for wavelengths between 400 and 500 nm (Lythgoe, 1979; Kröger, chapter 8 in this volume), thus producing a successively more pronounced blue shift of the spectrum of the remaining downwelling light. Most deep-sea fish have rod pigments absorbing maximally in the 470 to 495 nm region (Denton and Warren, 1956; Wald et al., 1957; Partridge et al., 1989), and similar rod visual pigments have been found in deep-diving cetaceans and the deep-diving elephant seal (Fig. 10.2; Table 10.1) (Lythgoe and Dartnall, 1970; McFarland, 1971; Fasick and Robinson, 1998, 2000; Southall et al., 2002; Levenson et al., 2006). These pigments thus seem to illustrate an evolutionary adaptation to the light conditions in the deep sea. In other pinnipeds, the rod λ_{max} is close to 500 nm, as in terrestrial carnivores (see Levenson et al., 2006). It can be argued that maintaining effective terrestrial vision exerted some pressure on rod opsin tuning in these amphibious animals.

The blue shift observed in cetacean rod pigments, from λ_{max} at 496 to 500 nm in terrestrial artiodactyls and coastal shallow-water odontocetes, to the 484 nm λ_{max} in deep-diving cetaceans such as *Mesoplodon bidens*, are partly obtained by amino acid substitutions identical to those observed in deep-sea marine fish and deepwater

Lake Baikal cottid fish. This is a striking convergence in molecular evolution (Hunt et al., 1996, 2001; Fasick and Robinson, 1998, 2000; Fasick et al., 1998). Using *Mesoplodon* as an example, we can estimate how deep this odontocete whale has to dive before its 484 nm pigment becomes more efficient in absorbing downwelling light than a 498 nm artiodactyl rod pigment. By integrating the product of pigment absorption and spectral intensity over wavelength (cf. Reuter, 1969:48–50), we find that the critical depth when both pigments are equally effective is about 20 meters for the Sargasso Sea spectra (measured by Lundgren and Højerslev [1971]), and 50 meters for the extremely pure Crater Lake water (for the depth spectra, see Lythgoe [1979:26]). Thus, at relatively shallow depths the blue-shifted pigments become more efficient than the rod pigments of terrestrial mammals. With increasing depth the calculated relative advantage of a 484 nm pigment increases, but a further blue shift of the pigment would have no dramatic effect.

At depths below 1000 meters no significant downwelling light remains, and bluish bioluminescence becomes the dominant source of light (Warrant and Locket, 2004). Elephant seals, beaked whales, and sperm whales have rod visual pigments peaking at about 485 nm (Fasick and Robinson, 2000; Southhall et al., 2002; Levenson et al., 2006). All three species regularly dive to depths of 800 to 1200 meters (Nowak, 1999) and seem to forage in the "bioluminescence zone."

The above discussion may give the erroneous impression that rod pigments always are adapted to the prevailing illumination, thus maximizing the absorption of available photons. Such adaptations are indeed seen when the spectral distribution of light is narrow, for instance in bluish marine environments and some yellowish lake waters (Lythgoe, 1979). Under terrestrial conditions, and in freshwaters with spectrally broad illumination, the rod pigment spectra lie at the short-wavelength end of the available light range, often far from the position predicted by a simple photon catch hypothesis (Fig. 10 of Reuter, 1969; Fig. 1.3 of Lythgoe, 1991).

Barlow (1957) first foresaw this discrepancy between pigments and ambient light. He pointed out that randomly occurring thermal activations of visual pigment molecules (the "dark isomerizations" described above) constitute an irreducible noise that sets an ultimate limit to the detection of weak light, and that the frequency of these noise events necessarily increases with a red shift of the pigment and correspondingly decreases with a blue shift. Thus an ecological situation demanding extreme sensitivity will favor a blue shift of the visual pigment spectra, and a rod pigment observed in nature is always a compromise beween two demands, efficient photon collection and low noise stemming from thermal isomerizations (Baylor et al., 1980; Firsov and Govardovskii, 1990; Bowmaker et al., 1994; Ala-Laurila et al., 2004). Thus three factors may contribute to the evolution of the blue-shifted pigments in deep-sea tetrapods: bluish downwelling light, bluish bioluminescence, and reduced noise levels in rods containing blue-shifted pigments.

The retinae of pinnipeds and whales show high densities of tightly packed rods, and sparsely distributed cones (Fig. 10.3). Across species, the cone-to-rod ratios range from 1:50 to 1:250 (Peichl and Moutairou, 1998; Peichl et al., 2001), which is similar to the cone-to-rod ratios in nocturnal terrestrial mammals (Peichl, 2005). Obviously, the photic conditions during foraging in the mesopelagic zone (Warrant and Locket, 2004) exerted more evolutionary pressure modulating the cone-to-rod ratios than did the conditions of bright illumination during periods spent at the surface or on land. A further feature to improve photon capture and thus sensitivity is a reflecting tapetum lucidum behind the retina of pinnipeds and cetaceans (see Kröger and Katzir, chapter 9 in this volume). The presence of a tapetum lucidum is shared with terrestrial carnivores and artiodactyls.

CONE VISUAL PIGMENTS

Flat-mounted retinae of pinnipeds and whales, stained by cone opsin-specific antibodies, show sparsely distributed L cones, but no S1 cones at all (Fig. 10.3). Hence diving cetaceans and pinnipeds seem to be L cone monochromats, surprisingly lacking the S1 cones that would be most sensitive to the short-wavelength light dominating at those depths (Peichl and Moutairou, 1998; Peichl et al., 2001; reviewed in Griebel and Peichl, 2003). Cetacean and pinniped genomes, however, show remnants of the lost S1 opsin in the form of mutated, nonfunctional S1 opsin genes (Fasick et al., 1998; Levenson and Dizon, 2003; Newman and Robinson, 2005; Levenson et al., 2006). Thus, it is likely that the earliest cetaceans and pinnipeds or their terrestrial ancestors had violet- to blue-sensitive cones and were at least dichromats.

The closest extant relatives of pinnipeds and cetaceans are terrestrial carnivores and artiodactyls, which have S1 cones with λ_{max} in the blue (ca. 450 nm) and L cones with λ_{max} in the yellowish green (approximately 550 nm). Many cetaceans and pinnipeds dive to depths of 200 to 300 meters, where the downwelling photon flux in the 410 to 480 nm region is reduced to less than 1% of that close to the surface, and at other wavelengths reduced much more (see Kröger, chapter 8 in this volume). As an adaptation to these light conditions the L cones of pilot whales, bottlenose dolphins, and harbor porpoises have absorption peaks at 531 nm, 524 nm, and 522 nm, respectively, clearly blue shifted from the peak at 550 to 555 nm observed in terrestrial artiodactyls (Table 10.1) (Newman and Robinson, 2005). Thus, for the fully aquatic cetaceans, the blue shift of deep-sea light has affected the evolution of both rod and cone vision.

The L cones of all pinnipeds studied to date have their absorption peaks (measured or inferred from known spectral tuning sites) around 550 to 560 nm, similar to those of terrestrial carnivores (Levenson et al., 2006). Even the very deep-diving elephant seal has an inferred L cone λ_{max} of 552 nm. As done for the rods above, it can be argued that maintaining effective terrestrial vision also exerted some pressure on pinniped L opsin tuning. For both pinnipeds and cetaceans, it remains enigmatic why they have lost the S1 cone type that would appear best suited for brightness and contrast perception in their blue-shifted underwater world.

COLOR VISION

A few meters down in a typical freshwater lake, and at ocean depths below 200 meters, the photon fluxes become very sparse and, furthermore, have a limited spectral range. These are adverse conditions for chromatic vision (Vorobyev, 1997; Warrant and Locket, 2004), and one might suppose that color vision is unimportant for deeper-living taxa. The evidence for that is not, however, univocal.

Tetrachromatic turtles living in relatively turbid freshwater have four cone types with four pigments, and they have good wavelength discrimination in three spectral regions, around 400 nm, 500 nm, and 600 nm (Arnold and Neumeyer, 1987; Neumeyer, 1998). Some sea birds such as eider ducks (Anatidae) and two species of shearwaters (Puffinus) are found to regularly forage at depths of 40 to 60 meters (Brun, 1971; Weimerskirch and Sagar, 1996; Keitt et al., 2000). They have typical avian cones (Partridge, 1989; Hart, 2001b), and the wedge-tailed shearwater has been shown to possess four cone pigments (Table 10.1) (Hart, 2004). In these shallow divers, the organization of the retina is fairly similar to that of nonaquatic birds. However, a number of bird species (e.g., shag [Phalacrocorax aristotelis], razorbill [Alca torda], and Manx shearwater [Puffinus puffinus]) that follow their prey underwater have higher relative densities of short-wave-sensitive cones (Hart, 2001b), consistent with their blue-shifted environment.

Some species of penguins seldom dive deeper than 20 meters in search of prey, whereas others dive to depths of more than 500 meters to catch fish and squid (Martin, 1999; Kooyman, 2002). During the daytime, king penguins (Aptenodytes patagonicus) usually forage at depths of 100 to 300 meters, and Humboldt penguins (Spheniscus humboldti) at somewhat shallower depths (Martin, 1999). The latter species is the only penguin for which the retina has been studied, and it shows clear adaptations to deep-ocean vision. Bowmaker and Martin (1985) studied Humboldt penguins and measured a limited number of L cones, S2 cones, and S1 cones (Table 10.1). They may have overlooked possible M cones, and it is therefore not clear that these birds have fewer than the typical set of four cone types. However, the λ_{max} of the L cone is at 543 nm, clearly less than the 555 to 571 nm range encompassing the L cones in the 31 nonpenguin bird species studied (Hart, 2001a). Furthermore the majority of the cone oil droplets in the retina of the Humboldt penguin are pale and similar to those of owls, another clear adaptation to the low light levels in the deep sea (Bowmaker and Martin, 1985).

Penguins are not alone in using a multicone system for oceanic predation. Many mesopelagic bony fish, probably stemming from tetrachromatic ancestors, have large eyes and two or three cone types, although regularly foraging at depths of 200 to 600 meters. Swordfish, marlin, and tunas have rod visual pigments peaking at about 485 nm, certifying that they are true deep-sea predators, and the peak sensitivities of their two dominating cone types lie at about 430 and 480 nm, respectively. In some cases they have a third cone pigment above 500 nm (Loew et al., 2002; Fritsches et al., 2003b). Marlin and swordfish cones are very large (i.e., efficient photon collectors) and thus similar to the huge cones of Lake Baikal cottid fish found in the cold water at depths of 50 to 400 meters (Bowmaker et al., 1994). Large cones would be very noisy at high temperatures but may evolve in cold habitats.

Color vision is less important in diving mammals. All available data indicate that pinnipeds and cetaceans have a single cone and a single rod type. Surprisingly enough, some experiments indicate that dolphins, as well as sea lions, may behave as dichromats, possibly as a result of rod-cone interactions (reviewed in Griebel and Peichl, 2003; Jacobs, 1993). In mammals the sparse S1 cones serve only color vision, whereas the more numerous L cones

serve color vision as well as form vision and acuity. This may be an explanation why whales and seals lost the S1 cones rather than the L cones.

In contrast to cetaceans and pinnipeds, the herbivorous sirenians live in shallow coastal waters and have diurnal retinae with dichromatic color vision and a significant proportion of cones (reviewed in Griebel and Peichl, 2003; Newman and Robinson, 2005). Two cone types and hence the prerequisite for dichromatic color vision are also present in the pygmy hippopotamus, the river otter, the sea otter, and the polar bear (Peichl et al., 2001, 2005; Levenson et al., 2006). Where studied, the L opsin and S opsin spectral tuning in these species is simiar to that in their terrestrial relatives (Table 10.1) (Newman and Robinson, 2005; Levenson et al., 2006).

PHOTORECEPTOR SENSITIVITY AND TEMPORAL SUMMATION

The properties described in the general section also apply to the photoreceptors of aquatic species. With highly rod-dominated retinae, the consequences are good sensitivity by large temporal summation, but poor temporal resolution. Low water temperatures generally reduce thermal dark noise in rods, increasing rod sensitivity. But that is relevant only for poikilotherm animals, as discussed in the next section.

ROD NOISE AND BODY TEMPERATURE

Deep-diving seals, cetaceans, and penguins are homoiotherm, and in the mesopelagic zone (200 to 1000 meters) with water temperatures of 4 to 14°C (Defant, 1961), their eyes can hardly reach the sensitivity of the relatively noise-free eyes of poikilotherm fish adapted to low temperatures (see above). On the other hand, deep-diving mammals and penguins may benefit from a high body temperature and fast sensory-motor reactions, which are made possible by an atmospheric-oxygen-fueled metabolism. The partly homoiotherm deep-diving swordfish may be used to illustrate the trade-off between visual reaction speed and absolute sensitivity, the former favored by high and the latter by low temperature. Swordfish have a specialized heating system warming the eye and brain to 10 to 15°C above ambient water temperature, and dramatically increasing their temporal resolution. Thus, at least for this fast predator, speed is worth more than maximum dark-adapted sensitivity (Fritsches et al., 2005).

GANGLION CELLS

SPATIAL SUMMATION AND VISUAL ACUITY

The photoreceptor arrangements of aquatic mammals appear evolutionarily adapted to low light levels, and this also holds for the ganglion cells. The retinal structure is characterized by efficient spatial summation and low sampling density, as a result of the large and sparsely distributed ganglion cells (Supin et al., 2001). This is illustrated by the retinal ganglion cell topographies of a bottlenose dolphin and a harp seal compared to the corresponding topographies of a goat and a wolf (Fig. 10.5), and by comparison of the central areas in a seal and a dog (Fig. 10.4). Compared to terrestrial mammals of similar size, dolphins have huge ganglion cells with large dendritic trees, and even the areas with maximum acuity have a relatively low ganglion cell density (Table 10.2). This is consistent with the low light intensity of their environment. Spatial summation of a ganglion cell over many photoreceptors, achieved by a large receptive field, is a good strategy to improve the detection of low-level signals.

The ganglion cell densities and the ganglion cell soma sizes of sea otters are similar to those of terrestrial mustelids, but because the eye is larger the potential visual acuity of the sea otter is approximately twice that of terrestrial mustelids (Table 10.2; Fig. 10.6). Mass and Supin (2000) proposed that the sea otter eye represents an early stage of adaptation to an aquatic lifestyle. The Florida manatee and the cetaceans form a coherent aquatic group at the low-acuity edge of Figure 10.6. Amphibious species, the sea otter and the pinnipeds, occupy intermediate positions.

The African elephant deviates from the general pattern by being a large terrestrial mammal with low calculated visual acuity, although higher than that of the phylogenetically related Florida manatee (Table 10.2). Gaeth et al. (1999) have suggested an aquatic stage in elephant evolution. Possibly the low visual acuity of the elephant might then be explained as a remnant of its previous adaptation to aquatic life.

GANGLION CELL TOPOGRAPHIES

For fish that swim in shallow waters or close to the bottom, the sand-water horizon may be as relevant as the sky horizon is for prairie dwellers. In fact, a visual streak giving enhanced visual acuity of this horizon is found in the ganglion cell topographies of a number of fish (see Generalized Anatomy and Function section above). But for vertebrates that dive and move three-dimensionally in deeper open waters there is no well-defined horizon. Consistent with this, there are no prominent visual streaks in any whales or seals (Fig. 10.5) (Supin et al., 2001). It is likely that ancestral cetaceans retained a visual streak, given that it occurs in all extant artiodactyls. The streak may have been modified by obliteration of the center, resulting in two areas of high ganglion cell density, as found in the bottlenose dolphin (Fig. 10.5B) (Supin et al., 2001). A similar pair of high-density areas, one laterally and one medially (temporal and nasal areas), is described in three further odontocetes, the false killer whale, the pacific white-sided dolphin, and the beluga (Murayama and Somiya, 1998). One investigated mysticete, the gray whale *(Eschrichtius gibbosus),* has a high-density field in the temporal retina but only a weak density increase on the nasal side (Mass and Supin, 1997). Some behavioral observations indicate that the forward-looking temporal area may be used for vision above the water surface and the laterally looking nasal area for vision underwater (Supin et al., 2001). Another hypothesis relates the forward-looking area to echolocation (see below).

It is not clear whether the terrestrial ancestors of pinnipeds had a visual streak, but no such structure has been found in any extant pinniped. In contrast, the sea otters possess a distinct visual streak (Fig. 10.5F) (Mass and Supin, 2000), although it is not as prominent as that of some terrestrial mustelids (Fig. 10.5E). Visual scanning of the horizon may be more relevant in sea otters than it is in pinnipeds. As stated above, the sea otter eye may represent an early stage of adaptation to an aquatic lifestyle (Mass and Supin, 2000).

EVOLUTIONARY HYPOTHESES

Evolutionary scenarios can be construed based on the retinal differences described above. For instance, the fact that whales and seals lost their blue-sensitive cones, whereas penguins did not, may have evolutionary implications. It is possible that cetaceans went through an extended stage in yellowish coastal waters, an evolutionary bottleneck for the visual system where S1 cones were of no use (Peichl et al., 2001). Alternatively, it is possible that the avian visual system, which normally relies deeply on chromatic analysis (good color vision), is more conservative and less prone to evolve toward achromatic vision than the mammalian visual system. Possibly the basic cortical visual functions of dichromatic mammals may be less dependent on chromatic information for discerning contour and shapes, making it easier to give up chromatic vision.

Rod and L cone spectra of different odontocetes reflect evolutionary adaptations to different foraging depths exploited by different species (Fasick and Robinson, 2000; Southall et al., 2002; Newman and Robinson, 2005). This suggests that for deep-diving odontocetes vision is indeed an important complement to echolocation.

In contrast, the rod and L cone pigment spectra of pinnipeds are very similar to those of terrestrial carnivores, except for the blue-shifted rod pigment of the extremely deep-diving elephant seal. Levenson et al. (2006) argue that this might reflect the pinnipeds' amphibious lifestyle: the rod and L cone tuning match the requirements for reasonably good terrestrial vision, the S cone lack matches the situation in the fully aquatic cetaceans.

Given the cranial position of the eyes, and the regions of highest ganglion cell density, the direction of highest visual acuity can be determined. As described above, there are two areas of maximum acuity in the dolphin retina, one looking forward and the other to the side. These directions seem to overlap with the two acoustic windows of *Stenella* dolphins; one for high sound frequencies and forward-directed echolocation, and one laterally directed for lower-frequency communication (Fig. 15 of Bullock et al., 1968). The behavior of searching dolphins is described in the following way by Bullock et al. "As a dolphin, emitting a train of clicks, approaches a target, it invariably moves its head back and forth, up and down in short, jerky circular movements" (p. 153). In doing so the dolphin may combine vision and echolocation (cf. Supin et al., 2001).

Concerning spatial summation and sensitivity, the aquatic and amphibious mammals in Figure 10.6 summarize the main message. The Florida manatee and the cetaceans form a coherent aquatic group at the low-acuity edge of the graph. Their estimated acuity does not improve significantly with increasing eye size. The amphibious species, the sea otter and the pinnipeds, occupy intermediate positions, and their acuities clearly improve with increasing eye size. In order to extract useful information at light levels that are low even in comparison with nocturnal terrestrial conditions, the eyes of dolphins and whales have to give up high acuity; the sparse photon flux and the high temperature of the cetacean eyeballs make a high acuity physically impossible. Instead whale eyes have large and sparsely distributed ganglion cells with huge receptive fields collecting and integrating photon signals from large retinal areas. With this "large pixel strategy" the cells can collect statistically significant photon numbers per summation time, and the eye can do what is possible for detecting contrast and movement down at the ocean depths.

ACKNOWLEDGMENTS

We thank the editors, J. G. M. Thewissen and Sirpa Nummela, for inviting us to contribute to this volume, and Christa Neumeyer and Victor Govardovskii for helpful discussions and constructive comments on an earlier draft of the manuscript.

LITERATURE CITED

Aho, A.-C. 1997. The visual acuity of the frog *(Rana pipiens)*. Journal of Comparative Physiology A 180:19–24.

Aho, A.-C., K. Donner, C. Hydén, L. O. Larsen, and T. Reuter. 1988. Low retinal noise in animals with low body temperature allows high visual sensitivity. Nature 334:348–350.

Aho, A.-C., K. Donner, and T. Reuter. 1993a. Retinal origins of the temperature effect on absolute visual sensitivity in frogs. Journal of Physiology (London) 463:501–521.

Aho, A.-C., K. Donner, S. Helenius, L. O. Larsen, and T. Reuter. 1993b. Visual performance of the toad *(Bufo bufo)* at low light levels: retinal ganglion cell responses and prey-catching accuracy. Journal of Comparative Physiology A 172:671–682.

Ala-Laurila, P., K. Donner, and A. Koskelainen. 2004. Thermal activation and photoactivation of visual pigments. Biophysical Journal 86:3653–3662.

Archer, S. N., M. B. A. Djamgoz, E. R. Loew, J. C. Partridge, and S. Vallerga (eds.). 1999. Adaptive Mechanisms in the Ecology of Vision. Kluwer Academic Publishers, Dordrecht.

Arnold, K., and C. Neumeyer. 1987. Wavelength discrimination in the turtle *Pseudemys scripta elegans*. Vision Research 27:1501–1511.

Barbour, H. R., M. A. Archer, N. S. Hart, N. Thomas, S. A. Dunlop, L. D. Beazley, and J. Shand. 2002. Retinal characteristics of the ornate dragon lizard, *Ctenophorus ornatus*. Journal of Comparative Neurology 450:334–344.

Barlow, H. B. 1957. Purkinje shift and retinal noise. Nature 179:255–256.

Bartol, S. M., and J. A. Musick. 2001. Morphology and topographical organization of the retina of juvenile loggerhead sea turtles *(Caretta caretta)*. Copeia 2001:718–725.

Bauer, G. B., D. E. Colbert, J. C. Gaspard III, B. Littlefield, and W. Fellner. 2003. Underwater visual acuity of two Florida manatees *(Trichechus manatus latirostris)*. International Journal of Comparative Psychology 16:130–142.

Baylor, D. A., and A. L. Hodgkin. 1973. Detection and resolution of visual stimuli by turtle photoreceptors. Journal of Physiology (London) 234:163–198.

Baylor, D. A., G. Matthews, and K.-W. Yau. 1980. Two components of electrical dark noise in toad retinal rod outer segments. Journal of Physiology (London) 309:591–621.

Bowmaker, J. K., and D. M. Hunt. 1999. Molecular biology of photoreceptor spectral sensitivity; pp. 439–462 in S. N. Archer, M. B. D. Djamgoz, E. R. Loew, J. C. Partridge, and S. Vallerga (eds.), Adaptive Mechanisms in the Ecology of Vision. Kluwer Academic Publishers, Dordrecht.

Bowmaker, J. K., and G. R. Martin. 1985. Visual pigments and oil droplets in the penguin, *Spheniscus humboldti*. Journal of Comparative Physiology A 156:71–77.

Bowmaker, J. K., L. A. Heath, S. E. Wilkie, and D. M. Hunt. 1997. Visual pigments and oil droplets from six classes of photoreceptor in the retinas of birds. Vision Research 37:2183–2194.

Bowmaker, J. K., V. I. Govardovskii, S. A. Shukolyukov, L. V. Zueva, D. M. Hunt, V. G. Sideleva, and O. G. Smirnova. 1994. Visual pigments and photic environment: the cottid fish of Lake Baikal. Vision Research 34:591–605.

Bridges, C. D. B. 1972. The rhodopsin-porphyropsin visual system; pp. 417–480 in H. J. A. Dartnall (ed.), Handbook of Sensory Physiology, Vol. VII-1. Springer-Verlag, Berlin.

Brun, E. 1971. Predation of *Chlamys islandica* by eiders *Somateria* spp. Astarte 4:23–29.

Bullock, T. H., A. D. Grinnell, E. Ikezono, K. Kameda, Y. Katsuki, M. Nomoto, O. Sato, N. Suga, and K. Yanagisawa. 1968. Electrophysiological studies of central auditory mechanisms in cetaceans. Zeitschrift für vergleichende Physiologie 59:117–156.

Collin, S. P., and J. D. Pettigrew. 1988a. Retinal topography in reef teleosts. I. Some species with well-developed areae but poorly-developed streaks. Brain, Behavior, and Evolution 31:269–282.

Collin, S. P., and J. D. Pettigrew. 1988b. Retinal topography of reef teleosts. II. Some species with prominent horizontal streaks and high-density areae. Brain, Behavior, and Evolution 31:283–295.

Collin, S. P., and A. E. O. Trezise. 2004. The origins of colour vision in vertebrates. Clinical and Experimental Optometry 87:217–223.

Copenhagen, D. R., K. Donner, and T. Reuter. 1987. Ganglion cell performance at absolute threshold in toad retina: effects of dark events in rods. Journal of Physiology (London) 393:667–680.

Copenhagen, D. R., S. Hemilä, and T. Reuter. 1990. Signal transmission through the dark-adapted retina of the toad (*Bufo marinus*). Journal of General Physiology 95:717–732.

Dartnall, H. J. A. 1968. The photosensitivities of visual pigments in the presence of hydroxylamine. Vision Research 8:339–358.

Defant, A. 1961. Physical Oceanography, Vol. I. Pergamon Press, Oxford.

Denton, E. J., and F. J. Warren. 1956. Visual pigments of deep-sea fish. Nature 178:1059.

Donner, K. O. 1953. The spectral sensitivity of the pigeon's retinal elements. Journal of Physiology (London) 122:524–537.

Donner, K. 1989a. Visual latency and brightness: an interpretation based on the responses of rods and ganglion cells in the frog retina. Visual Neuroscience 3:39–51.

Donner, K. 1989b. The absolute sensitivity of vision: can a frog become a perfect detector of light-induced and dark rod events? Physica Scripta 39:133–140.

Donner, K., D. R. Copenhagen, and T. Reuter. 1990. Weber and noise adaptation in the retina of the toad *Bufo marinus*. Journal of General Physiology 95:733–753.

Ehrlich, D. 1981. Regional specialization of the chick retina as revealed by the size and density of neurons in the ganglion cell layer. Journal of Comparative Neurology 195:643–657.

El Hassni, M., S. Ba M'Hamed, J. Repérant, and M. Bennis. 1997. Quantitative and topographical study of retinal ganglion cells in the chameleon (*Chameleo chameleon*). Brain Research Bulletin 44:621–625.

Fasick, J. I., and P. R. Robinson. 1998. Mechanism of spectral tuning in the dolphin visual pigments. Biochemistry 37:433–438.

Fasick, J. I., and P. R. Robinson. 2000. Spectral-tuning mechanisms of marine mammal rhodopsins and correlations with foraging depth. Visual Neuroscience 17:781–788.

Fasick, J. I., T. W. Cronin, D. M. Hunt, and P. R. Robinson. 1998. The visual pigments of the bottlenose dolphin (*Tursiops truncatus*). Visual Neuroscience 15:643–651.

Firsov, M. L., and V. I. Govardovskii. 1990. Dark noise of visual pigments with different absorption maxima [in Russian]. Sensory Systems 4:25–34.

Fritsches, K. A., N. J. Marshall, and E. J. Warrant. 2003a. Retinal specializations in the blue marlin: eyes designed for sensitivity to low light levels. Marine and Freshwater Research 54:333–341.

Fritsches, K. A., L. Litherland, N. Thomas, and J. Shand. 2003b. Cone visual pigments and retinal mosaics in the striped marlin. Journal of Fish Biology 63:1347–1351.

Fritsches, K. A., R. W. Brill, and E. J. Warrant. 2005. Warm eyes provide superior vision in swordfishes. Current Biology 15:55–58.

Gaeth, A. P., R. V. Short, and M. B. Renfree. 1999. The developing renal, reproductive, and respiratory systems of the African elephant suggest an aquatic ancestry. Proceedings of the National Academy of Sciences USA 96:5555–5558.

Goldsmith, T. H. 1991. The evolution of visual pigments and colour vision; pp. 62–89 in P. Gouras (ed.), The Perception of Colour, Vol. 6 of J. R. Cronly-Dillon (ed.), Vision and Visual Dysfunction. Macmillan Press, Houndmills, UK.

Gonzalez-Soriano, J., S. Mayayo-Vicente, P. Martinez-Sainz, J. Contreras-Rodriguez, and E. Rodriguez-Veiga. 1997. A quantitative study of ganglion cells in the goat retina. Anatomia, Histologia, Embryologia 26:39–44.

Govardovskii, V. I., and M. V. Vorobyev. 1989. The role of colored oil drops of cones in color vision [in Russian]. Sensory Systems 3:150–159.

Govardovskii, V. I., and L. V. Zueva. 1977. Visual pigments of chicken and pigeon. Vision Research 17:537–543.

Govardovskii, V. I., N. Fyhrquist, T. Reuter, D. G. Kuzmin, and K. Donner. 2000. In search of the visual pigment template. Visual Neuroscience 17:509–528.

Granda, A. M., and C. A. Dvorak. 1977. Vision in turtles; pp. 451–495 in F. Crescitelli (ed.), Handbook of Sensory Physiology, Vol. VII-5, The Visual System of Vertebrates. Springer-Verlag, Berlin.

Griebel, U., and L. Peichl. 2003. Colour vision in aquatic mammals: facts and open questions. Aquatic Mammals 29:18–30.

Hárosi, F. I., and Y. Hashimoto. 1983. Ultraviolet visual pigment in a vertebrate: a tetrachromatic cone system in the dace. Science 222:1021–1023.

Hart, N. S. 2001a. The visual ecology of avian photoreceptors. Progress in Retinal and Eye Research 20:675–703.

Hart, N. S. 2001b. Variations in cone photoreceptor abundance and the visual ecology of birds. Journal of Comparative Physiology A 187:685–698.

Hart, N. S. 2004. Microspectrophotometry of visual pigments and oil droplets in a marine bird, the wedge-tailed shearwater *Puffinus pacificus*: topographic variations in photoreceptor spectral characteristics. Jounal of Experimental Biology 207:1229–1240.

Hart, N. S., and M. Vorobyev. 2005. Modelling oil droplet absorption spectra and spectral sensitivities of bird cone photoreceptors. Journal of Comparative Physiology A 191:381–392.

Hayes, B. P. 1982. The structural organization of the pigeon retina. Progress in Retinal Research 1:197–226.

Hayes, B. P., G. R. Martin, and M. de L. Brooke. 1991. Novel area serving binocular vision in the retinas of procellariiform seabirds. Brain, Behavior, and Evolution 37:79–84.

Hebel, R. 1976. Distribution of retinal ganglion cells in five mammalian species (pig, sheep, ox, horse, dog). Anatomy and Embryology 150:45–51.

Henderson, Z. 1985. Distribution of ganglion cells in the retina of adult pigmented ferret. Brain Research 358:221–228.

Hisatomi, O., and F. Tokunaga. 2002. Molecular evolution of proteins involved in vertebrate phototransduction. Comparative Biochemistry and Physiology Part B 133:509–522.

Hisatomi, O., Y. Takahashi, Y. Taniguchi, Y. Tsukahara, and F. Tokunaga. 1999. Primary structure of a visual pigment in bullfrog green rods. FEBS Letters 447:44–48.

Hodos, W., R. F. Miller, and K. V. Fite. 1991. Age-dependent changes in visual acuity and retinal morphology in pigeons. Vision Research 31:669–677.

Howland, H. C., S. Merola, and J. R. Basarab. 2004. The allometry and scaling of the size of vertebrate eyes. Vision Research 44:2043–2065.

Hughes, A. 1977. The topography of vision in mammals of contrasting lifstyle: comparative optics and retinal organisation; pp. 613–756 in F. Crescitelli (ed.), Handbook of Sensory Physiology, Vol. VII-5, The Visual System of Vertebrates. Springer-Verlag, Berlin.

Hunt, D. M., J. Fitzgibbon, S. J. Slobodyanyuk, and J. K. Bowmaker. 1996. Spectral tuning and molecular evolution of rod visual pigments in the species flock of cottoid fish in Lake Baikal. Vision Research 36:1217–1224.

Hunt, D. M., K. S. Dulai, J. C. Partridge, P. Cottrill, and J. K. Bowmaker. 2001. The molecular basis for spectral tuning of rod visual pigments in deep-sea fish. The Journal of Experimental Biology 204:3333–3344.

Jacobs, G. H. 1993. The distribution and nature of colour vision among the mammals. Biological Reviews 68:413–471.

Jacobs, G. H., J. F. Deegan II, and J. Neitz. 1998. Photopigment basis for dichromatic color vision in cows, goats, and sheep. Visual Neuroscience 15:581–584.

Jane, S. D., and J. K. Bowmaker. 1988. Tetrachromatic colour vision in the duck (Anas platyrhynchos L.): microspectrophotometry of visual pigments and oil droplets. Journal of Comparative Physiology A 162:225–235.

Kawamura, S., and S. Yokoyama. 1998. Functional characterization of visual and nonvisual pigments of American chameleon (Anolis carolinensis). Vision Research 38:37–44.

Keitt, B. S., D. A. Croll, and B. R. Tershy. 2000. Dive depth and diet of the black-vented shearwater (Puffinus opisthomelas). The Auk 117:507–510.

Kelber, A., M. Vorobyev, and D. Osorio. 2003. Animal colour vision: behavioural tests and physiological concepts. Biological Reviews 78:81–118.

Kooyman, G. L. 2002. Evolutionary and ecological aspects of some Antarctic and sub-Antarctic penguin distributions. Oecologia 130:485–495.

Koskelainen, A., S. Hemilä, and K. Donner. 1994. Spectral sensitivities of short- and long-wavelength sensitive cone mechanisms in the frog retina. Acta Physiologica Scandinavica 152:115–124.

Land, M. F., and D.-E. Nilsson. 2002. Animal Eyes. Oxford University Press, Oxford.

Levenson, D. H., and A. Dizon. 2003. Genetic evidence for the ancestral loss of short-wavelength-sensitive cone pigments in mysticete and odontocete cetaceans. Proceedings of the Royal Society of London B 270:673–679.

Levenson, D. H., P. J. Ponganis, M. A. Crognale, J. F. Deegan II, A. Dizon, and G. H. Jacobs. 2006. Visual pigments of marine carnivores: pinnipeds, polar bear, and sea otter. Journal of Comparative Physiology A 192:833–843.

Liebman, P. A., and G. Entine. 1968. Visual pigments of frog and tadpole (Rana pipiens). Vision Research 8:761–775.

Loew, E. R. 1994. A third, ultraviolet-sensitive, visual pigment in the Tokay gecko (Gekko gekko). Vision Research 34:1427–1431.

Loew, E. R., and V. I. Govardovskii. 2001. Photoreceptors and visual pigments in the red-eared turtle, Trachemys scripta elegans. Visual Neuroscience 18:753–757.

Loew, E. R., W. N. McFarland, and D. Margulies. 2002. Developmental changes in the visual pigments of the yellowfin tuna, Thunnus albacares. Marine and Freshwater Behavior and Physiology 35:235–246.

Lundgren, B., and N. Højerslev. 1971. Daylight Measurements in the Sargasso Sea. Results from the "Dana" Expedition, January–April 1966. Institut for Fysisk Oceanografi, Report No. 14:1–33. Københavns Universitet, Copenhagen.

Lythgoe, J. N. 1979. The Ecology of Vision. Clarendon Press, Oxford.

Lythgoe, J. N. 1991. Evolution of visual behaviour; pp. 3–14 in J. R. Cronly-Dillon and R. L. Gregory (eds.), Evolution of the Eye and Visual System, Vol. 2 of J. R. Cronly-Dillon (ed.), Vision and

Visual Dysfunction. Macmillan Press, Houndmills, UK.

Lythgoe, J. N., and H. J. A. Dartnall. 1970. A "deep sea rhodopsin" in a mammal. Nature 227: 955–956.

Ma, J.-X., S. Znoiko, K. L. Othersen, J. C. Ryan, J. Das, T. Isayama, M. Kono, D. D. Oprian, D. W. Corson, M. C. Cornwall, D. A. Cameron, F. I. Harosi, C. L. Makino, and R. K. Crouch. 2001. A visual pigment expressed in both rod and cone photoreceptors. Neuron 32:451–461.

Makino, C. L., W. R. Taylor, and D. A. Baylor. 1991. Rapid charge movements and photosensitivity of visual pigments in salamander rods and cones. Journal of Physiology (London) 442:761–780.

Martin, G. R. 1999. Eye structure and foraging in king penguins Aptenodytes patagonicus. Ibis 141:444–450.

Martin, G. R., and M. de L. Brooke. 1991. The eye of a procellariiform seabird, the Manx shearwater, Puffinus puffinus: visual fields and optical structure. Brain, Behavior, and Evolution 37:65–78.

Mass, A. M., and A. Ya. Supin. 1992. Peak density, size and regional distribution of ganglion cells in the retina of the fur seal Callorhinus ursinus. Brain, Behavior, and Evolution 39:69–76.

Mass, A. M., and A. Ya. Supin. 1997. Ocular anatomy, retinal ganglion cell distribution, and visual resolution in the gray whale, Eschrichtius gibbosus. Aquatic Mammals 23:17–28.

Mass, A. M., and A. Ya. Supin. 2000. Ganglion cells density and retinal resolution in the sea otter, Enhydra lutris. Brain, Behavior, and Evolution 55:111–119.

Mass, A. M., and A. Ya. Supin. 2003. Retinal topography of the harp seal Pagophilus groenlandicus. Brain, Behavior, and Evolution 62:212–222.

Mass, A. M., and A. Ya. Supin. 2005. Ganglion cells topography and retinal resolution of the Steller sea lion Eumetopias jubatus. Aquatic Mammals 31:393–402.

McFarland, W. N. 1971. Cetacean visual pigments. Vision Research 11:1065–1076.

Murayama, T., and H. Somiya. 1998. Distribution of ganglion cells and object localizing ability in the retina of three cetaceans. Fisheries Science 64:27–30.

Nathans, J., D. Thomas, and D. S. Hogness. 1986. Molecular genetics of human color vision: the genes encoding blue, green and red pigments. Science 232:193–202.

Neumeyer, C. 1986. Wavelength discrimination in the goldfish. Journal of Comparative Physiology A 158:203–213.

Neumeyer, C. 1991. Evolution of colour vision; pp. 284–305 in J. R. Cronly-Dillon and R. L. Gregory (eds.), Evolution of the Eye and Visual System, Vol. 2 of J. R. Cronly-Dillon (ed.), Vision and Visual Dysfunction. Macmillan Press, Houndmills, UK.

Neumeyer, C. 1998. Color vision in lower vertebrates; pp. 149–162 in W. G. K. Backhaus, R. Kliegl, and J. S. Werner (eds.), Color Vision: Perspectives from Different Disciplines. Walter de Gruyter and Co., Berlin.

Neumeyer, C. 2003. Wavelength dependence of visual acuity in goldfish. Journal of Comparative Physiology A 189:811–821.

Newman, L. A., and P. R. Robinson. 2005. Cone visual pigments of aquatic mammals. Visual Neuroscience 22:873–879.

Nguyen, V. S., and C. Straznicky. 1989. The development and the topographic organization of the retinal ganglion cell layer in Bufo marinus. Experimental Brain Research 75:345–353.

Nowak, R. M. 1999. Walker's Mammals of the World, 6th edition, Vol. II. Johns Hopkins University Press, Baltimore.

Ödeen, A., and O. Håstad. 2003. Complex distribution of avian color vision systems revealed by sequencing the SWS1 opsin from total DNA. Molecular Biology and Evolution 20:855–861.

Palacios, A. G., F. J. Varela, R. Srivastava, and T. H. Goldsmith. 1998. Spectral sensitivity of cones in the goldfish, Carassius auratus. Vision Research 38:2135–2146.

Partridge, J. C. 1989. The visual ecology of avian cone oil droplets. Journal of Comparative Physiology A 165:415–426.

Partridge, J. C., J. Shand, S. N. Archer, J. N. Lythgoe, and W. A. H. M. van Groningen-Luyben. 1989. Interspecific variation in the visual pigments of deep-sea fishes. Journal of Comparative Physiology A 164:513–529.

Peichl, L. 1992. Topography of ganglion cells in the dog and wolf retina. Journal of Comparative Neurology 324:603–620.

Peichl, L. 2005. Diversity of mammalian photoreceptor properties: adaptations to habitat and lifestyle? Anatomical Record A 287A:1001–1012.

Peichl, L., and K. Moutairou. 1998. Absence of short-wavelength sensitive cones in the retinae of seals (Carnivora) and African giant rats (Rodentia). European Journal of Neuroscience 10:2586–2594.

Peichl, L., and H. Wässle. 1979. Size, scatter and coverage of ganglion cell receptive field centres in the cat retina. Journal of Physiology (London) 291:117–141.

Peichl, L., G. Behrmann, and R. H. H. Kröger. 2001. For whales and seals the ocean is not blue: a visual pigment loss in marine mammals. European Journal of Neuroscience 13:1520–1528.

Peichl, L., R. R. Dubielzig, A. Kübber-Heiss, C. Schubert, and P. K. Ahnelt. 2005. Retinal cone types in brown bears and the polar bear indicate dichromatic color vision. Investigative Ophthalmology and Visual Science 46:E-Abstract 4539.

Peterson, E. H., and P. S. Ulinski. 1979. Quantitative studies of retinal ganglion cells in a turtle, *Pseudemys scripta elegans*. Journal of Comparative Neurology 186:17–42.

Pettigrew, J. D., B. Dreher, C. S. Hopkins, M. J. McCall, and M. Brown. 1988. Peak density and distribution of ganglion cells in the retinae of microchiropteran bats: implications for visual acuity. Brain, Behavior, and Evolution 32:39–56.

Rahmann, H. 1967. Die Sehschärfe bei Wirbeltieren. Naturwissenschaftliche Rundschau 20:8–14.

Reuter, T. 1969. Visual pigments and ganglion cell activity in the retinae of tadpoles and adult frogs (*Rana temporaria* L.). Acta Zoologica Fennica 122:1–64.

Reuter, T. E., R. H. White, and G. Wald. 1971. Rhodopsin and porphyropsin fields in the adult bullfrog retina. Journal of General Physiology 58:351–371.

Rodieck, R. W. 1998. The First Steps in Seeing. Sinauer Associates, Sunderland, MA.

Roth, L. S. V., and A. Kelber. 2004. Nocturnal colour vision in geckos. Proceedings of the Royal Society of London B (Suppl.) 271:S485–S487.

Sillman, A. J., S. J. Ronan, and E. R. Loew. 1991. Histology and microspectrophotometry of the photoreceptors of a crocodilian, *Alligator mississippiensis*. Proceedings of the Royal Society of London B 243:93–98.

Southall, K. D., G. W. Oliver, J. W. Lewis, B. J. LeBoeuf, D. H. Levenson, and B. L. Southall. 2002. Visual pigment sensitivity in three deep diving marine mammals. Marine Mammal Science 18:275–281.

Stone, J., and P. Halasz. 1989. Topography of the retina in the elephant *Loxodonta africana*. Brain, Behavior, and Evolution 34:84–95.

Supin, A. Ya., V. V. Popov, and A. M. Mass. 2001. The Sensory Physiology of Aquatic Mammals. Kluwer Academic Publishers, Boston.

Timney, B., and K. Keil. 1992. Visual acuity in the horse. Vision Research 32:2289–2293.

Vorobyev, M. 1997. Costs and benefits of increasing the dimensionality of colour vision systems; pp. 280–289 in C. Tadei-Ferretti (ed.), Biophysics of Photoreception: Molecular and Phototransductive Events. World Scientific, Singapore.

Wald, G., P. K. Brown, and P. Smith Brown. 1957. Visual pigments and depths of habitat of marine fishes. Nature 180:969–971.

Walls, G. L. 1942. The Vertebrate Eye and its Adaptive Radiation. Cranbrook Press, Bloomfield Hills, MI.

Warrant, E. J., and N. A. Locket. 2004. Vision in the deep sea. Biological Reviews 79:671–712.

Wässle, H. 2004. Parallel processing in the mammalian retina. Nature Reviews Neuroscience 5:747–757.

Weimerskirch, H., and P. M. Sagar. 1996. Diving depths of sooty shearwaters *Puffinus griseus*. Ibis 138:786–794.

Yokoyama, S. 2000. Molecular evolution of vertebrate visual pigments. Progress in Retinal and Eye Research 19:385–419.

Yokoyama, S., N. Takenaka, D. W. Agnew, and J. Shoshani. 2005. Elephants and human color-blind deuteranopes have identical sets of visual pigments. Genetics 170:335–344.

Zueva, L. V. 1982. Microspectrophotometric study of cone oil droplets in two tortoises, *Emys orbicularis* and *Testudo horsfieldi* [in Russian]. Journal of Evolutionary Biochemistry and Physiology 18:304–307.

Hearing

11

The Physics of Sound in Air and Water

Sirpa Nummela and J. G. M. Thewissen

A vibrating sound source sets molecules in the transmitting medium, air or water, into motion, and these movements of the molecules lead to deviations from the static pressure. The deviations then propagate as longitudinal waves in the medium and are called sound waves. Sound waves received by the ear cause specialized neurons to fire, and these neurons may respond to one of two properties of sound. First, actual displacement of the molecules that transmit the sounds are detected by the ears of some vertebrates, and these ears are called displacement ears. In nearly all tetrapods, however, the ears are sensitive to the slight pressure fluctuations (deviations from ambient pressure) as sound waves reach the eardrum, and these ears are referred to as pressure ears.

Hearing is an animal's ability to respond to sound using a receptor for which such vibrations are the most effective stimulus (Wever, 1978). The hearing frequency range and the hearing sensitivity are constrained by the type of the ear, its structures, and its size. The ability to detect and localize sound then uses the acoustic cues that this peripheral sound transmission mechanism can provide.

At some level, hearing is a mechanoreception system, where the stimulus is transduced into a neural impulse by the hair cells embedded in the inner ear fluid in the cochlea (Fig. 11.1), protected from the outside world. Before the stimulus reaches the hair cells it is transmitted by specialized parts of the peripheral hearing organ, the outer, middle, and inner ear. The transmission mechanism through these parts of the ear is affected by the acoustic properties of the surrounding medium.

This chapter first discusses the physics of sound and hearing in water and air, focusing on the differences between these media in their ability to transmit sounds and the opportunities and challenges that living in two media presents to

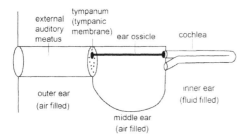

external auditory meatus | tympanum (tympanic membrane) | ear ossicle | cochlea

outer ear (air filled)

inner ear (fluid filled)

middle ear (air filled)

FIGURE 11.1. Diagram of the ear of a generalized tetrapod. The tympanic membrane (tympanum, eardrum) is on the boundary between outer and middle ear, and the deep part of the ear ossicle (footplate of the stapes as lodged in the oval window) is on the boundary between middle and inner ear.

the organism. These different aspects of sound all determine an animal's ability to detect sound. Then, this chapter addresses two important and often misunderstood subjects related to hearing. The first is sound localization, the mechanisms used by animals to determine the direction from which a sound arrives. The second is bone conduction, a particular mechanism of audition that is not of special importance in aerial hearing, but very significant in aquatic sound reception.

SOUND DETECTION

SOUND FREQUENCY, WAVELENGTH, AND VELOCITY

Sound can be characterized by its frequency (f, pitch), wavelength (λ), and amplitude (intensity, energy level). Sound velocity (c) is the velocity at which a sound wave travels in the transmitting medium, and it is medium dependent. Sound wavelength and sound frequency are inversely related, and sound velocity is the product of these two ($c = \lambda f$). In air, sound travels at a speed of approximately 340 m/s. In water, the sound velocity is nearly five times higher, approximately 1530 m/s; slight variation occurs due to such factors as temperature and salinity (Table 11.1). As a result, an object that vibrates at a constant frequency in water produces sound waves with nearly five

times longer wavelength than when vibrating at the same frequency in air.

The hearing frequency range of different tetrapods varies to a considerable degree (Table 11.2) (see, e.g., Fay, 1988; Heffner and Heffner, 1992). Of vertebrates, only mammals have a true, high-frequency hearing, exceeding well over 100 kHz in bats and odontocetes. Humans hear frequencies between 20 Hz and 20 kHz. The terms infrasound and ultrasound are based on the human frequency range; frequencies below 20 Hz are called infrasounds, and frequencies above 20 kHz are called ultrasounds.

High-frequency sounds have short wavelengths, and they attenuate quickly in both air and water. They are well suited for short-distance communication and may be very useful in determining the sound direction (see below). Low-frequency sounds have long wavelengths, and they attenuate more slowly than high-frequency sounds and can carry over long distances. Hence, low frequencies are more suitable for long-distance communication. In water, where the sound wavelength is five times longer than in air, low frequencies traveling in a certain water layer attenuate extremely little and are used, for example, by mysticetes (baleen whales) for their communication over hundreds and even thousands of kilometers (see, e.g., Richardson et al., 1995; Geisler, 1998).

SOUND PRESSURE LEVEL

The ears of most vertebrates are sensitive to sound pressure rather than to particle motion, detecting pressure fluctuations that indicate acoustic phenomena. The sound strength is expressed as sound pressure level (L_p) with decibel units, which gives the ratio p/p_{ref} in the logarithmic scale:

$$L_p = 20 \log (p/p_{ref}) \text{ dB},$$

where p is the sound pressure and p_{ref} is the generally agreed reference pressure, 20 μPa in air and 1 μPa in water. The sound pressure level

TABLE 11.1
Physical Parameters of Sound

	IN AIR	IN WATER
Sound velocity (c)	340 m/s	1530 m/s
Density (ρ)	1.3 kg/m^3	1030 kg/m^3
Acoustic impedance (Z)	0.4 kPa s/m	1500 kPa s/m
Frequencies used in echolocation	120 to 150 kHz (to 250 kHz, bats)	120 kHz to 150 kHz (toothed whales)
Reference pressure (p)	20 μPa	1 μPa

increases 20 dB when the sound pressure increases 10 times.

In the case of a propagated plane wave, the sound intensity (I) can be calculated with the help of the measured pressure, density of the medium (ρ), and sound velocity (c), using the equation $I = p^2/\rho c$.

The absolute hearing sensitivity, that is, the minimum hearing threshold at a particular frequency, is an important quantity in comparative auditory research. When hearing thresholds are being compared between air and water, sound intensity rather than sound pressure has been adopted as the measure for the sensitivities in these two media (see, e.g., Supin et al., 2001).

The density ratio of water and air, ρ_w/ρ_a, is about 800, and the sound velocity ratio c_w/c_a is about 4.5 (depending on the temperature, air pressure, and salinity in seawater). Thus, when the sound waves in water and air have same intensities $I = p^2/\rho c$, the sound pressure ratio is about 60, corresponding to 35.5 dB difference in the sound pressure levels. The ratio of the reference pressures, 20 μPa/1 μPa, corresponds to 26 dB. Taken together, if the threshold sound pressure level of an animal in water is 61.5 dB larger than the threshold sound pressure level of an animal in air, the sounds have equal intensities and the animals have equally sensitive hearing.

Hearing abilities of a particular species are often described with a behavioral audiogram, which shows the frequency range with hearing thresholds for pure tones through the animal's hearing range. Of special interest are four different features in the audiogram: the highest and the lowest frequencies the animal can hear at a sound pressure level of 60 dB (20 μPa in air), the frequency of best sensitivity, and the volume threshold sound pressure level at highest sensitivity (Fig. 11.2 threshold sound intensity level) (Masterton et al., 1969; Fay, 1988; Heffner and Heffner, 1998).

ACOUSTIC IMPEDANCE

In acoustics, impedance is a concept of central importance and perhaps can be best described as the intrinsic resistance of a substrate, or of a mechanism, against transmitting sounds. Impedance thus characterizes the sound propagation in a medium or through a sound-transmitting device (such as an amplifier or a middle ear). When sound moves from one medium to the other, the ratio between their impedances determines how much energy will be transmitted across the interface of these media, and how much will be reflected, resembling the situation of the refractive index for light (Kröger, chapter 8 in this volume). The more the impedance ratio differs from 1, the more energy will be reflected.

Each medium has its characteristic acoustic impedance, and each device (e.g., the inner ear) has its own specific acoustic impedance. Characteristic acoustic impedance of a medium is $Z = \rho c$, where ρ is the density of the medium and c is the sound velocity in that particular medium. The physical properties of water and

TABLE 11.2	
Upper frequency limits of hearing	
Amphibians	500 Hz to 1 kHz
Reptiles	1 kHz to 3 kHz
Birds	100 Hz to 12 kHz
Mammals	100 kHz to (180 to 250 kHz

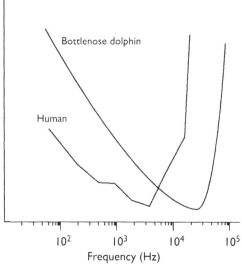

FIGURE 11.2. Audiograms for human (in air) and dolphin (in water). Note that frequency is usually plotted on a logarithmic scale and that the *y* axis (which shows threshold sound intensity levels in decibels) is not directly comparable in aquatic and airborne hearing.

air differ, and as a result sound propagation through each medium differs. The density and sound velocity are much higher in water than they are in air, and hence the characteristic acoustic impedance of water, Z_w, is about 1500 kPa s/m, much higher than the characteristic acoustic impedance of air, Z_a, which is about 0.4 kPa s/m (Table 11.1).

The vertebrate inner ear is a small, fluid-filled space housing hair cells functional in hearing and balance sensory systems. The inner ear is filled with watery liquids (perilymph and endolymph) and has an impedance that is somewhat smaller than Z_w. The specific acoustic impedance of the inner ear of humans is about 50 kPa s/m (Evans, 1982). Based on middle ear geometries, Hemilä et al. (1995, 1999) estimated that for most mammals this impedance is roughly a tenth of Z_w, that is, about 150 kPa s/m.

If sounds were transmitted from the surrounding medium directly to the inner ear, most sound energy would be lost. The tetrapod middle ear is a mechanical device between the outside world and the inner ear, and as a result of this the pressure at the oval window is increased compared to the air pressure at the tympanic membrane. By passing sounds from air through the middle ear to the inner ear, the impedance ratio is minimized, and less energy is lost by reflection. For this reason, the middle ear is necessary for efficient sound transfer in vertebrate hearing in air and is often referred to as an impedance matching device (e.g., Rosowsk; and Relkin 2001).

The physical properties of each of the parts of the peripheral ear structures vary, and each part has its own sound transmission properties.

The acoustic properties of air and water and the properties of the ear are tuned to each other, and understanding this is the key to understanding the adaptations for hearing in water, air, or both.

The impedance matching in a terrestrial vertebrate ear is accomplished by increasing the pressure p and by decreasing the particle velocity v between the tympanic membrane and the oval window ($Z = p/v$). In transmitting energy between these membranes, the pressure can be increased by the area ratio A_1/A_2, and a possible lever ratio between them, l_1/l_2. The pressure increase is thus $(A_1/A_2)(l_1/l_2)$; this factor is called the geometric transformer ratio of the vertebrate middle ear (Henson, 1974). Additionally, the lever mechanism affects the particle velocity between the tympanic membrane and the oval window by l_1/l_2. The particle velocity can be decreased by having the in-lever l_1 longer than the out-lever l_2, giving a lever ratio $l_1/l_2 > 1$. Hence, p/v at the oval window is multiplied by a factor $(A_1/A_2)(l_1/l_2)^2$ compared to p/v at the tympanic membrane. The impedance matching function of the tetrapod middle ear is also discussed by Hetherington (chapter 12 in this volume).

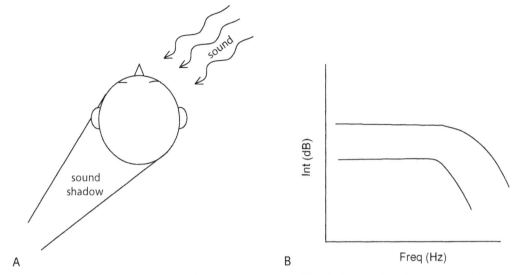

FIGURE 11.3. (A) Diagram of a mammalian head with sound coming obliquely from the right, showing that the left ear is in a sound shadow. (B) Approximate sound intensities at left and right ear for mammal in A. Upper curve refers to right ear.

SOUND LOCALIZATION

Sound localization is the ability to determine the direction of a sound source. Animals have two ways to determine the direction from which the sound is arriving: the interaural time difference and the interaural intensity difference between the two ears.

A sound arriving from the right leaves the left ear in a sound shadow created by the head (Fig. 11.3). This causes an intensity difference between the ears that depends on the sound frequency and the head size. In order to have a large intensity ratio, the interaural distance has to be larger than the wavelength. Thus the product of the interaural distance and the frequency should be larger than the sound velocity. In water this product has to be five times larger than in air.

Sound arriving not directly from front or from behind reaches one ear earlier than the other. This results in the interaural time difference between the signals to the ears, and this time difference can also be used to determine the direction from which a sound is coming. When the sound source is at the side, this time difference is the interaural distance divided by sound velocity. Hence, this time difference is

large in large animals and does not depend on frequency. Neural networks can determine surprisingly short time differences (0.0006 ms in owls) (Bradbury and Vehrencamp, 1998), but for very small animals, the interaural time difference may be too short a time to determine the sound direction.

In the case of tonal sounds, the interaural time difference is manifested in the phase difference of sine waves, which can also be determined by neural networks. This phase difference is proportional to both the interaural time difference and frequency. Thus it is large in large animals and increases with increasing frequency. However, phase differences larger than 90 degrees lead to ambiguity.

Only mammals are able to hear frequencies over 10 kHz, and their elaborate sound localization ability is dependent on their good high-frequency hearing (e.g., Heffner and Heffner, 1998; Heffner, 2004). It has been suggested that mammals have been under high selective pressures for good high-frequency hearing throughout their evolutionary history (Masterton et al., 1969; Heffner and Heffner, 1985, 1992; Rosowski, 1992). Heffner and Heffner (1985, 1992) showed that the high-frequency limit and the so-called functional interaural distance (the maximum interaural distance divided by the speed of sound) correlate well with each other for different species of mammals, including high-frequency hearing limits in air and water for terrestrial mammals, seals and sea lions, and odontocete whales. Further, such a correlation has not been found in the upper hearing limit of any other groups of vertebrates (see Heffner and Heffner, 1985, 1992, and references therein).

In air, even animals with small head size may have good directional hearing, provided they can hear very high frequencies, and indeed bats have excellent ability to locate sound sources. In water sound travels nearly five times faster than in air, reducing the interaural time difference in water to one-fifth of that in air, and also reducing interaural intensity differences (as this increases, when the ratio between the interaural distance and the wavelength increases, and in water the wavelength is fivefold). This fivefold sound velocity is no problem for whales and seals, as their large head size compensates for the increase of sound velocity.

In addition, the intensity difference between the sound reaching the two ears is reduced. The reason for this is that the impedance of the soft tissues of the head is similar to that of water, and thus most sound energy passes through the head instead of being reflected at the skin (bone conduction, see below). Sounds travel through the tissues of the head to both ears, creating less of an intensity difference between the ears.

ECHOLOCATION

Echolocation is the process in which the reflections of sounds emitted by an animal are perceived by that animal and are used to determine the shape of its surroundings and its prey. Sound localization is a major part of echolocation. Microbats and odontocete whales echolocate using high-frequency sounds well over 100 kHz, and thus wavelength ranges that vary by a factor of 5 (when the sound frequency is 100 kHz, the wavelength is 3.4 mm in air

and 15 mm in water). In echolocation, and sound localization in general, turning the head from side to side provides additional important cues.

BONE CONDUCTION

Bone conduction occurs when sound elicits skull vibrations that lead to hearing sensations. Three different bone conduction mechanisms can be distinguished (Tonndorf, 1968). Within the middle frequency range the inertial mechanism dominates. When sound sets the whole skull (including the inner ear), into vibration but the elastically coupled ossicles vibrate less, the relative movement of the stapes in the oval window leads to cochlear stimulation. In the inertial model a relative motion between the ossicular chain and the temporal bone leads to cochlear stimulation much the same way as in air-conducted hearing of terrestrial vertebrates (Bárány, 1938). This eventually leads to displacement of the basilar membrane, which will create a neural impulse.

Bone conduction is a relevant mechanism in bringing the stimulus into the inner ear in water where the density of the surrounding medium is close to the density of body tissues. In that case, the impedance ratio is nearly one because the densities of water and soft body tissues are similar, and sound will penetrate the body tissues with little attenuation (Heth et al., 1986). Thus, bone conduction is an unavoidable hearing mechanism in many aquatic animals. However, these sounds traverse the soft body tissues and reach the right and left cochlea almost simultaneously, causing disruption of sophisticated directional hearing. In aquatic taxa that need directional sensitivity, such as whales, the ears are acoustically isolated from the rest of the body in order to limit the reception of bone-conducted sounds.

ACKNOWLEDGMENTS

We thank Simo Hemilä for comments on the manuscript.

LITERATURE CITED

Bárány, E. 1938. A contribution to the physiology of bone conduction. Acta Oto-Laryngologica Stockholm Supplement 26:1–223.

Bradbury, J. W., and S. L. Vehrencamp. 1998. Principles of Animal Communication. Sinauer Associates, Sunderland, MA.

Evans, E. F. 1982. Functional anatomy of the auditory system; pp. 251–306 in H. B. Barlow and J. D. Mollon (eds.), The Senses. Cambridge University Press, Cambridge.

Fay, R. R. 1988. Hearing in Vertebrates: A Psychophysics Databook. Hill-Fay Associates, Winnetka, IL.

Geisler, C. D. 1998. From Sound to Synapse: Physiology of the Mammalian Ear. Oxford University Press, New York.

Heffner, H. E., and R. S. Heffner. 1992. Auditory perception; pp. 159–184 in C. Phillips and D. Piggins (eds.), Farm Animals and the Environment. CAB International, Wallingford, UK.

Heffner, H. E., and R. S. Heffner. 1998. Hearing; pp. 290–303 in G. Greenberg and M. M. Haraway (eds.), Comparative Psychology, A Handbook. Garland, New York.

Heffner, R. S. 2004. Primate hearing from a mammalian perspective. Anatomical Record 281A:1111–1122.

Heffner, R. S., and H. E. Heffner. 1985. Hearing in two cricetid rodents: wood rat (Neotoma floridana) and grasshopper mouse (Onychomys leucogaster). Journal of Comparative Psychology 99:275–288.

Hemilä, S., S. Nummela, and T. Reuter. 1995. What middle ear parameters tell about impedance matching and high-frequency hearing. Hearing Research 85:31–44.

Hemilä, S., S. Nummela, and T. Reuter. 1999. A model of the odontocete middle ear. Hearing Research 133:82–97.

Henson, O. W., Jr. 1974. Comparative anatomy of the middle ear; pp. 39–110 in W. D. Keidel and W. D. Neff (eds.), Handbook of Sensory Physiology, Vol. V-I, Auditory System. Springer-Verlag, New York.

Heth, G., E. Frankenberg, and E. Nevo. 1986. Adaptive optimal sound for vocal communication in tunnels of a subterranean mammal (Spalax ehrenbergi). Experientia 42:1287–1289.

Masterton, B., H. Heffner, and R. Ravizza. 1969. The evolution of human hearing. Journal of the Acoustical Society of America 45:966–985.

Richardson, W. J., C. R. Greene, Jr., C. I. Malme, and D. H. Thomson. 1995. Marine Mammals and Noise. Academic Press, San Diego.

Rosowski, J. J. 1992. Hearing in transitional mammals: predictions from the middle-ear anatomy and hearing capabilities of extant mammals; pp 615–631 in D. B. Webster, R. R. Pay, and A. N. Popper (eds.), The Evolutionary Biology of Hearing. Springer-Verlag, New York.

Rosowski, J. J., and E. M. Relkin. 2001. Introduction to the analysis of middle ear function; pp. 161–190 in A. F. Jahn and J. Santos-Sacchi (eds.), Physiology of the Ear, 2nd Ed. Thomson Learning, Hampshire, UK.

Supin, A. Ya., V. V. Popov, and A. M. Mass. 2001. The Sensory Physiology of Aquatic Mammals. Kluwer, Boston.

Tonndorf, J. 1968. A new concept of bone conduction. Archives of Otolaryngology 87:49–54.

Wever, E. G. 1978. The Reptile Ear: Its Structure and Function. Princeton University Press, Princeton, NJ.

12

Comparative Anatomy and Function of Hearing in Aquatic Amphibians, Reptiles, and Birds

Thomas Hetherington

This chapter describes patterns of evolutionary change in the auditory systems of amphibian, reptilian, and avian lineages that have returned to an aquatic lifestyle. It starts with a description of the basic features of the ancestral terrestrial auditory system, discusses the functional implications of placing such a system underwater, and moves on to descriptions of the auditory systems of different amphibious and aquatic tetrapod lineages. There are large gaps in our functional understanding of how the auditory systems of many aquatic tetrapods actually work underwater, but nonetheless some general patterns can still be discerned.

THE GENERALIZED TERRESTRIAL AUDITORY SYSTEM

OUTER EAR

Outer ear structures function to collect and transfer sound energy to the middle ear

(Rosowski, 1994). Outer ears are most elaborate in terrestrial mammals, many of which have large pinnae (ear flaps) to collect sound. The pinna may be important in improving directionality of hearing by modifying sound spectra depending on the position of the sound source (Heffner and Heffner, 1992; Brown, 1994), and many mammals have muscles that can move the pinna toward the source of a sound. Sounds then pass down a funnel-shaped space (concha) to a tubelike ear canal that ends at the tympanic membrane (tympanum, ear drum). Depending on their shape and dimensions, the concha and ear canal will have resonant properties that will selectively amplify certain sound frequencies before reaching the tympanic membrane (Rosowski, 1994). The outer ear therefore can aid in overcoming the problem of sound reflection at the air-tissue interface (see below). Birds have outer ears consisting of feathers arranged around an ear canal (Saito, 1980), and in some species the ear canal has a funnel shape (Schwartzkopff, 1973). The feathers surrounding the ear canal and the shape of the ear canal likely aid in sound collection and also can provide directional information (Norberg, 1978). Amphibians and many reptiles have no outer ear structures, although most lizards and crocodilians have short ear canals (Wever, 1978).

MIDDLE EAR

Vertebrates rely on hair cells, mechanoreceptors that are extraordinarily sensitive to displacements of their apical ends, to detect sound. Terrestrial vertebrates face a challenge in utilizing these receptors for hearing because they are located within an inner ear positioned deep inside an ossified skull, possibly the least likely location one would expect for auditory receptors. In air, most sound energy will be reflected off the surface of the body because of the mismatch between the acoustic impedances of air and tissue. Depending on the frequency of the sound and the mass, stiffness,

and friction associated with a particular body tissue, up to 99.9% of sound energy will be reflected. This dramatic loss in sound transmission between air and tissue traditionally is considered the functional explanation for the evolution of the tympanic middle ears of terrestrial vertebrates, although as discussed below, many other factors likely have shaped the evolution of tympanic ears.

The tympanic middle ear of tetrapods (Fig. 12.1) other than mammals typically consists of a tympanic membrane, an air-filled middle ear cavity, and auditory ossicle frequently referred to as the columella, but, because it almost certainly is homologous to the primitive hyomandibula, I call it the stapes. The tympanic membrane is supported around its margin, and often stiffened by a variety of skeletal elements and ligaments in different groups, and is connected to a cartilaginous element (extrastapes) that in turn is linked to the stapes. The air-filled middle ear cavity allows the tympanic membrane to move in response to pressure fluctuations acting on its outer surface. The extrastapes and stapes traverse the middle ear cavity toward the inner ear, and the presence of air also likely minimizes friction that would impede ossicular movement. The stapes usually has an expanded footplate that presses against the fluids of the inner ear at an opening in the otic capsule called the oval window. Sound-induced motion of the tympanic membrane transferred to the stapes eventually produces pressure waves in the inner ear fluids that stimulate hair cells (see below).

The middle ears of amphibians differ from those of reptiles and birds by having an additional ossicle, the operculum, lying in the oval window. The function of the operculum remains uncertain. There is evidence that it can function as an inertially sensitive element for detection of ground vibrations entering the body (Hetherington, 1988), and it may form part of a pathway for low-frequency sound reception (Lewis and Narins, 1999).

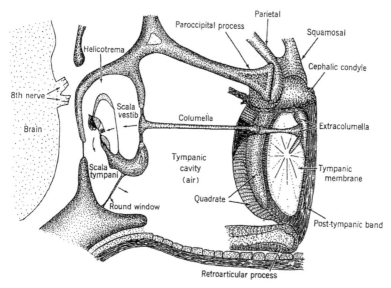

FIGURE 12.1. The right ear region of the lizard *Sceloporus magister* seen from posterior and left. Note the linkage of the tympanic membrane (tympanum) to the extracolumella (extrastapes) that in turn connects to the stapes (columella). The stapes extends to the inner ear. A tympanic cavity (middle ear cavity) lies beneath the tympanic membrane, which is supported around its margin by several structures. The tympanic cavity usually has a connection to the pharynx via a eustachian tube (not shown). Image taken from Wever (1985).

Tympanic middle ears employ two main strategies to facilitate the transmission of sound energy to the inner ear fluids. One strategy is to collect sound energy over a relatively large tympanic membrane and converge it onto a relatively small oval window, thereby amplifying the force acting on the inner ear fluids. Ratios of tympanic to oval window area appear to range from about 10 to 60 in vertebrates (Schwartzkopff, 1957; Kuhne and Lewis, 1985; Lombard and Hetherington, 1993). A second strategy is to configure the tympanic membrane and ossicle as a lever system to further amplify the force. The precise lever arrangement employed in middle ears varies, and ossicular motions can be complex and poorly understood, but the lever arm associated with the tympanic membrane is always longer than the lever arm driving the stapedial footplate. There also may be a mechanical lever arrangement within the tympanic membrane itself (Saunders et al., 2000). However, lever advantages are not dramatic, almost always being below a factor of 2 (Kuhne and Lewis, 1985; Lombard and Hetherington, 1993; Saunders et al., 2000). It is apparent that the convergence ratio of tympanic to oval window area is the most significant factor compensating for the impedance mismatch between air and body tissue. The maximum overall amplification resulting from both the tympanic–to–oval window area ratio and the lever advantage proposed for a vertebrate is 97 for a rodent (Webster and Webster, 1975). This degree of amplification does not match the loss in transmission of sound energy from air into body tissue at most sound frequencies but is sufficient to provide such high sensitivity that with additional amplification animals might hear a constant, dull hissing sound produced by the brownian motion of air molecules!

The tympanic middle ear often is described as an impedance-matching device that compensates for the transmission loss of sound from air to tissue (Wever, 1949; Lombard and Hetherington, 1993). A more accurate description is that it helps to compensate for the impedance mismatch between air and the inner ear fluids contained within the confined spaces of the otic capsule (Rosowski, 1994). The system also often is described as a signal transformer, transforming a pressure signal into displacement energy entering the inner ear (Rosowski, 1994). These are useful ways of understanding middle ear function, but the origin and evolution of tympanic middle ears may best be understood by appreciating more specific functional parameters. For example, tympanic middle ears are highly adaptable systems that can be modified to maximize selective transfer of acoustic power at certain frequencies (i.e., they can be "tuned" to sound frequencies of special importance). Many frogs, for example, have middle ears that are most effective in transmitting frequencies that match the dominant frequencies of intraspecific vocalizations (Wilczynski and Capranica, 1984; Jaslow et al., 1988). In general, tympanic middle ears are very effective for transferring power at relatively high sound frequencies (greater than about 1 kHz), and potentially selection for higher-frequency sensitivity may have been important in their evolution. Transmission of low-frequency sound is less impeded by mass than transmission of higher frequencies, so low-frequency sound can effectively penetrate many unspecialized body tissues (Lombard and Hetherington, 1993; Hetherington and Lindquist, 1999). Several reptiles and amphibians lack tympanic middle ears but nonetheless are quite sensitive to low-frequency sound (Wever, 1978, 1985; Lindquist et al., 1998; Hetherington and Lindquist, 1999). A system designed for coupling high-frequency sound to the inner ear should be characterized by light, stiff, and relatively small elements, as transmission of high-frequency signals is more impeded by mass but less impeded by stiffness and small dimensions than lower-frequency signals (Rosowski, 1994). In accordance with these expectations, the tympanic membrane of most tympanic ears is a thin membrane held in tension within a supporting

framework, and the middle ear ossicles typically are small, light, ossified elements with connections to surrounding bones that may further increase stiffness. The tympanic middle ears of some tetrapods, of course, may be adapted for low-frequency sound reception through a variety of modifications, typically increasing the mass and dimensions, and reducing the stiffness, of the different components.

Another important selective advantage of a tympanic middle ear may be its effectiveness for sound localization. Detecting a sound is one thing (and perhaps the simplest achievement of an auditory system), but it is directional information that allows an organism to most effectively respond to the signal. Tympanic middle ears appear to be very effective in allowing comparison of amplitude, timing, phase, and/or spectral differences between sound signals reaching the right and left ears (Knudsen, 1980; Brown, 1994). In small animals in which head dimensions minimize such differences, the left and right middle ears may be functionally linked to act as a pressure gradient receiver (Aertsen et al., 1986; Eggermont, 1988). Therefore, impedance matching and signal transformation do not provide a complete explanation for the evolution of the tympanic middle ear of terrestrial vertebrates, and because tympanic middle ears likely evolved multiple times among tetrapods (Lombard and Bolt, 1979; Presley, 1984; Clack and Allin, 2004), different selective pressures may have operated on different lineages.

INNER EAR

The inner ear basically consists of a vesicle (the membranous labyrinth) wrapped in a channel filled with perilymphatic fluid. Although the configuration of the channel differs among tetrapod groups (Fig. 12.2), one of its ends (near the stapedial footplate) is called the oval window and the other end is usually called the round window. Motion of the stapedial footplate in the oval window produces pressure waves in the perilymphatic fluid that pass along the channel, stimulate certain hair cell populations within the membranous labyrinth, and exit at the round

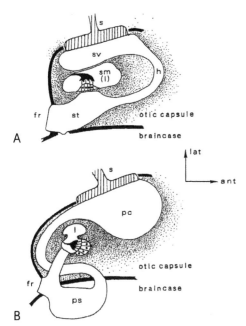

A

B

FIGURE 12.2. Diagrams of the inner ears of an amniote (A, reptile) and an amphibian (B) showing the configuration of the perilymphatic channels and the position of the basilar papilla, the auditory receptor of reptiles (and birds), and one of the auditory receptors of amphibians. Sound energy enters via the stapedial footplate (s) at the oval window and leaves at the round window (fr) and/or the intracranial cavity within the braincase. The sensory hair cells rest on a flexible membrane in amniotes and on dense connective tissue in amphibians. The sensory hairs are partly embedded in the overlying tectorial membrane. Skeletal elements are represented in black. Abbreviations: h, helicotrema; l, lagena; pc, perilymphaytic cistern; ps, perilymphatic sac; sm, scala media; st, scala tympani; sv, scala vestibuli. Image taken from Lombard and Bolt (1979).

window or some other pressure release point (Fig. 12.2, fr). Several distinct populations of hair cells are found in the membranous labyrinth, and certain ones can respond to acoustic (vibratory and sound) stimuli. For example, the otolithic end organs (utricular, saccular, and lagenar maculae) can serve as inertially sensitive detectors of vibrations and low-frequency sound penetrating the body. These acoustic signals produce differences in amplitude and/or phase between the dense, heavy otoliths and underlying hair cells, thereby shearing the tips of the latter. But all tetrapods also possess auditory papillae associated with detecting aerial sound. Amniotes have just one, called the basilar papilla

in reptiles and birds and the cochlear duct in mammals, whereas most amphibians have two, the basilar and amphibian papillae, that respond to relatively high and low sound frequencies, respectively. The basilar papilla often is considered to be homologous among tetrapods (Manley and Clack, 2004), and it appears to be present in the coelacanth *Latimeria chalumnae* (Fritzsch, 1992). However, the mechanism of stimulation of the basilar papilla of amniotes is different from that of both the basilar and amphibian papillae of amphibians. Waves in the perilymphatic fluids stimulate hair cells of the amniote basilar papillae by moving the membrane on which the hair cells are anchored, causing shearing action between the hair cells and an overlying fibrous, tectorial membrane (Fig. 12.2). In contrast, the hair cells of both the basilar and amphibian papillae of amphibians typically sit on dense connective tissue and are overlain by a more gelatinous tectorium (Lewis and Narins, 1999; Smotherman and Narins, 2004). The latter is better coupled to fluid motion and likely moves relative to the hair cells in response to fluid waves within the inner ear.

The key point in understanding effective middle and inner ear function in terrestrial tetrapods is to appreciate that acoustic energy enters at one point (the oval window) and exits at another (the round window). If acoustic energy were to enter simultaneously at both openings into the inner ear, no net fluid motion would occur and the hair cell receptors would not be stimulated. The auditory papillae require a differential transfer of acoustic power to one or the other of these windows. Presumably, they would be effectively stimulated if energy entered at the round window and exited at the oval window (i.e., the opposite of the normal pattern). Some tetrapods do not have a round window but have other pathways through which acoustic energy can exit the inner ear. The perilymphatic system of amphibians penetrates the cranial cavity (Fig. 12.2), and in some reptiles sound energy returns back to a large space outside the oval window (see turtle ears, below). In any case, a one-way flow system of acoustic energy is maintained. This pattern of energy flow into and out of the inner ear of terrestrial animals may have important implications for the evolution of the auditory systems of lineages that return to the water.

THE TERRESTRIAL AUDITORY SYSTEM UNDERWATER

OUTER EAR

The outer ear is largely superfluous underwater because of the match between the acoustical impedances of water and body tissue. The transparency of tissue to sound means that external structures are generally ineffective in collecting and guiding sound energy to a middle ear. It is not surprising therefore to find that external ears are absent or greatly reduced in secondarily aquatic forms. Loss of large external ears also likely helps to reduce drag during underwater locomotion. Aquatic birds retain external ear canals surrounded by feathers, but the feathers likely function to keep water out of the ear canal and prevent loss of aerial hearing sensitivity when not submerged (see below).

SOUND TRANSMISSION

If an animal with a generalized tympanic middle ear is submerged, a general match exists between the acoustic impedances of the medium and body tissues, allowing sound to easily penetrate the head. Ironically, this surfeit of acoustic energy penetrating the head could represent a hearing problem for a submerged animal with an auditory system adapted for reception of sound along particular paths. The ability of underwater sound to penetrate body tissues allows several possible strategies for hearing. These mechanisms need not be mutually exclusive; multiple mechanisms, effective over different frequency ranges, could be used. This situation produces a nightmare for morphological assessments of underwater hearing mechanisms but an outstanding opportunity to generate many clever but often untestable adaptive scenarios.

Several factors may effect the evolutionary transformation of a terrestrial auditory system into an aquatic one. The specific features of their auditory systems may cause different lineages to diverge along different paths specializing on one or the other of the strategies of underwater hearing described below. However, one factor that likely will influence the transformation in all fully aquatic lineages is the nature of the intermediate, amphibious phase through which they pass. This intermediate stage may require an auditory system that can function in both media, and maintenance of aerial sensitivity may come at the expense of underwater hearing sensitivity. Because we may not have a good understanding of the relative importance of hearing underwater versus hearing in air for any given species, study of amphibious forms is problematic for a full assessment of patterns of evolutionary transformations of terrestrial into aquatic ears. We can use information on how their auditory systems work in air and water, and perhaps view them as representative of intermediate stages in evolution, but they still likely represent an evolutionary compromise rather than a complete transformation.

Aquatic tetrapods have modified the generalized tympanic middle ear in a variety of ways corresponding to the utilization of new mechanisms of sound transfer to the inner ear. We discuss a variety of possible mechanisms, beginning with the potential retention of a functioning tympanic middle ear.

TYMPANIC MIDDLE EAR

The simplest expectation of how a terrestrial auditory system should function underwater would be to assume that the middle ear would continue as the pathway of sound transfer from the lateral head surface to the inner ear (Fig. 12.3A). Certain evolutionary modifications of this pathway would be expected because of the better impedance match between water and head tissues. For example, the stiffness that characterizes middle ears designed for sensitivity to high-frequency aerial sound might be reduced, and because force amplification by the

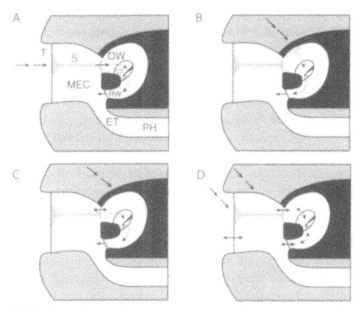

FIGURE 12.3.   Diagrams of mechanisms of underwater hearing in a generalized "terrestrial" tetrapod. (A) Sound transmission may occur along the standard tympanic middle ear pathway, producing fluid motion at the oval window that stimulates the auditory papilla and flows out at the round window. (B) Displacement energy may produce deformations of the otic capsule that in turn produce fluid waves in the perilymphatic channel that stimulate the auditory papilla. (C) Displacement energy of sound waves penetrating the head may (by inertial effects) produce relative motion between the stapedial footplate and inner ear fluids, generating fluid waves with the perilymphatic channels that stimulate the auditory papilla. (D) The middle ear cavity, mouth cavity, and/or lungs may transduce underwater sound pressure, producing pulsations of these cavities that cause tympanic and stapedial motion that in turn generates fluid motion at the oval window that stimulates the auditory papilla. Pulsations also could generate motion at the round window and produce fluid waves that travel in the opposite direction (note two-way arrows in the perilymphatic channel); pulsations of the pharynx, mouth cavity, and/or lungs also could produce high-amplitude displacements that could directly effect the inner ear. In all diagrams, the tympanic membrane (T) is connected to a stapes (S), which traverses an air-filled middle ear cavity (MEC). The stapedial footplate fits against the oval window (OW), and a perilymphatic channel connects the oval window to the round window (RW), wrapping around the part of the membranous labyrinth containing the auditory papilla with its hair cells and tectorial structure in the cochlear duct (CO) (scala media). The middle ear cavity is continuous with the pharynx (PH) via an eustachian tube (ET). *Solid arrows* represent sound-induced motion of structural elements, and *dashed arrows* represent fluid motion within the inner ear.

middle ear can now be reduced, both the ratio of tympanic–to–oval window area and the lever advantage of the system may be lower.

BONE CONDUCTION

Sound waves displace particles in the medium through which they travel and also compress the medium. Inner ear acoustic receptors of a submerged animal could respond to either or both of these components (i.e., displacement and/or pressure). Auditory stimulation based on displacement energy often is referred to as bone conduction (Tonndorf, 1972), and there are several ways that inner ear receptors could be stimulated by this component of sound waves. Displacement energy could directly stimulate hair cells of auditory papillae by moving them. Responses would be enhanced by large differences in density between the hair cells and overlying structures attached to their tips. This would produce an inertially sensitive system using the pronounced differences in amplitude and phase of motion between the hair cells and overlying structure to displace the tips of the hair cells. This hearing strategy is widely used in fishes that possess heavy otolithic structures overlying hair cells of the saccular, utricular, and/or lagenar maculae (Popper and Fay, 1999) and is directly comparable to the use of these same maculae by terrestrial vertebrates for detection of ground vibrations and low-frequency sound entering their bodies (Lombard and Hetherington, 1993). However, although underwater sound might stimulate macular organs in the inner ear of aquatic tetrapods in this manner, the auditory papillae have low-mass tectorial structures overlying them and are ill suited to act as part of an inertial system. Nonetheless, displacement energy also could stimulate the auditory papillae in more indirect ways, such as bone conduction through deformations of the otic capsule, and ossicular bone conduction.

In bone conduction through deformation of the otic capsule, penetrating sound waves produce slight deformations of the bones surrounding the inner ear that in turn produce motion of the inner ear fluids that stimulates the auditory papillae in the same manner as in aerial hearing (Fig. 12.3B).

Sound waves also could produce differences in the amplitude and phase of motion of the stapedial footplate relative to the perilymphatic fluid at the oval window (ossicular bone conduction). This would result in fluid motion within the inner ear that stimulates the papillae (Fig. 12.3C). Differences in motion between the stapedial footplate and inner ear fluids would be most pronounced if the footplate were especially heavy and loosely linked to the otic capsule, thereby maximixing inertial effects of the sound displacements.

PRESSURE TRANSDUCTION

Fluctuations in the pressure associated with underwater sound will cause pulsations of any gas cavity they encounter, thereby transducing the pressure component of the sound into high-amplitude displacement energy that then can be transferred to the hair cells (Fig. 12.3D). The presence of lungs, air-filled mouth cavities, and air-filled middle ear cavities could be employed in this way. For example, sound pressure–induced pulsations of an air-filled middle ear cavity may produce tympanic and ossicular motion that is relayed to the inner ear (Fig. 12.3D). The most pronounced resonations would be produced by sound frequencies close to the natural frequency of the air cavity, which is determined largely by its configuration and volume. Therefore, significant pulsations would tend to be restricted to relatively high frequencies in small animals and relatively low frequencies in large animals. However, the different air cavities of submerged animals would have different natural frequencies, so together they could contribute to pressure transduction over a wider range of frequencies. Hetherington and Lombard (1982) found that air cavities in the bodies of frogs produced significant pulsations over a fairly broad range of frequencies (200 to 3000 Hz), and Christensen-Dalsgard and Elepfandt (1995) found that the middle ear cavity and lungs of the African clawed frog *(Xenopus laevis)* produced significant pulsations from about 600 to 2200 Hz.

Pressure sensitivity is widespread among fishes that have air bladders and use the latter to transduce pressure fluctuations into displacement signals that stimulate inner ear receptors. Pressure transduction also appears to have occurred early on in the evolution of tetrapods. *Ichthyostega*, one of the earliest known tetrapods of the late Devonian, had an expanded, spoon-shaped middle ear ossicle (stapes) that formed part of the wall of an air chamber and may have transferred pressure-induced pulsations of the chamber to the inner ear (Clack et al., 2003).

Transduction of underwater sound pressure for hearing has several potential advantages over the use of displacement energy. This mechanism provides the most sensitive hearing in fishes (Popper and Fay, 1999) as, for a given sound intensity, pressure is greater in water than in air and the pulsations of an air cavity produced by pressure fluctuations can produce very high amplitude (often near-field) displacements. Pressure sensitivity therefore allows detection of sound signals at greater distances from the source. Pressure sensitivity also is better for hearing higher-frequency sounds because displacement energy decreases with increasing frequency. On the negative side, pressure effects on air cavities of a submerged animal may produce problems for normal inner ear function. For example, pulsations of the middle ear cavity could introduce acoustic power into both the oval window and round window (Fig. 12.3D), resulting in no net fluid motion available to stimulate the auditory papillae. Also, pressure is a scalar quantity that imparts no directional information as to the sound source. An air cavity likely would pulsate the same regardless of the direction of the sound waves penetrating the body. The displacement component of sound waves is an inherently directional vector quantity, and, in fishes, positioning hair cells in different orientations within a macula allows the extraction of this directional information from the movements generated between the otoliths and underlying hair cells (Popper and Fay, 1999). However, the more indirect mechanisms of displacement sensitivity described above (Figs. 12.3B and C) that would produce waves in the inner ear fluids to stimulate auditory hair cells (i.e., use of an inertially functioning stapedial footplate or otic capsule wall deformations) also would be problematic for sound localization. The same type of fluid motion could be produced by sound waves penetrating the body from a wide range of directions.

The effect of underwater sound pressure on air-filled spaces in the body of a submerged animal complicates attempts to understand precisely how middle ears that include an air-filled middle ear cavity function underwater. Above, for example, I described how a "terrestrial" middle ear may still serve as a low-attenuation channel of sound energy to the inner ear underwater. However, an air-filled middle ear cavity would likely pulsate in response to underwater sound pressure at least at certain frequencies, producing high-amplitude motions of the ossicular chain. Is sound energy funneled from the lateral head surface to the inner ear, or is tympanic and ossicular motion generated by pressure-induced pulsations, or both?

DIRECTIONAL HEARING

Besides the issue of the mechanism of sound transmission to the inner ear of submerged animals, there also is the issue of sound localization underwater. Terrestrial animals may use interaural differences in timing, phase, intensity, and frequency spectrum of sound signals reaching their left and right ears for sound localization. However, sound travels close to five times faster underwater than in air, making timing differences difficult except for those animals with very large heads. Wavelengths are much longer underwater, making interaural phase differences less pronounced and also decreasing acoustic shadow effects on the side of the head opposite a sound source. Amphibians, reptiles, and birds have inner ears tuned to lower frequencies than mammals, and the longer wavelengths associated with such lower frequencies exacerbate the problems of using interaural phase and intensity differences for determining sound direction. Use of the sound spectrum for determining directionality of aerial sound signals relies on external auditory structures that typically are lacking in aquatic tetrapods (although potentially, internal structures could serve the same role). Given these problems, one might expect auditory specializations for effective mechanisms of sound localization to be especially pronounced in aquatic lineages.

THE AUDITORY SYSTEMS OF AQUATIC AMPHIBIANS, REPTILES, AND BIRDS

AMPHIBIANS

Most amphibians are tied to water, or at least moisture, for most or part of their lives. Many salamanders are entirely aquatic. However, although it is generally assumed that salamanders evolved from amphibian ancestors that possessed a tympanic middle ear, thoroughly aquatic salamanders appear to retain a larval (i.e., aquatic) auditory system via paedomorphosis. Most aquatic salamanders have only one middle ear element, a stapes, that is basically an expanded footplate with a short process loosely attached to the quadrate or squamosal bones of the skull (Jaslow et al., 1988). The stapes likely acts as an inertial element that generates fluid waves in the inner ear in response to the displacement component of underwater sound. In any case, salamanders are not appropriate for the study of the evolutionary transformation of an aerial (i.e., postmetamorphic) auditory system into an aquatic one. In contrast, certain highly aquatic frogs likely evolved from forms that possessed a standard tympanic middle ear (Wever, 1985), and they provide a good opportunity to examine patterns of auditory evolution associated with a return to an aquatic existence.

Although the ears of frogs appear designed primarily for hearing aerial sounds (most vocalize in air), some studies have examined their relative effectiveness for hearing both aerial and underwater sound.

Lombard et al. (1981) analyzed hearing sensitivity in the American bullfrog (*Rana catesbeiana*, family Ranidae) and found that their sensitivity to sound pressure was significantly better in air at frequencies above about 200Hz. However, when sound intensity (in watts per square meter) is used, underwater hearing was considerably better below about 200 Hz and about the same as in air at higher frequencies. Sound intensity is probably the more appropriate measure because it represents the actual energy produced by a sound source. Lombard et al. (1981) suggested that below 200 Hz sound transmission to the low-frequency component of the amphibian papilla was more efficient in water than air and possibly the result of direct displacement effects on hair cells in that part of the papilla. These authors suggested that the low-frequency portion of the amphibian papilla, which also is its most primitive part (Lewis and Narins, 1999), may best be considered an ancestral receptor designed for underwater sound reception. Above 200 Hz

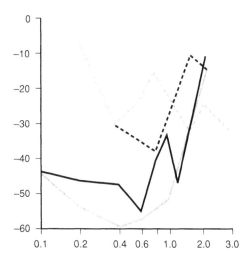

FIGURE 12.4. Underwater auditory threshold curves for the goldfish *(Carassius auratus) (pale solid line)*, the American bullfrog *(Rana catesbeiana) (dark solid line)*, a pipid frog *(Xenopus laevis) (pale dashed line)*, and the American alligator *(Alligator mississippiensis) (dark dashed line)*. The x axis shows frequency in kilohertz, and the y axis shows the auditory threshold in decibels (re 1 Pa). Data for the goldfish (Fay, 1988) and clawed frog (Elepfandt and Gunther, 1986) are based on behavioral measures of sensitivity, the data for the bullfrog are based on multicellular midbrain recordings (Lombard et al., 1981), and the data for the alligator are based on auditory brain stem responses (Higgs et al., 2002).

sound transfer to higher-frequency parts of the amphibian papilla and the basilar papilla was equally efficient in both air and water. This suggests a middle ear that is well suited for sound transfer in both air and water. Figure 12.4 compares underwater hearing sensitivity in the American bullfrog with other nonmammalian aquatic vertebrates for which there are data (although note that thresholds are determined by different techniques), showing that the bullfrog is only slightly less sensitive than the goldfish *(Carassius auratus)*, a fish hearing specialist. Although the bullfrog spends more time in the water than many other frogs, these findings suggest caution when assuming that frog ears are generally adapted for aerial hearing.

PIPID FROGS

Pipid frogs are largely aquatic frogs that only rarely come out on land and potentially are good models for an amphibian lineage reevolving an aquatic ear. It is useful to describe their auditory systems in some detail because, rather than having to rely mostly on speculation based on morphology, considerable experimental work has been done on underwater hearing in the pipid genus *Xenopus*. Pipid frogs produce a variety of vocalizations underwater, so there likely has been selection for effective underwater hearing. The African clawed frog *(Xenopus laevis)* retains a standard tympanic middle ear configuration but displays many unusual features (Wever, 1985; Elepfandt, 1996). There is no recognizable tympanic membrane; the tympanic area consists only of unspecialized skin overlying layers of loose connective tissue and adipose tissue (Fig. 12.5A). Beneath the adipose tissue there is a thick cartilaginous tympanic disk in which is embedded an ossified, relatively thick extrastapes. The latter connects to a relatively large stapes with a cartilaginous footplate. There are no reports of ratios of tympanic disk area to oval window area for *Xenopus*, but the figures presented by Wever (1985) show that the tympanic disk is significantly larger than the oval window. The middle ear cavity is a well-developed, trumpet-shaped space narrowing down to the oval window, and the left and right eustachian tubes are long tubes that extend to the midline before fusing and having a common connection to the mouth cavity. The left and right eustachian tubes of most frogs are short and connect independently to the left and right sides of the pharynx, respectively. The arrangement in *Xenopus* appears to allow the middle ear cavities, and eustachian tubes, to contain air while the mouth is filled with water in submerged animals (Elepfandt, 1996). The Surinam toad *(Pipa pipa)* also has a thick cartilaginous disk beneath its tympanic membrane. The opercular middle ear element, functionally linked to substrate vibration sensitivity in frogs (Hetherington, 1988), appears nonfunctional in pipids; it is very small and fused to the otic capsule in *Xenopus* and is absent in *Pipa* (Wever, 1985).

The inner ear of pipids contains both types of auditory papillae (basilar and amphibian). The basilar papilla is typical in morphology, and the

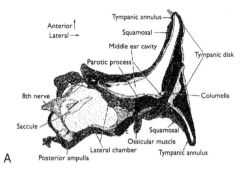

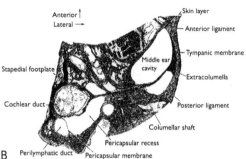

FIGURE 12.5. A frontal view of the ear of a pipid frog, *Xenopus laevis* (A), and the turtle *Trachemys scripta* (B). *Xenopus* is a completely aquatic frog lacking a tympanic membrane and instead having a cartilaginous tympanic disk underneath the lateral head surface. The stapes (columella) is attached to the disk and extends through the air-filled middle ear cavity toward the oval window. In *Trachemys*, the tympanic membrane consists of unspecialized skin. The shaft of the stapes passes through a narrow channel through the quadrate and a pericapsular recess to reach the oval window. Note that the middle ear cavity does not extend all the way inward to the inner ear. Illustration A is from Wever (1985) and illustration B is from Wever (1978).

larger amphibian papilla has the typical "primitive" configuration, lacking a posterior extension observed in more-derived frogs (Wever, 1985). This lack of the posterior projection of the amphibian papilla probably does not represent a secondarily reduced condition, because other closely related terrestrial groups also lack it. So the inner ear of pipids shows no significant difference from that of closely related, terrestrial frogs. Significant variation in the auditory system is largely restricted to the middle ear elements.

Several studies have examined underwater hearing in *X. laevis*. Behavioral studies have established the hearing range for this species to extend from about 200 to 4000 Hz (Elepfandt et al., 2000). *Xenopus laevis* appears to be less sensitive than the American bullfrog *(Rana catesbeiana)* and the goldfish *(Carassius auratus)*, a fish hearing specialist, to low-frequency underwater sound below about 1000 Hz but more sensitive at higher frequencies (Fig. 12.4). It is important to keep in mind that these thresholds were determined by different methods, but these findings suggest that *X. laevis* may have auditory specializations for reception of relatively high frequency underwater sound. Such sensitivity correlates well with the observation that most of the energy in the vocalizations of *Xenopus* is above 2000 Hz (Elepfandt, 1996).

Biophysical studies by Christensen-Dalsgaard et al. (1990) demonstrated that vibration of the tympanic disk produced by underwater sounds was about 30 dB greater than the particle displacement motion associated with those sounds in the experimental water chamber. The vibratory amplitudes of the tympanic disk matched expectations of how an air bubble would pulsate in the sound field, so these authors suggested that stimulation of the inner ear was based on pressure transduction by the air-filled middle ear cavities rather than action of sound impinging on the external tympanic surface. A later study (Christensen-Dalsgaard and Elepfandt, 1995) suggested that pulsations of the middle ear cavities drove tympanic motion at higher frequencies (peaking between 1600 and 2200 Hz) and that pulsations of the lungs (transferred to the middle ear cavity via respiratory passages) drove tympanic motion at lower frequencies (peaking between 600 and 1100 Hz). The middle ear of pipid frogs therefore works very differently underwater than in air. Rather than collecting sound energy at the lateral surface of the head and transferring it to the inner ear via the middle ear elements, sound penetrates the head and causes pulsations of the air-filled middle ear cavity and lungs. These pulsations subsequently produce movements of the tympanic disk that are then transferred via the stapes to the inner ear fluids. Hetherington and Lombard (1982) also found that the air-filled lungs, mouth, and middle ear cavities of frogs pulsate readily in response to underwater sound to the extent that their bod-ies effectively act as an air-water interface when placed in a tube filled with water. The surprisingly high sensitivity of *Xenopus* to the higher-frequency sounds that characterize its vocalizations appears to be based on the use of such pressure transduction mechanisms.

Behavioral studies have found that *X. laevis* is capable of localizing sound underwater about as accurately as terrestrial frogs can localize aerial sound sources (Elepfandt, 1996). This is surprising because, if these frogs are responding to the pressure component of underwater sound, localizing ability should be poor or absent. A middle ear cavity or lung should pulsate equally to sound entering the body from any direction, and the left and right middle ear cavities are interconnected. The ability of *Xenopus* to localize sound suggests they are using bilateral differences in tympanic response amplitudes, and indeed Christensen-Dalsgaard and Elepfandt (1995) found up to a 10 dB difference in tympanic disk vibration between ipsilateral and contralateral sound presentations. But how can these bilateral differences exist if the entire air cavity is pulsating? Perhaps resonation of the middle ear cavities is a complex phenomenon that somehow varies with respect to sound direction. In any case, the available information on underwater hearing in *Xenopus* raises some confounding issues and, as we will see below, stands in contrast to interpretations of underwater hearing in other aquatic tetrapods.

OTHER AQUATIC FROGS

Middle ear morphology also has been examined in *Telmatobius exsul*, an apparently mostly aquatic leptodactylid frog. However, little is known of the natural history of this species, and nothing is known of its acoustic behavior. This species has a reduced tympanic middle ear in which the tympanic membrane has been lost. There is a small tympanic annulus, but the head surface within it consists of unspecialized skin and there is no middle ear cavity (Jaslow et al., 1988). The stapes and extrastapes are slightly built and extend to the head surface. The lack of an air-filled middle ear cavity precludes the pressure transduction model of middle ear function described above for *Xenopus*, although resonations of the mouth cavity and/or lungs could stimulate the inner ear. Alternatively, sound energy could simply be collected at the head surface and channeled to the oval window along the stapes. In any case, the ear of *T. exsul* is sufficiently different from that of pipid frogs to suggest that different lineages of frogs, for whatever historical and functional reasons, may evolve different aquatic auditory systems. However underwater acoustic behavior, so well developed in *Xenopus*, may be minimal in species such as *T. exsul*.

TURTLES

Most turtles are amphibious creatures spending most of the time in water but also spending considerable time basking and moving about on land. There is no fossil evidence establishing that their terrestrial ancestors had well-developed tympanic middle ears, but as we shall see, the presence of a modified tympanic middle ear suggests this ancestral condition. Besides amphibious forms, living turtles include groups that are completely terrestrial (e.g., tortoises) and groups that are almost entirely aquatic in habits (e.g., sea turtles). The latter is the most appropriate group to examine for specializations for underwater hearing. These animals spend most of their lives in water, except for an initial mad dash from nesting site to sea and (for females) a return to a beach to lay eggs. In any case, turtles provide an opportunity to examine ears in a lineage with terrestrial, amphibious, and aquatic members that might provide insights into trends associated with a return to an aquatic lifestyle. However, our understanding of hearing in turtles is severely limited by the fact that almost all electrophysiological studies of turtle hearing have been conducted in air.

TYPICAL TURTLE EAR

The tympanic membrane of turtles typically is much thicker than that of most tetrapods (Fig. 12.5B). The extrastapes flattens into a broad plate beneath the tympanic membrane, and a thick layer of connective tissue lies between this plate and the outer layer of skin (Wever, 1978). The stapes is a thin element that extends medially from the middle ear cavity and through a narrow canal in the quadrate before reaching a large fluid-filled space (the pericapsular recess) lateral to the oval window (Wever, 1978). The stapes passes through the pericapsular recess to the oval window, where it expands into a large footplate (Fig. 12.5B). Turtles lack a round window, and the pericapsular recess appears to be part of the flow system for sound energy in the inner ear. Sound energy delivered at the oval window flows through the inner ear and eventually into the pericapsular recess, where it dissipates. Tympanic to oval window area ratios are relatively low in turtles (e.g., about 8.5 in a generalized pond turtle *Chrysemys picta* [Wever, 1978]). The stapedial footplate of turtles has the unique feature of being directly connected to the saccule by fibrous connective tissue strands (Wever, 1978). Presumably, movements of the footplate can stimulate hair cells within the saccular macula, an acoustic receptor of low-frequency sound and vibrations in other tetrapods (Lewis and Narins, 1999). Footplate movements also presumably produce fluid motion at the oval window that eventually stimulate hair cells of the basilar papilla. The middle ear cavity of turtles does not extend to the oval window as in most tetrapods and is largely encased by posterior extensions of the quadrate and/or squamosal bones of the skull (Fig. 12.6). This encasement of the middle ear cavity is not observed in the earliest turtles of the late Triassic (family Proganochelyidae) and other Mesozoic turtles, such as *Archelon* (Fig. 12.7). The habits of proganochelyids are presumed to be amphibious, and *Archelon* was a completely aquatic marine form. *Chisternon* from the Eocene shows a more developed, but still incomplete, wrapping of the cavity. The functional significance of this bony encasement of the middle ear cavity is not clear. It clearly could provide effective support for the margin of the tympanic area, but most tetrapods provide such support without encasing the entire middle ear

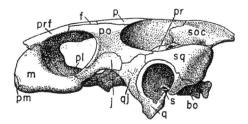

FIGURE 12.6. The skull of the turtle *Emys europaea*. Note the encasement of the middle ear cavity by the quadrate (q) and the stapes (s), extending through a narrow opening toward the otic capsule. The tympanic membrane (tympanum) lies over the conical shaped depression in the quadrate. Other abbreviations: bo, basioccipital; f, frontal; j, jugal; m, maxilla; p, parietal; pl, palatine; pm, premaxilla; po, postorbital; pr, prootic; prf, prefrontal; qj, quadratojugal; soc, supraoccipital; sq, squamosal. Illustration from Romer (1956).

cavity in bone. It could represent an adaptation for underwater hearing by somehow minimizing bone conduction of sound to the inner ear and restricting sound reception and transmission to the tympanic surface and middle ear. This could potentially improve sound localization by allowing interaural comparisons of sound cues. Enclosure of the middle ear cavity also might prevent its collapse in the face of high pressures associated with diving. The pat-

tern is observed in terrestrial tortoises as well (Romer, 1956), so it also could simply represent a trend of heavy ossification of the skull of turtles. However, a similar pattern of middle ear encasement is found in mosasaurs, thoroughly aquatic lizards of the Late Cretaceous (see below). Although the functional significance of encasement of the turtle middle ear cavity is not clear, some relationship to underwater hearing seems likely.

The basilar papilla of turtles is similar to that of other reptiles except that a significant number of hair cells rest on dense connective tissue (Fig. 12.8) rather than a thin basilar membrane (Wever, 1978). These hair cells, termed limbic hair cells, may be stimulated by movement of their overlying tectorial membrane rather than movement of the membrane on which they are anchored. This would make their pattern of stimulation comparable to that of hair cells of the basilar and amphibian papillae of amphibians. Given that the primitive part of the amphibian papilla has been hypothesized to represent an ancestral receptor for underwater hearing (Lombard et al., 1981), the limbic hair cells of turtles similarly may be related to underwater hearing. This possibility is further

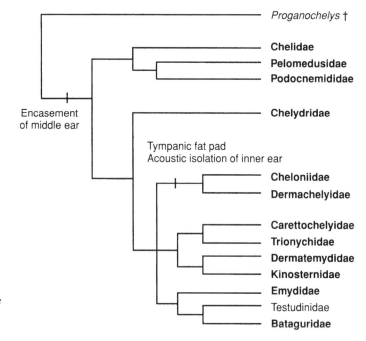

FIGURE 12.7. Evolutionary hypothesis for hearing characters in turtles. Aquatic turtles *(bold)* all share encasement of the middle ear, although this is a plesiomorphy for all modern turtles. Seaturtles (Cheloniidae and Dermatochelyidae) display additional aquatic features. The family names Bataguridae and Geoemydidae are synonyms.

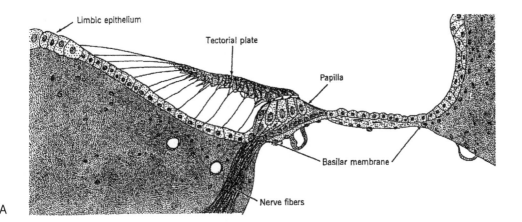

A

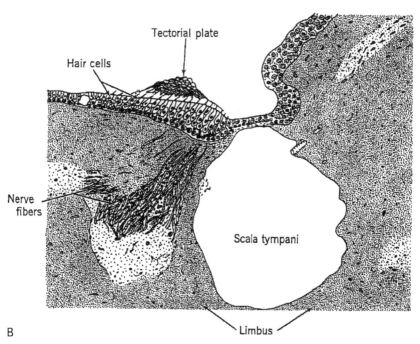

B

FIGURE 12.8. Two sections through different points along the basilar papilla of the turtle *Trachemys scripta* showing (A) hair cells of the papilla resting on the basilar membrane (the typical condition in tetrapods) and (B) hair cells of the papilla sitting instead on dense (limbic) connective tissues to one side of the basilar membrane. Illustration from Wever (1978).

discussed below in relation to the inner ears of sea turtles.

The ears of completely terrestrial turtles (e.g., tortoises) are very similar to those of more amphibious ones. This finding is problematic for the typical view that the ears of turtles represent a compromise for both aerial and underwater hearing sensitivity. Wever (1978) suggested that the thicker, more massive tympanic membrane would provide better coupling to water, although it would come at the expense of aerial hearing sensitivity. Wever (1978) found that turtles are quite sensitive to lower-frequency aerial sound below about 1000 Hz but are quite insensitive to higher frequencies, and he emphasized the latter finding as evidence of a generally reduced aerial sensitivity. However, this pattern of sensitivity may only suggest that turtle ears are specifically adapted for low-frequency sensitivity. The heavier tympanic membrane, for example, should increase responsiveness to low-frequency signals at the expense of higher frequencies. High-frequency sensitivity may be especially important for detecting vocalizations that often consist of frequencies that stand out above low-frequency environmental noise (Wiley and Richards, 1978). Turtles rarely vocalize, and those that do produce relatively low-frequency grunts (Gans and Maderson, 1973), so selection may favor sensitivity to low rather than high sound frequencies. Assertions about middle ear tuning characteristics based on auditory sensitivity must be made cautiously, however. Ruggero and Temchin (2002) measured sound-induced motion of the stapes of the red-eared turtle *Trachemys scripta* and found that the middle ear of this species vibrates quite effectively at high frequencies well beyond the range of hearing sensitivity measured by electrophysiological and behavioral methods. These workers suggested that the low-frequency sensitivity of turtles is a result of constraints within the inner ear. In other words, the middle ear of turtles can readily transmit higher-frequency sound energy, but the inner ear is not designed to respond to it. This finding demonstrates the limitations of assessing hearing sensitivity based solely on patterns of middle ear mor-

phology. Without physiological and/or behavioral measurements of hearing sensitivity of turtles conducted in both air and water, it is not possible to consider the generalized turtle ear as adapted for either aerial or aquatic hearing or both.

Lenhardt et al. (1983) suggested that the shell of turtles functions in bone conduction of sound to the inner ear. These researchers demonstrated that vibrations applied to the shell generate auditory brain stem responses. However, the amplitude of responses to shell vibration was not compared to the amplitude of responses to airbone sound, so it is not clear if bone conduction of sound from the shell is more or less effective than the middle ear as a route of sound conduction. Bone conduction of vibratory energy applied to the body or skull likely can be observed in any animal (Tonndorf, 1972), so it is important to directly compare the sensitivities of different putative routes of sound reception.

SEA TURTLES

The tympanic membrane of the sea turtles studied to date (genera *Chelonia*, *Caretta*, and *Lepidochelys*) is inconspicuous (Fig. 12.9), covered with normal skin and scales, and underlain by a thick layer of connective tissue filled with fatty tissue (Ridgway et al., 1969; Wever, 1978; Lenhardt et al., 1985). Computed tomography scans and magnetic resonance imaging of this fat layer suggest a density similar to fat bundles found in the mandible of odontocete

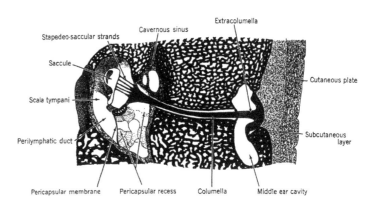

FIGURE 12.9. Semischematic representation of the ear of the green sea turtle *Chelonia mydas*. Note the thick tissue layers of the tympanic area (including layers of subcutaneous fat) and the expanded extracolumella (extrastapes) that forms a cartilaginous disk. The middle ear cavity is small, and the columella (stapes) extends through thick layers of cancellous bone to reach the inner ear. Illustration from Wever (1985).

whales (Ketten et al., 1999). It is likely that this fat layer acts as a low-impedance channel for conduction of sound to the inner ear (Ketten et al., 1999), although Lenhardt et al. (1985) also propose that when submerged, water pressure would cause the fat to push air out of the middle ear cavity. The outer surface of the tympanic membrane of sea turtles is hard and horny, also reminiscent of the thin bone layer over the mandibular fat pad of odontocetes. The convergence ratio of tympanic to oval-window area in *Chelonia mydas* is reported be about 3 (Lenhardt et al., 1985), although these authors considered it difficult to estimate the functional area of the inconspicuous tympanic region. However, it appears that the convergence ratio is quite low compared to that of less aquatic turtles, whose ratios were typically greater than 8. The low ratio possibly reflects the general impedance match between tissue and seawater. The middle ear cavity is small (Fig. 12.9), and Lenhardt et al. (1985) proposed that underwater it may be compressed such that there is little air remaining. As in most turtles, the middle ear cavity is completely encased by bone (mostly the quadrate). The inner ear of sea turtles also is surrounded by bone, although it is cancellous rather than dense in nature. Lenhardt et al. (1985) suggested that this pattern could acoustically isolate the inner ears in a manner analogous to the foam-filled sinuses surrounding the otic capsules of odontocete whales (Ketten, 2000). They suggested that bone-conducted sound passing through other denser skull bones would be reflected at the interface of the blood-filled cavities of the cancellous bone. Therefore, the pattern of ossification of the otic region, and bony encasement of the middle ear cavity, may limit sound reception in sea turtles to the tympanic area and sound transmission to the middle ear pathway. The pericapsular recess of sea turtles is heavily invaded by connective tissue, and this may further impede sound transfer to the inner ear other than through the stapes.

The basilar papilla of the sea turtle *Chelonia mydas* is relatively large, but there has been no allometric analysis to determine if it scales differently from that of other turtles. Otherwise, the auditory papilla is quite comparable to that of other turtles, except for one feature. The papilla has a disproportionately large number of limbic hair cells situated on dense connective tissue rather than a thin basilar membrane (Wever, 1978). Interestingly, Wever (1978) found very few limbic hair cells in the basilar papillae of mostly terrestrial turtles (e.g., the box turtle *Terrapene carolina* and thoroughly terrestrial tortoises *Testudo horsfieldi* and *Kinixys belliana*). Limbic hair cells also have been found in the basilar papillae of varanid (monitor) lizards that often show amphibious tendencies, and in the thoroughly aquatic snake *Acrochordus javanicus* (Wever, 1978). These observations suggest that limbic hair cells may be especially important for underwater hearing, and they may represent the only obvious specialization of the inner ear of sea turtles for aquatic hearing. The correlation of limbic hair cells with aquatic habits is not perfect, as they also are found in the basilar papillae of birds (Gleich and Manley, 2000), but a functional relationship of limbic hair cells with underwater hearing should be further examined.

From the review of the morphology above, it seems likely that sound reception in sea turtles involves the standard tympanic middle ear path. The dense fat layer beneath the tympanic membrane suggests that sound would be conducted along the stapes and into the oval window, and the encasement of the middle ear cavity and inner ear in bone further suggests restriction of sound input to the oval window. Other hypotheses of hearing, such as bone conduction, have been proposed for sea turtles (Lenhardt et al., 1985), but the potential acoustic isolation of the otic capsules and the lack of a heavy, inertially sensitive stapedial footplate suggest sound transmission occurs via the middle ear. Restriction of sound reception to the tympanic region may be important for localizing underwater sound. Although sound localization is more difficult underwater, the relatively large heads of sea turtles might allow the utilization of directional cues provided by sampling sound signals at the two ears. It also is possible, as demonstrated in the pipid frog *Xenopus laevis* (Elepfandt, 1996) and other frogs (Hetherington and Lombard, 1982), that the air-filled spaces of the middle ear cavity, mouth, and/or lungs could act as pressure transducers to generate fluid motion within the inner ear. The retention of an air-filled middle ear cavity in a thoroughly aquatic group may suggest that pressure transduction plays some role in hearing. Sea turtles therefore appear to retain a tympanic middle ear system for hearing underwater, either as a pathway of sound transfer from the lateral head surface or as a pressure transducer of underwater sound.

LEPIDOSAURS

MODERN LIZARDS

No living lizards are completely aquatic, although several forms are amphibious in their habits. However, little is known about the ears of these amphibious species. The marine iguana (*Amblyrhynchus cristatus*, family Iguanidae) of the Galapagos Islands submerges to feed for long periods of time, but there has been no study of its auditory structures. Many other lizards, such as monitor lizards (family Varanidae) frequently swim in water but nonetheless spend most of their time on land. Their middle and inner ears are well developed and display no obvious adaptation for underwater hearing (Wever, 1978).

MOSASAURS

Mosasaurs were large marine varanid lizards of the Late Cretaceous that probably were completely aquatic forms incapable of terrestrial locomotion. The mosasaurs' ear appears to be similar to that of their close relatives, the monitor lizards of the family Varanidae (Rieppel and Zehar, 2000). Pronounced suprastapedial and infrastapedial processes of the quadrate curved over the top and bottom of the stapedial shaft respectively (Fig. 12.10) and likely enclosed an air-filled middle ear cavity, as observed in turtles (Rieppel and Zehar, 2000). The quadrate bears a deep pit into which an internal process of the

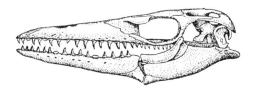

FIGURE 12.10.   Lateral view of the skull of the mosasaur *Clidastes* showing the encircling of the middle ear cavity by extensions of the quadrate bone. Modified after Romer (1956).

stapes likely inserted. The functional significance of the bony encasement of the middle ear cavity is unclear (see discussion above concerning turtle ears). Nonetheless, it suggests the presence of an air-filled middle ear cavity beneath a tympanic structure of some sort, and there is evidence that the tympanic membrane was ossified (Caldwell, pers. comm.). Potentially, the quadrate configuration helped to restrict sound reception to the tympanic area and sound transmission to the middle ear pathway. This could potentially improve sound localization by allowing interaural comparisons of sound cues. Mosasaurs therefore possessed a tympanic middle ear system, although it was probably modified in various ways. The mosasaur skull shows no evidence of any isolation of the otic region from other areas of the skull that might impede bone conduction of sound directly to the inner ear. The oval window is relatively larger than that of varanid lizards, and a groove that probably held the stapedial shaft extends lateroventrally from it. The large size of the oval window raises the possibility that the stapedial footplate could act as an inertial element in underwater hearing. If the tympanic element was ossified, this would increase the mass of the middle ear system and improve inertial sensitivity to bone-conducted sound. However, given the presence of a tympanic middle ear configuration, it seems more likely that mosasaurs utilized the ancestral middle ear pathway for sound reception underwater.

SNAKES

Snakes lack a standard middle ear, and their inner ears respond mostly to very low frequency sound (Hartline, 1969; Wever, 1978). The middle ear of snakes consists only of a large stapes, loosely connected to the quadrate, and often with an expanded, platelike footplate fitting against the oval window. This element likely is adapted to function as an inertial element responding to low-frequency acoustic energy (vibrations) entering the body from the substrate, but it also seems to function to hear low-frequency sounds that penetrate the body tissues (Hartline, 1969). This arrangement also should be effective for detecting underwater sound signals directly penetrating the head of submerged snakes. Although snakes traditionally are considered to have evolved from fossorial lizard ancestors (Zehar and Rieppel, 2000), it also is possible that their ancestors were marine forms (Caprette et al., 2004). Therefore, it is not clear if the snake middle ear originally was adapted for fossorial or underwater hearing. If the immediate ancestors of snakes were indeed marine forms, the snake middle ear is a good example of the evolutionary transformation of a standard terrestrial middle ear into an inertially sensitive aquatic middle ear. The latter subsequently was retained by later terrestrial snakes for detection of substrate vibrations and low-frequency aerial sound.

The ears of snakes that are fully aquatic, such as the sea snake *Pelamis platurus* (family Elapidae) or elephant trunk snake *Acrochordus javanicus* (family Colubridae), have middle and inner ears that are comparable to terrestrial snakes (Wever, 1978). This is not surprising, as the inertially sensitive middle ear system should work well on a substrate, underground, and underwater.

CROCODILIANS

Crocodilians are amphibious animals, spending much time floating at the water surface but also spending considerable time basking on land or guarding nest sites. Some marine forms were completely aquatic and possess flippers rather than limbs, but detailed descriptions of their auditory structures are limited (see below). Extant crocodilians are vocal animals, producing a variety of vocalizations involved in

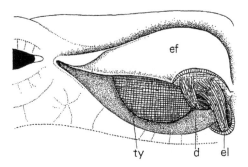

FIGURE 12.11. Left side of the head of an American alligator *(Alligator mississippiensis)* showing the tympanic membrane (ty), the superior ear flap (ef), and the muscles that elevate (el) or depress (d) it. A smaller inferior earlid that also can be elevated and lowered is not shown. Modified after Shute and Bellairs (1955).

territorial and mating behavior (Garrick and Lang, 1977; Garrick et al., 1978; Vliet, 1989). They have specialized external muscular flaps (Fig. 12.11) both above and below the opening of the external auditory canal that reflexively seal off the external auditory meatus upon submergence (Shute and Bellairs, 1955; Garrick and Saiff, 1974). The evolutionary origin of these derived structures has not been studied. Extant crocodilians have relatively normal middle ears, very comparable to those of their fellow archosaurs, birds (Gleich and Manley, 2000). The stapes of a marine dryosaurid crocodilian from the Cretaceous has been described as a slender rod about 2 mm in diameter, flaring distally to likely articulate with a cartilaginous extrastapes (Brochu et al., 2002). Therefore, the only middle ear evidence from a completely aquatic crocodilian suggests retention of a relatively light, rodlike stapes that likely did not function as a heavy inertial element but rather was involved in sound transfer to the inner ear. The middle ear cavity of crocodilians is a complicated space largely encased in bone, and the left and right cavities have numerous connections across the skull. Several channels pass medially to join their counterparts from the other side and form a common median tube that then opens into the pharynx. Other channels extend across the roof of the skull to connect to their counterparts from the other side.

Other extensions of the middle ear cavity penetrate deep into the quadrate and even the articular bone. These remarkably complex channels have long puzzled morphologists. Even the great comparative anatomist Richard Owen speculated about their function, suggesting that these channels may have allowed pressure equalization between the cavities during diving (Owen, 1850). However, this is generally dismissed because it does not explain the need for such a complex system of tubular connections (Wever, 1978). The eustachian tubes of the aquatic frog *Xenopus laevis* described above are similar in certain respects to those of crocodilians. In *Xenopus* the left and right eustachian tubes join in the midline and enter the pharynx via a single median opening (Wever, 1985). Might this pattern have some functional significance for underwater hearing? This possibility is weakened by the fact that these complicated channels are an ancient feature of crocodilians, found in some of the earliest members of the group, such as the Triassic protosuchians (Langston, 1973). These early crocodilians were not aquatic animals, so these middle ear channels presumably are not adaptations for aquatic hearing. The middle ear cavities of many birds also have numerous extensions into surrounding bones, so this general pattern may be an ancestral archosaurian feature. The functional basis of these complicated middle ear channels remains uncertain and a good candidate for study. Study of the eustachian channels of completely aquatic fossil groups of crocodilians may provide comparative information relevant to formulating hypotheses of functions in hearing. For example, the eustachian channels of a marine dryosaurid crocodilian do differ from living crocodilians in certain respects, although the differences provide no obvious insight into their function (Brochu et al., 2002).

The inner ear of crocodilians is surrounded by a thoroughly ossified otic capsule that is not isolated from other parts of the skull (Wever, 1978). Potentially, bone-conducted sound could penetrate directly to the inner ear through various skull pathways. The crocodilian basilar papilla is relatively long compared to that of other reptiles, and very similar to that of birds. Its size is comparable to that of large birds such as the Ostrich *(Struthio camelus)* (Gleich and Manley, 2000). Klinke and Pause (1980) studied aerial hearing in the spectacled caiman *Caiman crocodilus* by recording from single nerve fibers in the auditory nerve and found that hearing sensitivity was comparable to that of most birds except that sensitivity was reduced at relatively high frequencies (over about 5000 Hz). Higgs et al. (2002) studied hearing sensitivity of the American alligator *(Alligator mississippiensis)* in both air and water using auditory brain stem responses. The frequency range of hearing was greater in air (100 to 8000 Hz) than underwater (100 to 2000 Hz), and peak sensitivity occurred at 1000 Hz in air and 800 Hz underwater. Aerial hearing sensitivity was better than that of many birds at frequencies below about 1000 Hz, but less sensitive at higher frequencies, suggesting an ear tuned to lower frequencies. Vocalizations of adult crocodilians tend to be in this lower frequency range, and the energy in distress calls made by juvenile alligators is centered around 900 Hz and matches well with peak aerial sensitivity in the adults. Underwater hearing sensitivity of the alligator is compared to that of some other vertebrates in Figure 12.4. The alligator is generally less sensitive than the goldfish *(Carassius auratus),* a fish hearing specialist, the American bullfrog *(Rana catesbeiana),* and African clawed frog *(Xenopus laevis).* However, thresholds determined by the auditory brain stem response technique used for the alligator often are 10 to 30 dB higher than behavioral thresholds (Gorga et al., 1988) used for the goldfish and the clawed frog, so the alligator potentially could be comparably sensitive to these species over a wide range of frequencies. The alligator is most sensitive to lower-frequency underwater sound below about 1000 Hz. Overall, alligators appear to hear quite well in air but comparably less so underwater, suggesting that their auditory systems are designed more for aerial rather than underwater hearing.

What mechanism(s) are alligators using for hearing underwater? Higgs et al. (2002) found that hearing thresholds were the same whether or not the auditory meatus was filled with air. Because an air bubble trapped against the tympanic membrane should effect sound transfer through the middle ear, this observation prompted them to argue that the middle ear of alligators does not act as a pathway for underwater sound reception. They argued instead that alligators hear by bone conduction (i.e., sound directly passes through the head tissues and skull to reach the inner ear and stimulate the auditory hair cells via the displacement component of the sound energy). Higgs et al. (2002) therefore considered the ear flap to be an adaptation to prevent water from entering the auditory meatus (rather than an adaptation for underwater hearing), and the alligator middle ear to be adapted primarily for aerial hearing. However, although they removed air from the external auditory meatus, the middle ear cavity (and the mouth cavity and lungs as well) retained air and potentially could transduce underwater sound pressure signals, as suggested by studies on *X. laevis* (Elepfandt, 1996) and other frogs (Hetherington and Lombard, 1982). Removal of air from the middle ear cavity would be required to adequately test the precise role of the middle ear in underwater hearing in alligators. The basilar papilla of crocodilians is basically the same as that of terrestrial birds and may respond best to fluid motion entering the inner ear via middle ear elements. The middle ear could simply transfer sound energy from the lateral head surface or, perhaps more likely, could function as part of a pressure-transducing system of air-filled spaces. The precise mechanism of underwater hearing in crocodilians remains an open question.

Although not related to the auditory system, all living crocodilians also have unique sensory receptors for detecting water disturbances at the air-water interface called dome pressure receptors (Soares, 2002). These receptors lie in the skin of the snout and sides of the jaws and respond to fluctuations in water pressure caused by surface movements of water (Soares, 2002). Crocodilians can use these receptors to orient to water drops in complete darkness. The skulls of crocodilians show foramina matching the branches of the trigeminal nerve that innervates these receptors. Crocodilians initially were a terrestrial lineage, but by the Early Jurassic had evolved semiaquatic habits. By comparing the distribution of these foramina, Soares (2002) was able to show that the origin of these receptors occurred along with the evolution of aquatic habits. These findings serve as a reminder that aquatic tetrapod lineages may evolve sensory systems to detect acoustic signals that do not involve the inner ear. Hair cells do not hold a monopoly on the detection of acoustic signals by vertebrates.

ICHTHYOSAURS

Ichthyosaurs were completely aquatic mesozoic reptiles that show no evidence of a tympanic middle ear (Romer, 1956; McGowan, 1973). It is not clear if their terrestrial ancestors possessed a tympanic middle ear, so ichthyosaurs may not be good models for the evolutionary transformation of a well-developed terrestrial ear into an aquatic one. The stapes (hyomandibula) is a stout structural element bracing the braincase (Fig. 12.12).

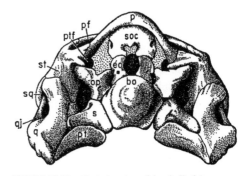

FIGURE 12.12. Posterior view of the skull of the ichthyosaur *Ophthalmosaurus*. Note the massive stapes. Abbreviations: bo, basioccipital; eo, exoccipital; op, opisthotic; p, parietal; pf, postfrontal; pt, pterygoid; ptf, posttemporal fenestra; q, quadrate; qj, quadratojugal; s, stapes; soc supraoccipital; sq, squamosal; and st, supratemporal. Illustration from Romer (1956).

The large stapes potentially could function as an inertial element for detection of displacement energy. Hearing could depend on bone conduction of sound to generate different motion between the heavy stapes and inner ear fluids. However, the mouth cavity and lung also could function as pressure transducers at low frequencies, introducing high-amplitude displacements directly into the inner ear or causing inertial movements of the stapes at the oval window.

SAUROPTERYGIANS

Sauropterygians include such extinct groups as the nothosaurs, which likely were amphibious in nature, and plesiosaurs, which were obligate aquatic forms. Some nothosaurs show evidence of a tympanic notch in the back of the quadrate, and in others there is a pronounced concavity of the quadrate and squamosal bones, suggesting a relatively large middle ear cavity (Carroll and Gaskill, 1985; Lin and Rieppel, 2000). A stapes has been found in only one nothosaur *(Neusticosaurus)*, and it was described as a short, cylindrical rod (Carroll and Gaskill, 1985). Regarding plesiosaurs, although Williston (1907) reported a short, stout stapes in a pliosauroid plesiosaur, this element was not found in situ. Taylor (1992) described a very different type of stapes found in situ in the plesiosaur *Rhomaleosaurus*. The stapes is remarkably delicate, a long, curved, and thin element that was found extending from the oval window toward the quadrate. Its total length, in this animal with a skull about 85 cm long, was close to 10 cm. Medially the stapes was a narrow rod and distally it became leaflike. The medial end of the stapes displayed no expanded footplate, although bone fragments nearby potentially could represent part of one. The leaflike distal end of the stapes likely articulated with a cartilaginous extrastapes (Taylor, 1992). There is no otic notch at the back of the skull of *Rhomaleosaurus* that would support a tympanic membrane. Taylor (1992) suggested that a large depressor mandibulae muscle may have left little room for a tympanic membrane, and he proposed that a tympanic membrane was absent

and that the stapes-extrastapes attached instead to the quadrate or hyoid. Although it is not clear whether the terrestrial ancestor of plesiosaurs had a tympanic middle ear, the presence of a lightly built stapes suggests this. If the stapes had been a stout structural element, it seems unlikely to have evolved into the type of delicate structure observed in *Rhomaleosaurus*. The relatively lightly built stapes clearly is ill suited to function as an inertial element for underwater sound reception, and its retention as a long, thin element suggests a continued role in sound transmission. Even if a tympanic membrane was absent, an extrastapes may have been linked to the lateral head surface and transferred sound energy from the head surface to the oval window. Alternatively, if a middle ear cavity existed, it could produce stapedial motion via pressure transduction of underwater sound. There is no evidence of acoustic isolation of the otic capsule from the rest of skull, and in fact the otic region of *Rhomaleosaurus* forms part of the load-bearing structure of the skull (Taylor, 1992). However, the heads of plesiosaurs potentially were large enough that even underwater, interaural cues related to differential sound transfer through the left and right middle ears could allow sound localization.

BIRDS

The auditory systems of aquatic birds appear to be very similar to those of terrestrial birds (Schwartzkopff, 1957; Saiff, 1976, 1978; Kuhne and Lewis, 1985). It is likely that hearing in air is more important than hearing underwater, as reproduction and communication occur on land and feeding underwater seems largely mediated by vision. Some specializations, however, to prevent water from entering the external auditory canal are present in aquatic forms. The external ear opening is generally small (e.g., only 0.1 mm in diameter in the cormorant *Phalacrocorax bougainvillei*) (Saiff, 1978), and feathers that extend posteriorly over the external ear opening are dense and overlap, so that upon submergence water pressure will seal off the ear opening. The external ear opening also may be offset from the tympanic membrane, leaving a space in the external ear canal to accommodate the flattening of feathers (Kartaschew and Iljitschew, 1964).

These modifications to prevent the entrance of water into the external auditory canal are adaptations to preserve aerial hearing sensitivity. The only report of a modification of the auditory system to improve underwater hearing is from Ketten et al. (1999) who found that the external auditory canals of two species of sea birds (taxa unspecified) were sheathed in fat with a density similar to that of seawater. This fat sheath could function as a low-impedance path for underwater sound to the tympanic membrane.

Aquatic birds also tend to have lower ratios of tympanic to oval-window area. Schwartzkopff (1957) observed that whereas most birds have ratios from about 20 to 30, ratios in grebes were about 15 and ratios in ducks were not greater than 20. Kartaschew and Iljitschew (1964) found ratios of about 5 to 13 in diving birds of the alcid family (auks). These lower ratios suggest some compromise for hearing underwater, where the impedance match between the body and surrounding water does not require as much amplification of acoustic power. The inner ears of aquatic birds do not appear to be significantly different from those of terrestrial forms (Kartaschew and Iljitschew, 1964).

Overall, aquatic birds seem to show few auditory modifications for underwater hearing. The specializations of their external auditory canals serve to protect aerial hearing sensitivity rather than facilitate underwater hearing. Underwater sound conduction along the middle ear may be facilitated by fat channels and other modifications of middle ear elements, but the middle ear appears to function in basically the same manner underwater as in air.

TRENDS IN EAR TRANSFORMATIONS

Understanding the evolutionary transformations of the ears of secondarily aquatic tetrapods is complicated by a general lack of experimental studies on the actual mechanisms used for underwater hearing in the various lineages. As described in this chapter, a multitude of possibilities exist for hearing underwater, and different mechanisms may act in the reception of different sound frequencies. Examination of morphology alone usually is not sufficient to determine the specifics of auditory function. Nonetheless, certain trends can be discerned. One general pattern is the retention of the tympanic middle ear even in forms that are completely aquatic (e.g., pipid frogs and sea turtles). This may seem surprising given that the impedance-matching function of a tympanic middle ear is much less significant underwater. Certain modifications (e.g., presence of a cartilaginous tympanic disk in lieu of a typical tympanic membrane, a decrease in the ratio of tympanic to oval-window area, etc.) are observed, and these indeed may be related to the close match between the acoustic impedances of water and tissue. However, the basic design of the middle ear remains the same. Maintenance of a tympanic middle ear is expected in amphibious forms that likely need to maintain effective aerial hearing, but the retention of such an ear in completely aquatic forms would not necessarily be expected.

The other general trend is the apparent retention of the basic mechanism of stimulation of inner ear auditory receptors. The general morphology of the auditory papillae is conserved in aquatic tetrapods, suggesting that stimulation relies on fluid waves passing through the perilymphatic channels. Increases in the density of the tectorial structures overlying the auditory hair cells, which could increase sensitivity to displacement energy reaching the inner ear via bone conduction, are not observed. The actual mechanisms producing stimulatory fluid waves remains unclear, although stapedial footplate motion in the oval window is likely involved. Footplate motion may be produced by sound transmission along the ossicular pathway from the tympanic membrane or lateral head surface, by pressure-induced pulsations of the middle ear cavity and/or other air-filled spaces in the body, or by inertial effects of bone-conducted sound on the stapes. The former two are most likely, because there would be little selective value in retaining a tympanic structure and air-filled middle ear cavity if these elements were not involved in sound reception. Reduction or complete loss of tympanic middle ears is observed in many lineages of terrestrial amphibians and reptiles (Jaslow et al., 1988; Hetherington and Lindquist, 1999), so the evolutionary loss of these elements does not seem to be constrained. Their presence in fully aquatic tetrapods suggests that all of the elements are functionally involved in sound reception. Bone conduction of sound directly to the inner ear, therefore, while possibly involved, is not the only mechanism of underwater sound detection.

Studies of underwater hearing in amphibians suggest that the retention of an air-filled middle ear cavity in aquatic forms may be related to pressure transduction of underwater sound signals. Pressure transduction, in which sound pressure waves penetrating the body cause high-amplitude pulsations of an air cavity, may extend underwater hearing sensitivity to relatively high frequencies. The African clawed frog, *Xenopus laevis*, appears to use its middle ear cavity to transduce higher-frequency sound up to 4000 Hz, beyond the effective hearing range of many other amphibians and reptiles. This high-frequency sensitivity in this species may be especially important for detection of intraspecific vocalizations that contain significant energy above 2000 Hz. Other air cavities within the body (e.g., mouth cavity, lungs) may, depending on their volumes and other characteristics, transduce lower-frequency sounds. Therefore, a functional tympanic middle ear, combined with a system of interconnected air cavities, can effectively serve in the reception of a wide range of sound frequencies underwater. It should be noted, however, that the hearing range of nonmammalian aquatic tetrapods, including *Xenopus*, is restricted to much lower frequencies than that of cetaceans, which utilize high sound frequencies (to over 100 kHz) for echolocation (Fay, 1988).

Crocodilians and aquatic birds show relatively little modification of the tympanic middle ear, likely because of selection for aerial hearing sensitivity. The American alligator shows no pronounced sensitivity to underwater sound, suggesting little specialization for underwater hearing. Both the crocodilian and avian lineages have evolved different strategies for keeping water out of the external auditory canal and thereby maintaining acute aerial hearing after emergence from the water. Crocodilians have dorsal and ventral muscular flaps that seal off the canal, whereas sea birds use elongate feathers that collapse under the effect of water pressure to the same effect.

The issue of sound localization by aquatic amphibians, reptiles, and birds remains problematic. Only sea turtles display any potential acoustic isolation of the inner ears from bone-conducted sound penetrating the skull, and there is no experimental evidence establishing that such isolation exists. Nonetheless, sound energy may be preferentially transmitted along the middle ear pathway of most aquatic tetrapods, and in large animals this may provide sufficient interaural cues for localization. Pipid frogs display behavioral abilities to localize sounds, and their small size makes this a remarkable achievement. This feat is even more remarkable given that pressure transduction by their middle ear cavity and lungs appear to be responsible for their underwater hearing. Pressure is a scalar quantity that should contain no directional information about a sound's source, so this case serves as another reminder of how good animals are at confounding biologists. Future studies may find that aquatic tetrapods are far better at sound localization than we would expect.

Overall, the auditory systems of aquatic tetrapods provide a good example of the role of contingency in evolution. These lineages began with terrestrial auditory systems and retain many of the elements in the new medium. The configuration of the inner ear and mechanism of hair cell stimulation appear highly conserved, and the middle ear elements are retained to input acoustic energy in a manner comparable to that in air. The conservation of the terrestrial design may have been facilitated by a long "amphibious" phase during which sensitivity to aerial sound had to be maintained. However, some aquatic lineages are ancient (e.g., sea turtles), and hearing aerial sound likely has been of no real selective value. Despite their terrestrial hearing heritage, some aquatic tetrapods that have been studied are nonetheless quite good at hearing underwater sound and display a sensitivity comparable to that of the most sensitive fishes studied to date.

LITERATURE CITED

Aertsen, A. M. H. J., M. S. M. G. Vlaming, J. J. Eggermont, and P. I. M. Johannesma. 1986. Directional hearing in the grassfrog *(Rana temporaria)*. II. Acoustics and modelling of the auditory periphery. Hearing Research 21:17–40.

Brochu, C. A., M. L. Boouare, F. Sissoko, E. M. Roberts, and M. A. O'Leary. 2002. A dryosaurid crocodyliform braincase from Mali. Journal of Paleontology 76:1060–1071.

Brown, C. H. 1994. Sound localization; pp. 57–96 in R. R. Fay and A. N. Popper (eds.), Comparative Hearing: Mammals. Springer-Verlag, New York.

Caprette, C. L., M. S. Y. Lee, R. Shine, A. Mokany, and J. F. Downhower. 2004. The origin of snakes (Serpentes) as seen through eye anatomy. Biological Journal of the Linnean Society 81:469–482.

Carroll, R. L., and P. Gaskill. 1985. The nothosaur *Pachypleurosaurus* and the origin of plesiosaurs. Philosophical Transactions of the Royal Society of London B 309:343–393.

Christensen-Dalsgaard J., and A. Elepfandt. 1995. Biophysics of underwater hearing in the clawed frog, *Xenopus laevis*. Journal of Comparative Physiology A 176:317–324.

Christensen-Dalsgaard J., T. Breithaupt, and A. Elepfandt. 1990. Underwater hearing in the clawed frog, *Xenopus laevis*, tympanic motion studied with laser vibrometry. Naturwissenschaften 77:135–137

Clack, J. A., and E. Allin. 2004. The evolution of single- and multiple-ossicle ears in fishes and tetrapods; pp. 128–163 in G. A. Manley, A. N. Popper, and R. R. Fay (eds.), Evolution of the vertebrate auditory system. Springer-Verlag, New York

Clack, J. A., P. E. Ahlberg, S. S. Finney, A. P. Dominguez, J. Robinson, and R. A. Ketcham.

2003. A uniquely specialized ear in a very early tetrapod. Nature 425:65–69.

Eggermont, J. J. 1988. Mechanisms of sound localization in anurans; pp. 307–336 in B. Fritzsch. M. J. Ryan, W. Wilczynski, T. E. Hetherington, and W. Walkowiak W (eds.), The Evolution of the Amphibian Auditory System. John Wiley and Sons, New York.

Elepfandt, A. 1996. Underwater acoustic and hearing in the clawed frog, Xenopus; pp. 177–193 in R. C. Tinsley and H. R. Kobel (eds.), The Biology of *Xenopus*. Oxford Science Publications, Oxford.

Elepfandt, A., I. Eistetter, A. Fleig, E. Gunther, M. Hainich, S. Hepperle, and B. Traub. 2000. Hearing thresholds and frequency discrimination in the purely aquatic frog *Xenopus laevis laevis* (Pipidae): measurement by means of conditioning. Journal of Experimental Biology 203:3621–3629

Fay, R. R. 1988. Hearing in Vertebrates: A Psychophysics Databook. Fay-Hill, Winnetka, IL.

Fritzsch, B. 1992. The water-to-land transition: evolution of the tetrapod basilar papilla, middle ear, and auditory nuclei; pp. 351–376 in D. B. Webster, R. R. Fay, and A. N. Popper (eds.), The Evolutionary Biology of Hearing. Springer-Verlag, New York.

Gans, C., and P. F. A. Maderson. 1973. Sound producing mechanism in recent reptiles: review and comment. American Zoologist 13:1195–1203.

Garrick, L. D., and J. W. Lang. 1977. Social signals and behavior of adult alligators and crocodiles. American Zoologist 13:1195–1203.

Garrick, L. D., and E. I. Saiff. 1974. Observations on submergence reflexes of *Caiman scleros*. Journal of Herpetology 8:231–235.

Garrick, L. D., J. W. Lang, and H. A. Herzog. 1978. Social signals of adult American alligators. Bulletin of the American Museum of Natural History 160:153–192.

Gleich, O., and G. A. Manley. 2000. The hearing organ of birds and Crocodilia; pp. 70–138 in R. J. Dooling, R. R. Fay, and A. N. Popper (eds.), Comparative Hearing: Birds and Reptiles. Springer-Verlag, Berlin.

Gorga, M., J. Kaminski, K. Beauchaine, W. Jesteadt. 1988 Auditory brainstem response to tone bursts in normally hearing subjects. Journal of Speech and Hearing Research 31:89–97.

Hartline, P. H. 1969. Auditory and vibratory responses in the midbrains of snakes. Science 163:1221–1223.

Heffner, R. S., and H. E. Heffner. 1992. Evolution of sound localization in mammals; pp. 691–716 in D. B. Webster, R. R. Fay, and A. N. Popper (eds.), The Evolutionary Biology of Hearing. Springer-Verlag, New York.

Hetherington, T. E. 1988. Biomechanics of vibration reception in the bullfrog *Rana catesbeiana*. Journal of Comparative Physiology 163:43–52.

Hetherington, T. E., and E. D. Lindquist. 1999. Lung-based hearing in an "earless" anuran amphibian. Journal of Comparative Physiology 184:395–401.

Hetherington, T. E., and R. E. Lombard. 1982. Biophysics of underwater hearing in anuran amphibians. Journal of Experimental Biology 98:49–66.

Higgs, D. M., E. F. Brittan-Powell, D. Soares, M. J. Souza, C. E. Carr, R. J. Dooling, and A. N. Popper. 2002. Amphibious auditory responses of the American alligator *(Alligator mississippiensis)*. Journal of Comparative Physiology 188:217–223.

Jaslow, A. P., T. E. Hetherington, and R. E. Lombard. 1988. Structure and function of the amphibian middle ear; pp. 69–92 in B. Fritzsch, M. Ryan, W. Wilczynski, T. E. Hetherington, and W. Walkowiak (eds.), Evolution of the Amphibian Auditory System. Wiley and Sons, New York.

Kartaschew, N., and W. D. Iljitschew. 1964. Über das Gehörorgan der Alkenvögel. Journal of Ornithology 105:113–136.

Ketten, D. R. 2000. Cetacean ears; pp. 43–108 in W. W. L. Au, A. N. Popper, and R. R. Fay (eds.), Hearing by Whales and Dolphins. Springer-Verlag, New York.

Ketten, D. R., C. Merigo, E. Chiddick, and H. Krum. 1999. Acoustic fatheads: parallel evolution of underwater sound reception mechanisms in dolphins, seals, turtles, and sea birds. Journal of the Acoustical Society of America 105:1110.

Klinke, R., and M. Pause. 1980. Discharge properties of primary auditory fibers in *Caiman crocodilus*: comparisons and contrasts to the mammalian auditory nerve. Experimental Brain Research 38:137–150.

Knudsen, E. I. 1980. Sound localization in birds; pp. 289–322 in A. N. Popper and R. R. Fay (eds.), Comparative Studies of Hearing in Vertebrates. Springer-Verlag, New York.

Kuhne, R., and B. Lewis. 1985. External and middle ears; pp. 227–272 in A. S. King and J. McClelland (eds.), Form and Function in Birds, Vol. 3. Academic Press, New York.

Langston, W. 1973. The crocodilian skull in historical perspective; pp. 263–284 in C. Gans and T. S. Parsons (eds.), Biology of the Reptilia, Vol. 4. Academic Press, New York.

Lenhardt, M. L., S. Bellmund, R. A. Byles, S. W. Harkins, and J. A. Musick. 1983. Marine turtle reception of bone-conducted sound. Journal of Auditory Research 23:119–125.

Lenhardt, M. L., R. C. Klinger, and J. A. Musick. 1985. Middle ear anatomy of the marine turtle. Journal of Auditory Research 25:66–72.

Lewis, E. R., and P. M. Narins. 1999. The acoustic periphery of amphibians: anatomy and physiology; pp. 101–154 in R. R. Fay and A. N. Popper (eds.), Comparative Hearing: Fish and Amphibians. Springer-Verlag, New York.

Lin, K., and O. Rieppel. 2000. Functional morphology and ontogeny of *Keichosaurus hui* (Reptilia, Sauropterygia). Fieldiana Geology 39:1–110.

Lindquist, E. D., T. E. Hetherington, and S. Volman. 1998. Biomechanical and neurophysiological studies on audition in eared and earless harlequin frogs *(Atelopus)*. Journal of Comparative Physiology 183:265–271.

Lombard, R. E., and J. Bolt. 1979. Evolution of the tetrapod ear: an analysis and reinterpretation. Biological Journal of the Linnean Society 11:19–76.

Lombard, R. E., and T. E. Hetherington. 1993. The structural basis of hearing and sound transmission; pp. 241–301 in J. Hanken and B. Hall (eds.), The Vertebrate Skull. University of Chicago Press, Chicago.

Lombard, R. E., R. R. Fay, and Y. L. Werner. 1981. Underwater hearing in the frog, *Rana catesbeiana*. Journal of Experimental Biology 91:57–71.

Manley, G. A., and J. A. Clack. 2004. An outline of the evolution of vertebrate hearing organs; pp. 1–27 in G. A. Manley, A. N. Popper, and R. R. Fay (eds.), Evolution of the Vertebrate Auditory System. Springer-Verlag, New York.

McGowan, C. 1973. The cranial morphology of the Lower Liassic latipinnate ichthyosaurs of England. Bulletin of the British Museum of Natural History A 24:1–109.

Norberg, R. A. 1978. Skull asymmetry, ear structure and function, and auditory localization in Tengmalm's owl, *Aegolius funereus* (Linne). Philosophical Transactions of the Royal Society of London 282:325–410.

Owen, R. 1850. On the communication between the cavity of the tympanum and the palate in the Crocodilia (gavials, alligators and crocodiles). Philosophical Transactions of the Royal Society of London 140:521–527.

Popper A. N., and R. R. Fay. 1999. The auditory periphery in fishes; pp.43–100 in R. R. Fay and A. N. Popper (eds.), Comparative Hearing: Fish and Amphibians. Springer-Verlag, New York.

Presley, R. 1984. Lizards, mammals and the primitive tetrapod tympanic membrane. Symposium of the Zoological Society of London 52:127–152.

Ridgway, S. H., E. G. Wever, J. G. McCormick, J. Palin, and J. H. Anderson. 1969. Hearing in the giant sea turtle, *Chelonia mydas*. Proceedings of the National Academy of Sciences USA 64: 884–890.

Rieppel, O., and H. Zehar. 2000. The braincase of mosasaurs and *Varanus*, and the relationship of snakes. Zoological Journal of the Linnean Society 129:489–514.

Romer, A. S. 1956. Osteology of the Reptiles. University of Chicago Press, Chicago.

Rosowski, J. J. 1994. Outer and middle ears; pp. 172–248 in R. R. Fay and A. N. Popper (eds.), Comparative Hearing: Mammals. Springer-Verlag, New York.

Ruggero, M. A., and A. N. Temchin. 2002. The roles of the external, middle, and inner ear in determining the bandwidth of hearing. Proceedings of the National Academy of Sciences USA 99: 13206–13210.

Saiff, E. I. 1976. Anatomy of the middle ear region of the avian skull: Sphenisciformes. Auk 93:749–759.

Saiff, E. I. 1978. The middle ear of the skull of birds: Pelecaniformes and Ciconiiformes. Zoological Journal of the Linnean Society 63:315–370.

Saito, N. 1980. Structure and function of the avian ear; pp. 241–260 in A. N. Popper and R. R. Fay (eds.), Comparative Studies of Hearing in Vertebrates. Springer-Verlag, New York.

Saunders, J. C., R. K. Duncan, D. D. Doan, and Y. L. Werner. 2000. The middle ear of reptiles and birds; pp. 13–69 in R. J. Dooling, R. R. Fay, and A. N. Popper (eds.), Comparative Hearing: Birds and Reptiles. Springer-Verlag, Berlin.

Schwartzkopff, J. 1957. Die Grössenverhältnisse von Trommelfell, Columella-Fussplatte und Schnecke bei Vögeln verschiedenen Gewichts. Zeitschrift für Morphologie und Ökologie der Tiere 45:365–378.

Schwartzkopff, J. 1973. Mechanoreception; pp. 417–477 in D. S. Farner, J. R. King, and K. C. Parkes (eds.), Avian Biology, III. Academic Press, New York.

Shute, C. C. D., and A. d'A. Bellairs. 1955. The external ear in Crocodilia. Proceedings of the Zoological Society of London 124:741–749.

Smotherman, M., and P. Narins. 2004. Evolution of the amphibian ear; pp. 164–199 in G. A. Manley, A. N. Popper, and R. R. Fay (eds.), Evolution of the Vertebrate Auditory System. Springer-Verlag, New York.

Soares, D. 2002. An ancient sensory organ in crocodilians. Nature 417:241–242.

Taylor, M. A. 1992. Functional anatomy of the head of the large aquatic predator *Rhomaleosaurus zetlandicus* (Plesiosauria, Reptilia) from the Toarcian (Lower Jurassic) of Yorkshire, England. Philosophical Transactions of the Royal Society of London B 335:247–280.

Tonndorf, J. 1972. Bone conduction; pp. 197–237 in J. V. Tobias (ed.), Foundations of Modern Auditory Theory, Vol. II. Academic Press, New York.

Vliet, K. A. 1989. Social displays of the American alligator *(Alligator mississippiensis)*. American Zoologist 29:1019–1031.

Webster, D. B., and M. Webster. 1975. Auditory system of Heteromyidae: functional morphology and evolution of the middle ear. Journal of Morphology 146:343–376.

Wever, E. G. 1949. Theory of Hearing. John Wiley and Sons, New York.

Wever E. G. 1978. The Reptile Ear. Princeton University Press, Princeton, NJ.

Wever E. G. 1985. The Amphibian Ear. Princeton University Press, Princeton, NJ.

Wilczynski, W., and R. R. Capranica. 1984. The auditory system of anuran amphibians. Progress in Neurobiology 22:1–38.

Wiley, R. H., and D. G. Richards. 1978. Physical constraints on acoustic communication in the atmosphere: implications for the evolution of animal vocalizations. Behavioral Ecology and Sociobiology 3:69–94.

Williston, S. W. 1907. The skull of *Brachauchenius*, with special observations on the relationships of the plesiosaurs. Proceedings of the U.S. National Museum 32:477–489.

Zehar, H., and O. Rieppel. 2000. A brief history of snakes. Herpetological Review 31:73–76.

Hearing in Aquatic Mammals

Sirpa Nummela

The mammalian auditory system is a complex biological structure of well-understood function and with parts that easily fossilize, thus providing exciting opportunities to study evolution.

Mammals have inherited their hearing mechanism from mammal-like reptiles, synapsids. The organization of the ear is explained by Hetherington (chapter 12 in this volume) and Nummela and Thewissen (Fig. 11.1 in this volume). Primitively, mammals have an outer ear pinna, three small bones in their middle ear, and, in therian mammals, an elongated and coiled cochlea. The evolution of the auditory system in synapsids is well documented by fossils, as well as by embryology: bones contributing to the lower jaw gradually move caudally, and the morphology of the jaw joint and the middle ear structrures is reorganized (Hopson, 1966; Allin, 1975; Allin and Hopson, 1992; Clack and Allin, 2004). Inner ear evolution in vertebrates is reviewed by Manley and Clack (2004). With fossils it is also possible to study the early evolution of mammalian hearing (Rosowski and Graybeal, 1991; Rosowski, 1992). The first semiaquatic mammals are known from the Mesozoic (Ji et al., 2006), although these have not left modern descendants.

Although the land mammal hearing mechanism works well in air, it is not suitable for hearing in water (see Nummela and Thewissen, chapter 11 in this volume). Air and water are acoustically different media, and good hearing in water is thus dependent on the morphological evolution of the ear providing adaptations to the new habitat. These adaptations can be analyzed in a functional context, and they can thus provide information about the most important evolutionary changes during the history of the ear.

Experimental data on hearing of different species are also an important source for understanding ear evolution: they document hearing frequency range, hearing sensitivity, and directional hearing ability; such data are available for several species of mammals (e.g., Heffner and Heffner, 1985, 1990; Fay, 1988). These data together with the anatomical data on ears of different species are very useful for the comparative study of hearing.

GENERALIZED ANATOMY AND FUNCTION OF HEARING IN MAMMALS

THE LAND MAMMAL EAR

The outer, middle, and inner ear form the peripheral part of the mammalian auditory system; the auditory path beyond that is part of the central nervous system. Sound in air passes through the outer and middle ear to the inner ear. In the inner ear, sound energy is transformed into a neural impulse that causes a hearing sensation in the brain. The basic components and function of the ear are similar among different mammalian species, but they do vary considerably in their more detailed morphology, size, and function. Most of these differences are adaptations. Outer, middle, and inner ear structures all affect hearing: frequency range, sensitivity, the frequency of best sensitivity, and directional and spatial abilities.

The mammalian ear structures have been described for many different species, and the functional significance of these structures is well understood (e.g., Hyrtl, 1845; Doran, 1879; Lay, 1972; Fleischer, 1973, 1978; Henson, 1974; Webster and Webster, 1975; Heffner and Heffner,

1985, 1992; Fay, 1988; Rosowski, 1992; Hemilä et al., 1995; Nummela, 1995, 1997; Geisler, 1998; Rosowski and Relkin, 2001; Ruggero and Temchin, 2002; Nummela and Sanchez-Villagra, 2006).

OUTER EAR

The outer ear of land mammals consists of a pinna and an outer ear canal, the external auditory meatus. The pinna, and sometimes the pinna and external auditory meatus are missing in some species. The pinna, a flange, collects sound and directs it into the external acoustic meatus. In it, sound waves travel to the tympanic membrane. The size and shape of the pinna and the length of the meatus determine which sound frequencies are collected best and amplified before reaching the tympanic membrane and the middle ear ossicles (Shaw, 1974); the shape of the pinna was found to correlate with hearing sensitivity measures in a large sample of primates (Coleman and Ross, 2004). The functional significance of the pinna as a sound collector is also obvious in many bats whose pinna is very large and of extraordinary shape. Carnivores and also desert mammals have large pinnae, probably used in the perception of low-frequency (long-wavelength) sounds in long-distance communication. Some fossorial mammals do not have a pinna at all, and their outer ear canal may also have degenerated (Lay, 1972; Webster and Webster, 1975; Webster and Plassmann, 1992).

MIDDLE EAR

Sound waves cause vibrations of the tympanic membrane, and these vibrations move through the middle ear ossicles: the malleus, incus, and stapes in the middle ear cavity. The outermost ossicle, the malleus, is attached to the tympanic membrane through its manubrium, a thin handle. The head of the malleus shares a joint with the incus, and together these two ossicles act somewhat like a swing. The incus shares a joint with the stapes, passing on the vibrations of the tympanic membrane to the distal end of the stapes at the oval window.

The size and morphology of the tympanic membrane and the ossicles vary across different groups of mammals and are related to their hearing frequency range. The stiffness and the mass of the middle ear mechanism are the main factors limiting high-frequency hearing, whereas the elasticity of the system and the volume of the middle ear cavity are the main factors allowing or impeding the low-frequency hearing. The size of the middle ear ossicles has been shown to limit the high-frequency hearing of mammals when experimental data from audiometric results have been combined with morphometric data (Hemilä et al., 1995; Nummela, 1995; Nummela and Sanchez-Villagra, 2006). An inverse correlation between the interaural distance and the high-frequency hearing limit has been identified by Heffner and Heffner (e.g., 1985, 1992). The audible frequency range is significantly broader in mammals than in other vertebrates (Fay, 1988).

INNER EAR

The mammalian cochlea is housed inside the petrosal bone, which is composed of very compact bone and protects the inner ear. Except for the cochlea, the inner ear also comprises the balance system (see Sipla and Spoor, chapter 14 in this volume; Spoor and Thewissen, chapter 16 in this volume). The cochlear part of the mammalian inner ear is an elongated and coiled structure (except in monotremes) and filled with endolymph and perilymph. The organ of Corti is located on the basilar membrane, a fibrous structure that divides the scala media and the scala tympani. The organ of Corti is a highly specialized structure, it contains hair cells that are the receptor cells, as well as their nerve endings and supporting cells. The hair cells consist of one row of inner hair cells on the modiolar side (toward the axis of the cochlea) of the arch of Corti, and between three and five rows of outer hair cells. When the stapes vibrates at the oval window, it sets the perilymph in the scala vestibuli into motion. This motion is passed on to the scala media and scala vestibuli and finally stimulates the hair cells. As a result, a neural impulse is created. The hair cells are arranged tonotopically along the organ of Corti: different frequencies are represented at different parts of the membrane. In this case hair cells at the beginning of the organ of Corti fire best at high frequencies, and the hair cells at the end of the basilar membrane, at the apex of the cochlea, fire most sensitively at low frequencies.

The number of coils in the cochlea varies and is not related to the body size of mammals. The number of coils is smaller in animals whose hearing is at high frequencies, and larger in animals whose hearing is at lower frequencies (Fleischer, 1973; Henson, 1974).

The inner ear functions no differently in water than in air and is therefore deemphasized in this chapter, which focuses on outer and middle ear adaptations in aquatic taxa. Aquatic mammals, in particular cetaceans, do display a number of pronounced inner ear specializations, as studied by Ketten and coworkers (reviewed in Wartzok and Ketten, 1999), however these are not adaptations for underwater hearing per se.

SOUND TRANSMISSION: AIRBORNE SOUND

Airborne sound is transmitted from the lower acoustic impedance of air (Z_{air} = 0.4 kPa s/m) to the much higher acoustic impedance of the inner ear fluid (Z_c = 150 kPa s/m, approximately) (see Nummela and Thewissen, chapter 11 in this volume, and references therein). Such a difference in impedance would lead to reflection of most sound energy. In a land mammal ear, reflection of sound is avoided by the insertion of the middle ear, acting as an impedance-matching device between the air and the inner ear.

Impedance matching in a land mammal ear is accomplished by increasing the pressure p and by decreasing the particle velocity v between the tympanic membrane and the oval window ($Z = p/v$, see above). The pressure can be increased by transmitting sound from a larger to a smaller area, from the tympanic membrane to the oval window. The area ratio between these two is the main impedance-matching mechanism in the ear. The particle velocity can be decreased by means of a malleus lever arm and the incus lever arm so that the former is longer than the latter. This lever ratio between the malleus and the incus decreases the particle velocity between the tympanic membrane and the oval window. The product of the area ratio and the lever ratio is called the geometric transformer ratio of the mammalian middle ear. With this mechanical transformer it is possible to minimize reflection of sound energy at the tympanic membrane; in other words, the middle ear input impedance is matched with the input impedance of the cochlea. Generally, the transformer ratio is large in animals that hear high frequencies, and it is small in animals that hear lower frequencies.

In addition to the impedance-matching function, the mammalian middle ear also functions as an intensity amplifier. The intensity amplification is accomplished when energy is transferred from a larger to a smaller area, in this case from the tympanic membrane to the oval window. In a land mammal ear, the area ratio between the tympanic membrane and the oval window serves two functions, the impedance matching and the intensity amplification accomplished by the ear.

SOUND TRANSMISSION: BONE CONDUCTION

The tympanic route to the inner ear is the most common way to perceive airborne sound. However, under certain conditions sound reaches the inner ear by passing through an animal's substrate and its body: bone conduction (Bárány, 1938; Tonndorf, 1968; see also Nummela and Thewissen, chapter 11 in this volume). In bone-conducted hearing, vibrations of the skull create a stimulus to the cochlear hair cells. Bone conduction takes place in all mammals (e.g., when chewing) but is especially significant when the skull is in physical contact with a dense medium such as water or soil, and some animals do display specific adaptations for this type of sound transmission. For instance, fossorial mammals, such as chrysochlorids (Afrotheria) are often adapted for increased bone conduction by having a heavy malleus.

FLEISCHER'S MIDDLE EAR TYPES

Gerald Fleischer (1978) grouped mammalian middle ears with different morphology into different categories and presented a hypothesis of the origin and evolutionary radiation of these different types. Additionally, he presented a qualitative model for the functional significance of the varying morphologies. Here, I briefly discuss the categories of ears found in land mammals. Below, I discuss the mammalian aquatic ear types and expand Fleischer's classification by identifying new types.

ANCESTRAL MIDDLE EAR TYPE

In the ancestral middle ear type, the malleus has a long processus gracilis, which is tightly connected to the anterior rim of the tympanic annular ring. This type is functionally stiff and thus most suitable for transmitting high frequencies to the inner ear. Most small mammals, rodents, and insectivores have an ancestral middle ear.

MICROTYPE MIDDLE EAR

The microtype middle ear resembles the ancestral middle ear very closely; the connection between the long and slim processus gracilis of the malleus and the tympanic ring is equally clear and stiff, and the incus is clearly smaller than the malleus. The structure that separates the microtype middle ear from the ancestral one is the orbicular apophysis, a bulbous structure in the posterior part of the malleus, dorsal to the manubrium. The orbicular apophysis increases the mass of the malleus, making it relatively even larger than the incus. This additional mass also changes the rotation axis of the malleus-incus unit from that of the ancestral middle ear. The ancestral and the microtype middle ears are most suitable for transmitting high frequencies, due to their stiffness, and this type of ear is found in small, echolocating bats.

TRANSITIONAL MIDDLE EAR TYPE

The transitional middle ear has a long processus gracilis of the malleus, as in the primitive type. The processus gracilis is attached to the tympanic ring, but not as tightly as in the two previous types. The incus has increased in size in comparison with the malleus, and the malleus-incus unit has rotated dorsally so that the processus gracilis does not lie directly in the line with the tympanic ring. The transitional middle ear is well suited to transmit somewhat lower frequencies than the two previous ones and is found in medium-sized mammals.

FREELY MOBILE MIDDLE EAR TYPE

In the freely mobile middle ear, the connection between ossicles and skull is loose. The freely mobile middle ear has lost the long processus gracilis, which appears only as a short, truncated process, not connected by bone to the tympanic ring. The incus is similar in size to the malleus, and the malleus-incus unit has moved even further dorsally. The malleus lacks a transversal part, and the malleus head and manubrium lie in the same line with each other and with the crus longum of incus. Together, these adaptations bring compliance to the ossicular chain, and it is well suited for transmitting low sound frequencies. This type occurs in large mammals, such as primates and some carnivores.

In his compilation, Fleischer (1978) suggests that the ancestral type gave origin to both the microtype and the transitional type, the latter then leading to the freely mobile middle ear, thus being transitional between the ancestral and the freely mobile type. The main changes have been that the processus gracilis has shortened and is unfused to the anterior rim of the tympanic ring, the malleus is relatively smaller than the incus, and the ossicular chain has rotated so that when the processus gracilis and manubrium were dorsoventral, and the malleus transversal part and the incus long process were oriented rostrocaudally, the manubrium and the incus long process came to lie dorsoventrally. Generalizing, the absolute ear size increased between the ancestral and the freely mobile type.

ANATOMY AND FUNCTION OF HEARING IN AQUATIC MAMMALS

Excellent general discussions of hearing in marine mammals can be found in works by Wartzok and Ketten (1999), Dehnhardt (2002), and Berta et al. (2006). In this chapter, I expand on this by focusing on the classification by Fleischer (1978), as it provides an excellent opportunity to study outer and middle ear morphology in an evolutionary context.

Fleischer (1978) distinguished two middle ear types unique to marine mammals, the *Tursiops* type, and the *Kogia* type. Both *Tursiops* and *Kogia* are odontocete cetaceans, the *Tursiops* middle ear type is found in most odontocetes and is therefore here renamed as such, whereas the *Kogia* type is found in physeterids (sperm whales) and ziphiids (beaked whales) and is here renamed the physeteroid type. In addition, I characterize three more ear types of marine mammals, found in mysticetes, phocids, and sirenians (Table 13.1). I characterize these on the basis of outer and middle ear morphology, because, in marine mammals, the outer ear is commonly significantly altered. Pinnipeds differ from cetaceans and sirenians in being amphibious, and hence they need to hear in both air and water. As a result, the pinnipeds are more similar to land mammals, and pinnipeds hear well in air and water. Based on the outer and middle ear types discussed below, some inferences can be made about hearing evolution in cetaceans and pinnipeds (Fig. 13.1). For cetaceans, these inferences build on the work of Nummela et al. (2004b).

ODONTOCETE EAR TYPE

The morphology of the odontocete ear is radically different from the generalized mammalian ear (Doran, 1879; Reysenbach de Haan, 1957; Oelschläger, 1986, 1990; Ketten, 1992; Ketten et al., 1992; Nummela et al., 1999a, 1999b). In the odontocete outer ear, there is no pinna, and the external auditory meatus runs through a fatty blubber tissue and is degenerated and partly occluded. The lower jaw of

TABLE 13.1
Aquatic Mammals with Ear Specializations Probably at Least in Part Related to Underwater Hearing

EAR	FOUND IN	SOUND-TRANSMITTING MEDIUM	SOUND-TRANSMITTING MECHANISM IN WATER	SOUND-RECEIVING AREA OF MIDDLE EAR
Odontocete type	Most odontocetes	Water	Odontocete mechanism	Mandible/tympanic plate
Physeteroid type	Physeteridae Kogiidae Ziphiidae	Water	Odontocete mechanism?	Tympanic plate?
Mysticete type	Mysticeti	Water	Bone conduction/ unique mysticete mechanism	Tympanic plate?
Sirenian type	Sirenia	Water	Bone conduction/ unique sirenian mechanism	Zygomatic process
Phocid type	Phocidae Odobenidae Hippopotamus	Air/water	Bone conduction/ unique phocid mechanism	Skin surface/ tympanic membrane

odontocetes has a wide mandibular canal, which houses a fatty tissue, a fat pad, reaching posteriorly to the middle ear wall.

The tympanic and periotic bones together form a tympanoperiotic complex, the ear complex in cetaceans. The tympanic bone (ectotympanic) forms the auditory bulla, also called tympanic bulla, and contains the middle ear cavity, where the middle ear ossicles are situated. The middle ear cavity is covered inside with cavernous tissue, which can swell and decrease the volume of the cavity, increasing the air pressure. The lateral wall of the tympanic bone is very thin and is called the tympanic plate, the medial part of the tympanic bone is very thick and is called the involucrum. A tympanic bone with a thick involucrum is typical of all modern cetaceans. Inside the tympanic bone, at the dorsocaudal part of the lateral wall, is the tympanic membrane, an elongated and relatively small conical structure, sometimes also called the tympanic conus or the tympanic ligament. Its medial tip attaches to the outermost middle ear

ossicle, the malleus. The three middle ear bones form an ossicular chain in the cetacean middle ear, like they do in terrestrial mammals. The malleus has a thin processus gracilis (gonial process, anterior process), which is fused to the anterior rim of the tympanic bone. The malleus lacks a manubrium, and the contact with the tympanic membrane is only through the tip of the tympanic conus. The incus is smaller than the malleus, and the two articulate. As in land mammals, the incus has two crura, crus breve and crus longum; however, they are inverse in length: the cetacean crus breve is longer than the crus longum. The crus breve is attached to the periotic bone through ligaments, and the crus longum makes a joint with the third ossicle, the stapes. The tympanoperiotic complex is surrounded by air sinuses and is thus acoustically isolated from the rest of the skull.

The function of the modern odontocete ear is fairly well understood. The lateral wall of the lower jaw is of very thin bone (Nummela et al.,

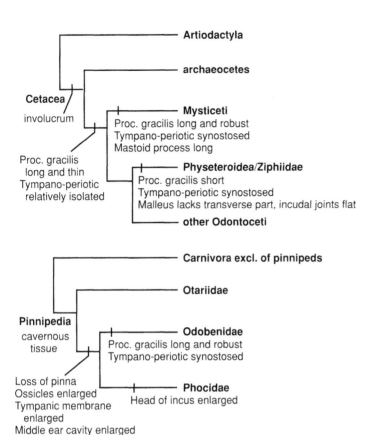

FIGURE 13.1. Evolutionary hypothesis for the origin of some ear features in modern cetaceans and pinniped carnivores. Base phylogeny of pinnipeds differs from that in chapter 1, both interpretations are supported by part of the data.

2004a; 2007). The mandibular canal is filled with a fat pad that extends posteriorly to the tympanic bone. The density of the fatty substance is such that it matches well with the density of water, and the impedance difference between water and the fat pad is minimal (Varanasi and Malins, 1971, 1972). It has been shown that the sound velocity in this substance is very high (Møhl et al., 1999). The lower jaw is a functional analog to the outer ear pinna of terrestrial mammals, as suggested originally by Norris (1964, 1968). When sound arrives at the mandible, it penetrates easily through it and is then conducted by the mandibular fat pad further posteriorly to the tympanic plate. When the tympanic plate starts to vibrate, the vibration is transmitted further to the middle ear ossicles.

Middle ear function in odontocetes was modeled by Hemilä et al. (1999, 2001). The tympanic bone is thin laterally and heavy medially (involucrum) and is not synostosed to the periotic, although the two are tightly connected. Additionally, anteriorly and posteriorly, there are two bony pedicles between the tympanic and periotic. These have areas of very thin bone that form "hinges," allowing the vibration of the tympanic plate in relation to the periotic. The involucrum is nearly immobile, and tympanic plate vibration is transmitted via the processus gracilis of malleus, to the ossicular chain. The tympanic plate thus functionally corresponds to the tympanic membrane of land mammals. The cone-shaped tympanic membrane of odontocetes most likely functions in pressure regulation, rather than sound reception.

The ossicular chain of odontocetes functions as a lever-arm system, as in land mammals, but unlike in other mammals, the tympanic bone and the ossicles form two levers. The first lever is made by the tympanic plate, and the second one by the malleus-incus unit. Both these levers contribute to impedance matching between the surrounding water and the inner ear by increasing particle velocity between the tympanic plate and the cochlea.

The density of the tympanic, the periotic, and the ossicles is very high, much higher than that of the skull. This high density ratio between the ear and the rest of the skull creates a barrier for sound penetration and helps in isolating the ear acoustically. The high bone density is advantageous for the odontocete hearing mechanism; it makes it possible for the tympanic plate to function as one single unit. If a land mammal tympanic membrane had the size of an odontocete tympanic plate, the membrane would not vibrate as one single unit, rather, the vibration would break up into several smaller units, each one vibrating independently. This would lead to a tremendous loss of energy, and hearing would not be possible.

The odontocete cochlea inside the periotic bone is relatively similar to the generalized mammalian cochlea, with its membranous labyrinths of scala media, scala tympani, and scala vestibuli. Based on the cochlear morphology, Ketten (1992, 2000) divided the modern cetacean cochleae into three different acoustic categories: type I odontocetes (upper-range ultrasonics), type II odontocetes (lower-range ultrasonics), and mysticetes (potentially infrasonic). According to her work, these types differ in their cochlear construction and support of the basilar membrane. In odontocetes, the basilar membrane is very narrow, the spiral ganglion cell density is high, and the bony outer spiral lamina is extensive. The basilar membrane thickness-to-width ratio at the basal turn of the cochlea is higher in odontocete ears than in any other mammals.

The model presented by Hemilä et al. (1999, 2001) for middle ear sound transmission in odontocetes explains experimental data from audiograms, although it still needs to be further tested experimentally. That model was based on the qualitative model of Fleischer (1978), with many modifications. Ketten and Wartzok (1990) and Ketten (1992, 1997, 2000) have presented interpretations of the cetacean

hearing mechanisms and frequency ranges mainly based on cetacean inner ear morphology. Odontocetes use high frequencies that may well exceed 100 kHz. These very high frequencies are used for communication and for echolocation.

PHYSETEROID EAR TYPE

Fleischer (1978) found that the pygmy sperm whale *Kogia* had some middle ear structures clearly different from those of other odontocetes. A similar morphology is also found in ziphiids and *Physeter*. In this middle ear type there is no tympanic membrane, but a small bony plate (the tympanic plate), the malleus lacks a transversal part, the malleus-incus unit has no rotation axis, and the crus breve makes a flat contact with the periotic. The processus gracilis of the malleus is short or missing altogether, and the connection to the tympanic bone is wide: no bending can occur between the processus gracilis and malleus. The surface joints in the malleus and incus are more flat and wide than in other odontocetes, and the tympanic and periotic are tightly fused to each other through a synostosis laterally, unlike in other odontocetes.

Fleischer (1978) used the term tympanic plate to refer to a small area of very thin bone in the dorsal part of the lateral tympanic wall, close to the sigmoid process. The term tympanic plate is also used by Hemilä et al. (1999) and Nummela et al. (1999a, 1999b) to indicate a much larger area of the lateral tympanic wall. In this functional redefinition, *Kogia* does have a tympanic plate, and this plate can vibrate similarly to other odontocetes. There are no functional models for physeteroid ears.

MYSTICETE EAR TYPE

The mysticete ear structures resemble the odontocete ear to a large extent, but there are also several clear differences. Fleisher (1978) did not describe this as a separate middle ear type, but the differences are significant, and I here do identify it as such. Mysticetes do not have an outer ear pinna, and their external auditory meatus is very thin and partially occluded. At the proximal end of the meatus, attached to the tympanic membrane on its lateral side, is a wax plug, often called the glove finger. The tympanic membrane is conical in its shape and situated as in odontocetes on the inner side of the lateral wall of the tympanic bone. The tympanic and periotic bones together form a tympanoperiotic complex, but unlike in the odontocete ear type, they are actually fused to each other anteriorly and posteriorly, resembling the physeteroid ear type. The tympanic has a thick medial part, the involucrum, and a thinner lateral part, the lateral wall. It is doubtful if this wall can be called a "tympanic plate" since that is a functional term, and it is still unclear whether the tympanic lateral wall has a similar function in sound reception as it has in odontocetes. The periotic has a long mastoid process extending posteriorly and situated in a groove in the mysticete skull. This mastoid process is the contact of the tympanoperiotic complex to the skull. The whole ear complex is situated not on the lateral side of the skull, as in odontocetes, but more proximally, close to the midline. The malleus is connected with its processus gracilis to the lateral wall of the tympanic bone; it is a strong connection. The middle ear ossicles of mysticetes resemble those of odontocetes in their shape but are much larger. The malleus does not have a manubrium, and the conical tympanic membrane is attached at its tip to the malleus. The incus makes contact with the periotic with its crus breve, while its crus longum joins with the stapes.

The ear morphology of mysticetes is in many respects similar to that of odontocetes, but there are also clear differences in the anatomy, the size range of the ear, and the hearing frequency ranges of these two cetacean groups. Mysticetes vocalize in a low-frequency sound range (Mednis, 1991). Low frequencies of only 100 or 200 Hz attenuate very little in water and can carry over long distances—hundreds or thousands of kilometers (Urick, 1975).

Mysticetes use these low-frequency sounds for communication over large distances.

The mysticete lower jaw is not the thin sound-receiving structure it is in odontocetes. The tympanoperiotic complex is very large and heavy and lacks the elastic springs present in the odontocete ear, and it is situated close to the midline of the skull, and not laterally.

The differences in the mysticete ear morphology indicate that the function of the mysticete ear is in many ways very different from the odontocete ear function. At present, it is unclear how sound travels from the surrounding water to the mysticete inner ear. Given that sounds are a considerable part of the life of these animals, it is clear that the hearing mechanism of mysticetes is extremely elaborate, and that they can also determine the sound direction.

SIRENIAN EAR TYPE

Sirenians do not have a pinna, but a very small opening, hardly visible on the surface of the head. The external auditory meatus is very thin and apparently occluded (Bullock et al., 1980; Ketten et al., 1992; Chapla et al., 2007). The tympanic bone forms a U-shaped oval ring, not a bulla as in whales, indicating that the sirenian middle ear cavity is mostly not surrounded by bony walls. This bony ring is fused to the periotic bone, which is much more massive than the tympanic, being thus relatively immobile. The sirenian tympanoperiotic complex is acoustically well isolated from the skull. The tympanic membrane is large, and unlike in other mammals, it bulges outward. The middle ear ossicles are massive, and they have a high bone density (Robineau, 1969; Fleischer, 1978; Chapla et al., 2007). The malleus is strongly fused to the tympanic bone through its gonial process (processus gracilis), which is very sturdy to the degree that the tympanic and the malleus actually form a functional unit together. The malleus also has a long and massive manubrium, which pushes the tympanic membrane outward, unlike in any other mammals. The incus is fused to the periotic through its crus breve. The joint surfaces between the malleus and incus are very small compared to the size of the ossicles, and of uncommon shape among mammals in general. The periotic is situated in a concave cavity of the squamosal, and further with the zygomatic process of the squamosal, this process forms part of the zygomatic arch, which consists of spongy bone and is inflated and filled with oil. The composition of lipids in the zygomatic process is not the same as in the mandibular fat pad of odontocetes. It is unclear how the underwater sound reaches the inner ear. It has been suggested that sound could reach the middle ear through the fatty zygomatic process (Bullock et al., 1980; Ketten et al., 1992; Chapla et al., 2007).

Despite very clear anatomical differences between cetaceans and sirenians, Fleischer (1978) claims that many functional similarities can be seen between these two groups. The hinges between the tympanic (ring) and the periotic function as a rotational axis allowing vibrational movements between tympanic and periotic. Other similarities with the cetaceans include that the bony ear structures have a very high density, the ear complex is decoupled from the rest of the skull, and due to the compliant connections between the tympanic and the periotic, the former can vibrate in relation to the latter. The sirenian middle ear anatomy indicates that the massiveness and compactness are important for sound transmission, but for the moment there are no models for the sound transmission mechanism in these animals.

PHOCID EAR TYPE

Of the three families of pinnipeds (phocids, otariids, and odobenids), otariid middle ears are similar to those of the freely mobile type. This type occurs widely in the terrestrial relatives of the pinnipeds, bears and mustelids, and was probably primitive for pinnipeds. Otariids have a small outer ear pinna, which they can also use to close their ear opening when diving. The middle ear in general resem-

bles that of fissiped carnivores, and the middle ear ossicles are similar in size to those of generalized land mammals; they are not hypertrophied.

Phocids lack an outer ear pinna and instead only have a small opening for the ear canal on the lateral side of their head, which they can close underwater. The external acoustic meatus is not wide but leads to the tympanic membrane. The middle ear is formed by a large tympanic bulla, which surrounds the middle ear cavity and the middle ear ossicles. The middle ear cavity is covered by cavernous tissue, which is richly supplied by blood vessels and through swelling can cause diminishing of the cavity volume, thus enhancing the air pressure in the cavity and equalizing it with the pressure outside the animal when diving. The ossicles are hypertrophied to various degrees (relatively little in *Phoca*, very much in *Mirounga*), and the tympanic membrane is enlarged. Malleus, incus, and stapes are enlarged, although the head of the incus is enlarged disproportionally. The enlarged ossicular mass is clear (Fig 13.2) and is most likely an adaptation to underwater hearing; a large mass increases the ossicular inertia, leading to displacement of mass center point from the rotation axis; this improves the inertial bone conduction (see above; Bárány, 1938; Tonndorf, 1968; Repenning, 1972). Hence, bone conduction may play an important role in phocid hearing, although it is very likely that at some frequencies, the ear functions like the freely mobile middle ear type described by Fleischer.

In air, the hypertrophied ossicles of phocids probably function like those of a normal land mammal ear, but the increased mass limits the high-frequency hearing and shifts the hearing frequency range of phocids and odobenids toward lower frequencies. This is consistent with experimental results (Møhl, 1968; Moore and Schusterman, 1987; Kastak and Schusterman, 1999). Although directional hearing is usually a problem with bone-conducted hearing, partial detachment of the ear complex from the skull helps in directional

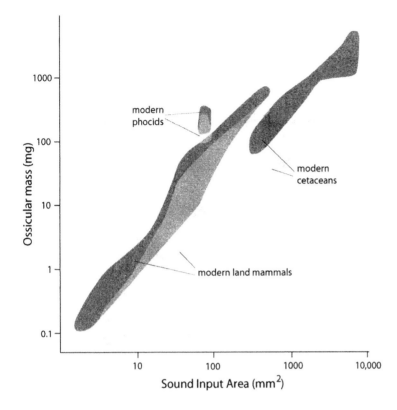

FIGURE 13.2. Bivariate plot of ossicular mass versus sound input area. Outline envelopes represent ossicular masses of a sample of land mammals, phocid seals, and modern cetaceans. Weights for malleus plus incus (dark gray) and incus only (light gray) are plotted. This plot shows that phocids have ossicles that are heavier than those of land mammals. Cetacean ossicles are also heavy; however, the cetacean sound input area (the tympanic plate) is greater in area than the tympanic membrane of many land mammals. Original modified from Nummela et al. (2004b).

hearing (Møhl, 1968). Interestingly, in otariids the high-frequency hearing limits in air and water coincide, whereas in phocids the hearing limit in air is at much lower frequencies than in water. This has been pointed out by Hemilä et al. (2006), suggesting that cochlear constraints dominate the high-frequency hearing in otariids in both air and water, whereas inertial constraints clearly dominate the phocid hearing in air. Sound transmission is phocids is poorly understood, and it is possible that phocids have still another mechanism for underwater hearing.

The third family of pinnipeds, odobenids (walrus), assumes an intermediate place morphologically. It lacks an outer ear pinna, the ear opening is small, but the external acoustic meatus is functional. The middle ear ossicles are hypertrophied, being about 10 times heavier than in land carnivores of the same skull size. Repenning (1972) pointed out that the odobenid ossicles are more otariidlike in their shape, but larger. The tympanic membrane is nonetheless not especially large (Repenning, 1972; Nummela, 1995).

CONCLUSION

Much remains to be learned about hearing underwater in mammals. The inner ear works similarly to that of land mammals, but there are fascinating frequency specializations, probably independent of the aquatic specializations. The outer and middle ear are highly specialized in some (cetaceans, phocids, and sirenians), and barely specialized in others (otariids). Some understanding of middle ear transmission in odontocetes has been reached, but this is not the case for mysticetes and sirenians. It appears that phocids and odobenids have a freely mobile middle ear type in air and use bone conduction in water; their sound transmission mechanism is likely to be more complicated, including other modes of transmission not known at the present time. These issues can be addressed by future generations of students.

LITERATURE CITED

Allin, E. F. 1975. Evolution of the mammalian middle ear. Journal of Morphology 147:403–438.

Allin, E. F, and J. A. Hopson. 1992. Evolution of the auditory system in Synapsida ("mammal-like reptiles" and primitive mammals) as seen in the fossil record; pp. 587–614 in D. B. Webster, R. R. Fay, and A. N. Popper (eds.), The Evolutionary Biology of Hearing. Springer-Verlag, New York.

Bárány, E. 1938. A contribution to the physiology of bone conduction. Acta Oto-Laryngologica Stockholm Supplement 26:1–223.

Berta, A., J. L. Sumich, and K. M. Kovacs. 2006. Marine Mammals: Evolutionary Biology, 2nd Ed. Elsevier, Amsterdam.

Bullock, T. H., D. P. Domning, and R. C. Best. 1980. Evoked brain potentials demonstrate hearing in a manatee *(Trichechus inunguis)*. Journal of Mammalogy 61:130–133.

Chapla, M. E., D. P. Nowacek, S. A. Rommel, and V. A. Sadler. 2007. CT scans and 3D reconstructions of Florida manatee *(Trichechus manatus latirostris)* heads and ear bones. Hearing Research 228:123–135.

Clack, J. A., and E. Allin. 2004. The evolution of single- and multiple-ossicle ears in fishes and tetrapods; pp. 128–163 in G. A. Manley, A. N. Popper, and R. R. Fay (eds.), Evolution of the Vertebrate Auditory System. Springer-Verlag, New York.

Coleman, M. N., and C. F. Ross. 2004. Primate auditory diversity and its influence on hearing performance. Anatomical Record A 281:1123–1137.

Dehnhardt, G. 2002. Sensory systems; pp. 116–141 in A. R. Hoelzel (ed.), Marine Mammal Biology: An Evolutionary Approach. Blackwell Publishing, Oxford.

Doran, A. H. G. 1879. Morphology of the mammalian ossicula auditus. Transactions of the Linnean Society Series 2, Zoology 1:371–497.

Fay, R. R. 1988. Hearing in Vertebrates: A Psychophysics Databook. Hill-Fay Associates, Winnetka, IL.

Fleischer, G. 1973. Studien am Skelett des Gehörorgans der Säugetiere, einschliesslich des Menschen. Säugetierkundliche Mitteilungen 21:131–239.

Fleischer, G. 1978. Evolutionary principles of the mammalian middle ear. Advances in Anatomy, Embryology, and Cell Biology 55:1–70.

Geisler, C. D. 1998. From Sound to Synapse: Physiology of the Mammalian Ear. Oxford University Press, New York.

Heffner, R. S., and H. E. Heffner. 1985. Hearing in mammals: the least weasel. Journal of Mammalogy 66:745–755.

Heffner, R. S., and H. E. Heffner. 1990. Vestigial hearing in a fossorial mammal, *Geomys bursarius*. Hearing Research 46:239–252.

Heffner, R. S., and H. E. Heffner. 1992. Hearing and sound localization in blind mole rats *(Spalax ehrenbergi)*. Hearing Research 62:206–216.

Hemilä, S., S. Nummela, and T. Reuter. 1995. What middle ear structures tell about impedance matching and high-frequency hearing. Hearing Research 85:31–44.

Hemilä, S., S. Nummela, and T. Reuter. 1999. A model of the odontocete middle ear. Hearing Research 133:82–97.

Hemilä, S., S. Nummela, and T. Reuter. 2001. Modeling whale audiograms: effects of bone mass on high-frequency hearing. Hearing Research 151:221–226.

Hemilä, S., S. Nummela, A. Berta, and T. Reuter. 2006. High-frequency hearing in phocid and otariid pinnipeds: inertial and cochlear constraints. Journal of the Acoustical Society of America 120:3463–3466.

Henson, O. W., Jr. 1974. Comparative anatomy of the middle ear; pp. 39–110 in W. D. Keidel and W. D. Neff (eds.), Handbook of Sensory Physiology, Vol. V-1, Auditory System. Springer-Verlag, Berlin.

Hopson, J. A. 1966. The origin of the mammalian middle ear. American Zoologist 6:437–450.

Hyrtl, J. 1845. Vergleichend-anatomische Untersuchungen über das innere Gehörorgan des Menschen und der Säugethiere. Verlag von Friedrich Ehrlich, Prague.

Ji, Q., Z.-X. Luo, C.-X. Yuan, and A. R. Tabrum. 2006. A swimming mammaliaform from the Middle Jurassic and ecomorphological diversification of early mammals. Science 311:1123–127.

Kastak, D., and R. J. Schusterman. 1999. In-air and underwater hearing sensitivity of a northern elephant seal *Mirounga angustirostris*. Canadian Journal of Zoology 77:1751–1758.

Ketten, D. R. 1992. The marine mammal ear: specializations for aquatic audition and echolocation; pp. 717–750 in D. B. Webster, R. R. Fay, and A. N. Popper (eds.), The Evolutionary Biology of Hearing. Springer-Verlag, New York.

Ketten, D. R. 1997. Structure and function in whale ears. Bioacoustics 8:103–135.

Ketten, D. R. 2000. Cetacean ears; pp. 43–108 in W. W. L. Au, A. N. Popper, and R. R. Fay (eds.), Hearing by Whales and Dolphins. Springer-Verlag, New York.

Ketten, D. R., and D. Wartzok. 1990. Three-dimensional reconstructions of the dolphin ear; pp. 81–105 in J. A. Thomas and R. A. Kastelein (eds.), Sensory Abilities of Cetaceans. NATO ASI Series A: Life Sciences, Vol. 196. Plenum Press, New York.

Ketten, D. R., D. K. Odell, and D. P. Domning. 1992. Structure, function, and adaptation of the manatee ear; pp. 77–95 in J. A. Thomas, R. A. Kastelein, and A. Ya. Supin (eds.), Marine Mammal Sensory Systems. Plenum Press, New York.

Lay, D. M. 1972. The anatomy, physiology, functional significance and evolution of specialized hearing organs of gerbilline rodents. Journal of Morphology 138:41–120.

Manley, G. A., and J. A. Clack. 2004. An outline of the evolution of vertebrate hearing organs; pp. 1–26 in G. A. Manley, A. N. Popper, and R. R. Fay (eds.), Evolution of the Vertebrate Auditory System. Springer-Verlag, New York.

Mednis, A. 1991. An acoustic analysis of the 1988 song of the Humpback whale, *Megaptera novaeangliae*, off eastern Australia. Memoirs of the Queensland Museum 30:323–332.

Møhl, B. 1968. Auditory sensitivity of the common seal in air and water. Journal of Auditory Research 8:27–38.

Møhl, B., W. W. L. Au, J. Pawloski, and P. E. Nachtigall. 1999. Dolphin hearing: relative sensitivity as a function of point of application of a contact sound source in the jaw and head region. Journal of the Acoustical Society of America 105:3421–3424.

Moore, P. W. B., and R. J. Schusterman. 1987. Audiometric assessment of northern fur seals, *Callorhinus ursinus*. Marine Mammal Science 3:31–53.

Norris, K. S. 1964. Some problems of echolocation in cetaceans; pp. 317–336 in W. N. Tavolga (ed.), Marine Bio-Acoustics. Pergamon Press, Oxford.

Norris, K. S. 1968. The evolution of the acoustic mechanisms in odontocete cetaceans; pp. 297–324 in E. T. Drake (ed.), Evolution and Environment. Yale University Press, New Haven, CT.

Nummela, S. 1995. Scaling of the mammalian middle ear. Hearing Research 85:18–30.

Nummela, S. 1997. Scaling and modeling the mammalian middle ear. Comments on Theoretical Biology 4:387–412.

Nummela, S., and M. R. Sanchez-Villagra. 2006. Scaling of the marsupial middle ear and its functional significance. Journal of Zoology 270: 256–267.

Nummela, S., T. Wägar, S. Hemilä, and T. Reuter. 1999a. Scaling of the cetacean middle ear. Hearing Research 133:71–81.

Nummela, S., T. Reuter, S. Hemilä, P. Holmberg, and P. Paukku. 1999b. The anatomy of the killer whale middle ear *(Orcinus orca)*. Hearing Research 133:61–70.

Nummela, S., J. E. Kosove, T. E. Lancaster, and J. G. M. Thewissen. 2004a. Lateral mandibular wall thickness in *Tursiops truncatus*: variation due to sex and age. Marine Mammal Science 20:491–497.

Nummela, S., J. G. M. Thewissen, S. Bajpai, S. T. Hussain, and K. Kumar. 2004b. Eocene evolution of whale hearing. Nature 430:776–778.

Nummela, S., J. G. M. Thewissen, S. Bajpai, S. T. Hussain, and K. Kumar. 2007. Sound transmission in archaic and modern whales: anatomical adaptations for underwater hearing. Anatomical Record A 290:716–733.

Oelschläger, H. A. 1986. Comparative morphology and evolution of the otic region in toothed whales (Cetacea, Mammalia). American Journal of Anatomy 177:353–368.

Oelschläger, H. A. 1990. Evolutionary morphology and acoustics in the dolphin skull; pp 137–162 in J. A. Thomas and R. A. Kastelein (eds.), Sensory Abilities of Cetaceans. NATO ASI Series A: Life Sciences, Vol. 196. Plenum Press, New York.

Repenning, C. A. 1972. Underwater hearing in seals: functional morphology; pp. 307–331 in R. J. Harrison (ed.), Functional Anatomy of Marine Mammals. Academic Press, London.

Reysenbach de Haan, F. W. 1957. Hearing in whales. Acta Oto-Laryngologica Supplement 134:1–114.

Robineau, D. 1969. Les osselets de l'ouie de la Rhytine. Mammalia 29:412–425.

Rosowski, J. J. 1992. Hearing in transitional mammals: predictions from the middle-ear anatomy and hearing capabilities of extant mammals; pp. 615–631 in D. B. Webster, R. R. Fay, and A. N. Popper (eds.), The Evolutionary Biology of Hearing. Springer-Verlag, New York.

Rosowski, J. J., and A. Graybeal. 1991. What did *Morganucodon* hear? Zoological Journal of the Linnean Society 101:131–168.

Rosowski, J. J., and E. M. Relkin. 2001. Introduction to the analysis of middle-ear function; pp. 161–190 in A. F. Jahn and J. Santos-Sacchi (eds.), Physiology of the Ear, 2nd Ed. Singular, New York.

Ruggero, M. A., and A. N. Temchin. 2002. The roles of external, middle, and inner ears in determining the bandwidth of hearing. Proceedings of the National Academy of Sciences USA 99: 13206–13210.

Shaw, E. A. G. 1974. The external ear; pp. 455–490 in W. D. Keidel and W. D. Neff (eds.), Handbook of Sensory Physiology, Vol. V-1, Auditory System. Springer-Verlag, Berlin.

Tonndorf, J. 1968. A new concept of bone conduction. Archives of Otolaryngology 87:49–54.

Urick, R. J. 1975. Principles of Underwater Sound. McGraw-Hill, New York.

Varanasi, U., and D. C. Malins. 1971. Unique lipids of the porpoise *Tursiops gilli*: differences in triacylglycerols and wax esters of acoustic (mandibular canal and melon) and blubber tissues. Biochimica Biophysica Acta 231:415–418.

Varanasi, U., and D. C. Malins. 1972. Triacylglycerols characteristic of porpoise acoustic tissues: molecular structures of diisovaleroylglycerolides. Science 176:926–928.

Wartzok, D., and D. R. Ketten. 1999. Marine mammal sensory systems; pp. 117–175 in J. E. Reynolds III and S. A. Rommel (eds.), Biology of Marine Mammals. Smithsonian Institution, Washington, DC.

Webster, D. B., and W. Plassmann. 1992. Parallel evolution of low-frequency sensitivity in Old World and New World desert rodents; pp. 633–636 in D. B. Webster, R. R. Fay, and A. N. Popper (eds.), The Evolutionary Biology of Hearing. Springer-Verlag, New York.

Webster, D. B., and M. Webster. 1975. Auditory systems of Heteromyidae: functional morphology and evolution of the middle ear. Journal of Morphology 146:343–376.

Balance

14

The Physics and Physiology of Balance

Justin S. Sipla and Fred Spoor

Functional Morphology
Function on Land and in Water

The sensory organ of balance, or vestibular system, is located in the inner ear of all vertebrates and provides an organism with accurate information on its body movements. This information is transmitted to the brain for participation in muscular reflex arcs that function to stabilize the trunk, head, and visual field during postural and locomotor activities. The focus in this brief review is on part of the organ of balance, the semicircular canal system, because it is the comparative, functional, and evolutionary morphology of this part that is increasingly well understood.

The vestibular system was one of the first sensory systems established in vertebrate evolution and arose early in the evolution of the lineage as a modification of the basic neuromast organ (Walker and Liem, 1994). A neuromast is a population of hair cells with cilia. The cilia are embedded in a gelatinous matrix called a cupula. The neuromast produces electrical signals in response to mechanical stimulation of the cupu-

lar mass. Vestibular neuromasts are located in a "labyrinth" of membranous ducts and compartments filled with endolymph, encased in the hard, avascular bony otic capsule of the cranium. Each hair cell of the vertebrate neuromast bears a bundle of cilia, which cannot be moved actively by the cell. These are arranged in ranks of stereocilia of increasing length, and headed by a single, substantially longer kinocilium. Bending of the stereocilia causes cell excitation when the bending is toward the kinocilium, and inhibition when away from the kinocilium.

In gnathostome vertebrates, each side of the bilateral vestibular system consists of two types of motion sensors: two otolith organs, in the membranous utricle and saccule, which inform the brain about changes in gravitational and other forms of linear acceleration, and three toroidal (more or less circular) semicircular ducts—anterior, posterior, and lateral—which sense rotational movement in three planes (Fig. 14.1). Following the *Terminologia Anatomica* (Federal Committee on Anatomical Terminology, 1998), we use the term semicircular canal to refer to the bony morphology, and the term semicircular duct to the membranous duct inside the

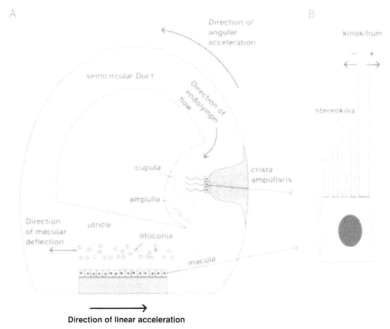

Direction of linear acceleration

FIGURE 14.1. Schematic representation of the vestibular system, showing the major components and properties of the system (A), and detail of a sensory neuron (B) in the macula and crista ampullaris. Linear acceleration of the head causes the otoconia to deflect off the stereocilia and kinocilium of the macula, and angular acceleration of the head causes the endolymph of the semicircular canals to deflect the cupula and its embedded cilia.

canal. The term semicircular canal system covers the entire functional unit, including both bony and soft-tissue aspects.

The functional basis of the otolith system is that linear acceleration of the head moves dense crystals (otoliths or otoconia) by inertia, and that these movements are detected by an associated bed of neuromast hair cells called a macula. The semicircular canal system functions on the principle that when the head experiences angular acceleration endolymph inside the semicircular ducts undergoes relative movement with respect to the duct wall, because it lags behind due to inertia. This fluid motion is resisted by the viscoelastic properties of the cupula, which spans the base of each duct in a dilated region called the ampulla and prevents endolymph from flowing around it. Displacement or deformation of the cupula activates hair cells in the ampulla, which are situated on a raised area of supporting cells called the crista ampullaris (Hillman and McLaren, 1979; Dickman et al., 1988).

Stereocilia within the ampullary regions of the semicircular canals are oriented in one direction, such that all hair cells in a given ampulla are maximally excited by angular rotation in a single direction (Wersäll and Bagger-Sjöbäck, 1974).

The semicircular ducts are aligned in approximately orthogonal planes in three-dimensional space so that rotational acceleration stimulates each canal to a different degree, thereby informing the brain about the nature of the motion. Pairs of canals on opposite sides of the skull tend to parallel one another, and when activated generate outputs of opposite sign (referred to as a "push-pull" operational mode [Graf, 1988]). The functional pairs are right lateral–left lateral, right anterior–left posterior, and left anterior–right posterior. By combining input from each of the canal pairs, the brain creates a representation of the vector describing the instantaneous speed of head rotation relative to three-dimensional space (Graf, 1988). Because the canals are not perfectly toroidal and do not lie in perfectly flat planes (Oman et al., 1987), and because endolymph conduits are hydrodynamically shared by all three canals in the utricular vestibule and common crus (Muller and Verhagen, 1988a, 1988b), interactions between the canals are complicated and driven by pressure gradients (Rabbitt et al., 1998). Although the physical stimulus of the semicircular canals is angular acceleration, the neural output from the sensory cells in the ampulla instantaneously represents the velocity of the head; that is, deflection of the cupula is proportional to angular head velocity (Goldberg and Fernández, 1975).

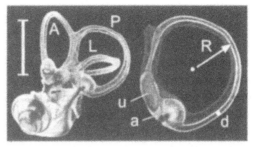

FIGURE 14.2. Lateral view of a left dissected baboon inner ear, showing the anterior (A), posterior (P) and lateral (L) semicircular canals and the enclosed membranous ducts (*left*). The whorl-shaped organ near the *bottom left* is the cochlea. On the *right*, the posterior duct has been isolated, demonstrating the ampulla (a) with the crista and cupula inside, and the utricle (u). The radius of curvature (R) and the lumen diameter (d) of the duct are indicated. Scale is 5 mm and pertains to *left image*. Modified after Gray (1907).

FUNCTIONAL MORPHOLOGY

When a duct is rotated, the endolymph carries out an excursion with respect to the duct wall, followed by a restoring movement to its original position. The excursion and restoring movements are characterized by two time constants (see work by Wilson and Melvill Jones [1979] and Rabbitt et al. [2004] for reviews of the biomechanics of the canal system). The long time constant is a measure of the recovery time of the system—that is, the time it takes the cupula to return to its original position after a deflection. The short time constant is a measure of the response speed of the system, or how long it takes for the cupula to displace. The maximum displacement of the endolymph relative to angular head velocity is a measure of the sensitivity of the semicircular canal system. Sensitivity represents the ability to resolve differences in angular velocity and acceleration (Muller, 1990, 1994, 1999; Rabbitt et al., 2004).

The physical dimensions of the duct parts influence the response time and sensitivity of the semicircular canal system. Response time increases with the lumen size of the duct, and sensitivity increases with both the lumen size and the circuit length of the duct, the latter often expressed by its radius of curvature (Fig. 14.2, *d* and *R*, respectively). Taking reliable measurements of these fragile structures requires fixa-

tion of the inner ear soon after death by either dissection or serial sectioning (Curthoys and Oman, 1986, 1987; Lindenlaub et al., 1995). However, the bony semicircular canals that enclose these organs can be investigated easily using computed tomography and other techniques (Spoor et al., 2000). These techniques enable researchers to reconstruct and measure two of the most important properties of semicircular canal function: arc size and planar orientation. Arc size, in particular, is important in the context of evolutionary studies, as this property has been linked to locomotor behavior (Spoor et al., 2002; Spoor et al., 2007; Sipla et al., 2003, 2004). In contrast, the lumen size of the bony canal is of limited functional significance as it does not adequately reflect the lumen size of the enclosed membranous duct in many groups of vertebrates (e.g., Gray, 1907, 1908; Ramprashad et al., 1984).

Numerous comparative studies have suggested an association between the size of the semicircular canals and locomotor behavior (reviewed in Spoor, 2003). Jones and Spells (1963) observed that among a comparative sample of 87 vertebrate species, duct dimensions scaled to body mass with negative allometry. Since the semicircular canal system functions as an angular velocity transducer over a limited range of sinusoidal frequencies only, they inferred that this "middle"-frequency bandwidth of the canal system would be attuned to the natural frequency content of an animal's normal head movements, with a corresponding shift in the range of cutoff frequencies. They speculated that interspecific differences in duct dimensions were due to functional attunements of the semicircular duct system "to match the likely patterns of head movements . . . according to size, shape and habitat of the animal" (p. 416). In larger animals, where physical constraints impose slower head movements, Jones and Spells (1963) argued that a more sensitive duct organ was needed to maintain the dynamic response characteristics of the system. They hypothesized, on biophysical grounds established by Steinhausen (1931, 1933) and Van Egmond et al. (1949), that this was achieved by increasing the internal radius of the duct lumen (d) and the circumferential length (proportional to R) of the semicircular duct.

Even though the body size of larger animals likely corresponds to reduced angular head accelerations, Spoor (2003) questioned whether the allometric increase in canal arc size represents a functional compensation for increased sluggishness, as suggested by Jones and Spells (1963). There is considerable variation in the size and shape of the semicircular canal system among species with the same body mass. Among these, it is the species with a more acrobatic behavior that shows canals with a larger arc size, not the sluggish ones (Spoor et al., 2007). This, it was argued, is because a larger arc size increases the sensitivity, thus providing the accurate canal response that is required when navigating a three-dimensional environment in an agile and acrobatic way (Spoor et al., 2002; Spoor, 2003; Spoor et al., 2007). This approach emphasizes that it is the interspecific deviations from the general scaling trend of vertebrates that are of interest, rather than the slope itself, as they represent attunements of the system to the movement criteria of individual species.

The semicircular canals exert reflexive control over eye position and function to maintain a stable retinal image despite changes in head position. In response to a given head rotation, the vestibulo-ocular reflex is activated to keep eye orientation invariant relative to space by generating an involuntary eye movement having the same amplitude but opposite direction as the perceived head motion (Minor et al., 1999). This is accomplished by transmission of sensory signals to the vestibular ganglion, which projects to vestibular brain stem nuclei via the vestibulo-cochlear nerve (cranial nerve VIII). Projections from the vestibular nuclei interact with motor nuclei of the extraocular muscles to generate reflexive movements of the eyes.

The vestibular nuclei also project to motor nuclei of the neck and trunk musculature, and to the cerebellum. These projections mediate a variety of reflexes of the neck, limbs and vertebral column. These reflexes contribute to head and trunk stabilization during locomotion and are variably integrated with visual and proprioceptive feedback (e.g., Money and Scott, 1962; Suzuki and Cohen, 1964; Dietz, 1992; Ito et al., 1992). Due to the multisensory integration of the vestibular system with multiple reflexive motor systems, the overall function of the semicircular canals should not be viewed in an isolated context, but as part of a feedback system incorporating all of these components.

FUNCTION ON LAND AND IN WATER

Unlike most other sense organs, the biophysical and physiological aspects of the vestibular organ do not differ functionally in air and in water. However, habitual aquatic behavior places different demands on vestibular input than do forms of terrestrial locomotion. Freed in water from the need for intermittent contact with a surface substrate, and with strongly reduced effects of gravity, a wider repertoire of body movements is potentially possible, including forms of rotations that would not be feasible on land. On the other hand, legged animals that walk, run, or leap experience a footfall-related pattern of discrete mechanical pulses of acceleration and deceleration, whereas aquatic animals typically remain in constant contact with a high-viscosity medium, and this tends to dampen such pulses. Aquatic animals move in a medium that itself moves, generally at a lower velocity and frequency than the animal itself (water currents). In principle the vestibular system will sense the sum of all movements, whereas an animal would primarily require feedback only regarding its own motion relative to the direct environment. Aquatic behavior shares some of these characteristics with airborne flight, but the differences in viscosity and density of water and air create a fundamentally different set of circumstances.

Varied approaches to swimming place different demands on the vestibular sensory apparatus. Differences in the hydrodynamics of swimming movements inherent to different vertebrate body plans and fin morphologies further modify the sensory environment. Many secondarily aquatic tetrapods use forms of lateral undulation to generate propulsive forces, such as most mosasaurs (Lingham-Soliar, 1991) and crocodiles, or lateral hindfoot movements, as in true seals (phocids) and the walrus. Penguins, sea turtles, and eared seals (otariids) employ a birdlike forelimb flight stroke when swimming, often characterized as subaqueous flying, in which the upstroke and downstroke both generate thrust (Walker, 1971; Clark and Bemis, 1979; Fish, 1996; Fish et al., 2003). Cetaceans and sirenians undulate the trunk dorsoventrally during swimming, rather than laterally, and have evolved horizontal tail flukes for thrust generation. Other specializations may be held in common among groups of aquatic vertebrates (e.g., neck reduction due to streamlining), owing to the high viscosity of water and the general slowing effect of the medium on accelerations, and the need to sense them. In addition, it must be noted that body temperature affects the corresponding temperature of labyrinthine fluids (i.e., endolymph) and may be a factor in determining the sensitivity of the canal organs. Ten Kate and Kuiper (1970), for instance, suggested that the larger semicircular canal size of fishes relative to land vertebrates was due in part to their lower body temperature, and the same could apply to other ectotherm vertebrates. Differences in the functional parameters of the canal system (size, shape, and planar orientation of the canals) are therefore expected to correspond in some way with preferred movements across the range of secondarily aquatic taxa, and possibly with a wider range of physiological adaptations associated with aquatic habits.

LITERATURE CITED

Clark, D. B., and W. Bemis. 1979. Kinematics of swimming of penguins at the Detroit Zoo. Journal of Zoology (London) 188:411–428.

Curthoys, I. S., and C. M. Oman. 1986. Dimensions of the horizontal semicircular duct, ampulla and utricle in rat and guinea pig. Acta Oto-Laryngologica 101:1–10.

Curthoys, I. S., and C. M. Oman. 1987. Dimensions of the horizontal semicircular duct, ampulla and utricle in human. Acta Oto-Laryngologica 103:254–261.

Dickman, J. D., P. A. Reder, and M. J. Correia. 1988. A method for controlled mechanical stimulation of single semicircular canals. Journal of Neuroscience Methods 25:111–119.

Dietz, V. 1992. Human neuronal control of automatic functional movements: interaction between central programs and afferent input. Physiological Reviews 72:33–69.

Federal Committee on Anatomical Terminology. 1998. Terminologia Anatomica. Thieme, Stuttgart.

Fish, F. E. 1996. Transitions from drag-based to lift-based propulsion in mammalian swimming. American Zoologist 36:628–641.

Fish, F. E., J. Hurley, and D. P. Costa. 2003. Maneuverability by the sea lion Zalophus californianus: turning performance of an unstable body design. Journal of Experimental Biology 206:667–674.

Goldberg, J. M., and C. Fernández. 1975. Vestibular mechanisms. Annual Review of Physiology 37:129–162.

Graf, W. 1988. Motion detection in physical space and its peripheral and central representation. Annals of the New York Academy of Sciences 545:154–169.

Gray, A. A. 1907. The Labyrinth of Animals, Vol. 1. Churchill, London.

Gray, A. A. 1908. The Labyrinth of Animals, Vol. 2. Churchill, London.

Hillman, D. E., and J. W. McLaren. 1979. Displacement configuration of semicircular canal cupulae. Neuroscience 4:1989–2000.

Ito, S., S. Odahara, and M. Hirando. 1992. Cristospinal reflex in circular walking. Acta Oto-Laryngologica 112:170–173.

Jones, G. M., and K. E. Spells. 1963. A theoretical and comparative study of the functional dependence of the semicircular canal upon its physical dimensions. Proceedings of the Royal Society of London B 157:403–419.

Lindenlaub, T., H. Burda, and E. Nevo. 1995. Convergent evolution of the vestibular organ in the subterranean mole-rats, *Cryptomys* and *Spalax*, as compared with the aboveground rat, *Rattus*. Journal of Morphology 224:303–311.

Lingham-Soliar, T. 1991. Locomotion in mosasaurs. Modern Geology 16:229–248.

Minor, L. B., D. M. Lasker, D. D. Backous, and T. E. Hullar. 1999. Horizontal vestibuloocular reflex evoked by high-acceleration rotations in the squirrel monkey. I. Normal responses. Journal of Neurophysiology 82:1254–1270.

Money, K. E., and J. W. Scott. 1962. Functions of separate sensory receptors of nonauditory labyrinth of the cat. American Journal of Physiology 202:1211–1220.

Muller, M. 1990. Relationship between semicircular duct radii with some implications for time constants. Netherlands Journal of Zoology 40:173–202.

Muller, M. 1994. Semicircular duct dimensions and sensitivity of the vertebrate vestibular system. Journal of Theoretical Biology 167:239–256.

Muller, M. 1999. Size limitations in semicircular duct systems. Journal of Theoretical Biology 198:405–437.

Muller, M., and J. H. G. Verhagen. 1988a. A new quantitative model of total endolymph flow in the system of semicircular ducts. Journal of Theoretical Biology 134:473–501.

Muller, M., and J. H. G. Verhagen. 1988b. A mathematical approach enabling calculation of the total endolymph flow in the semicircular ducts. Journal of Theoretical Biology 134:503–529.

Oman, C. M., E. M. Marcus, and I. S. Curthoys. 1987. The influence of semicircular canal morphology on endolymph flow dynamics. Acta Otolaryngologica (Stockholm) 103:1–13.

Rabbitt, R. D., R. Boyle, and S. M. Highstein. 1998. Responses of patent and plugged semicircular canals to linear and angular acceleration. Society for Neuroscience Abstracts 24:1653.

Rabbitt, R. D., E. R. Damiano, and J. Wallace Grant. 2004. Biomechanics of the semicircular canals and otolith organs; pp. 153–201 in S. M. Highstein, R. R. Fay, and A. N. Popper (eds.), The Vestibular System. Springer-Verlag, New York.

Ramprashad, F., J. P. Landolt, K. E. Money, and J. Laufer. 1984. Dimensional analysis and dynamic response characterization of mammalian peripheral vestibular structures. American Journal of Anatomy 169:295–313.

Sipla, J. S., J. A. Georgi, and C. A. Forster. 2003. The semicircular canal dimensions of birds and crocodilians: implications for the origin of flight. Journal of Vertebrate Paleontology 23 (3, Suppl.): 97A.

Sipla, J. S., J. A. Georgi, and C. A. Forster. 2004. The semicircular canals of dinosaurs: tracking major transitions in locomotion. Journal of Vertebrate Paleontology 24 (3, Suppl.): 113A.

Spoor, F. 2003. The semicircular canal system and locomotor behaviour, with special reference to hominin evolution. Courier Forschungsinstitut Senckenberg 243:93–104.

Spoor, F., N. Jeffery, and F. Zonneveld. 2000. Imaging skeletal growth and evolution; pp. 123–161 in P. O'Higgins and M. Cohn (eds.), Development, Growth and Evolution. Academic Press, London.

Spoor, F., S. Bajpai, S. T. Hussain, K. Kumar, and J. G. M. Thewissen. 2002. Vestibular evidence for the evolution of aquatic behaviour in early cetaceans. Nature 417:163–166.

Spoor, F., T. H. Garland, G. Krovitz, T. M. Ryan, M. T. Silcox, and A. Walker. 2007. The primate semicircular canal system and locomotion. Proc. Nat. Acad. Sci. 104:10808–10812.

Steinhausen, W. 1931. Über den Nachweis der Bewegung der Cupula in der intaktene Bogengangsampulle des Labyrinthes bei der naturlichen rotatorischen und calorischen Reizung. Pflugers Archiv 225:322–328.

Steinhausen, W. 1933. Über die Beobachtungen der Cupula in den Bogengangsampullen des Labyrinthes des Lebendes Hechts. Pflugers Archiv 232:500–512.

Suzuki, J.-I., and B. Cohen. 1964. Head, eye, body and limb movements from semicircular canal nerves. Experimental Neurology 10:393–405.

Ten Kate, J. H., and J. W. Kuiper. 1970. The viscosity of the pike's endolymph. Journal of Experimental Biology 53:495–500.

Van Egmond, A. A. J., J. J. Groen, and L. B. W. Jongkees. 1949. The mechanics of the semicircular canal. Journal of Physiology 110:1–17.

Walker, W. F. 1971. Swimming in sea turtles of the family Cheloniidae. Copeia 2:229–233.

Walker, W. F., and K. F. Liem. 1994. Functional Anatomy of the Vertebrates. Saunders College Publishing, New York.

Wersäll, J., and D. Bagger-Sjöbäck. 1974. Morphology of the vestibular sense organ; pp. 123–170 in H. H. Kornhuber (ed.), Handbook of Sensory Physiology: Vestibular System. Springer-Verlag, New York.

Wilson, V. J., and G. Melvill Jones. 1979. Mammalian Vestibular Physiology. Plenum Press, New York.

15

Comparative and Functional Anatomy of Balance in Aquatic Reptiles and Birds

Justin A. Georgi and Justin S. Sipla

The inner ear of vertebrates, which houses all of the vestibular endorgans responsible for the sensation of balance and orientation, is a structure of complex morphology and functional dynamics. The complexity of this system allows for variations in its morphology to affect changes in its functional response, and, in this way, the vestibular system may be fine-tuned to specific sensory demands imposed by an organism's behavior, including the specifics of locomotor function.

Accurate perception of balance and orientation is a key factor for successful locomotion. Prior research into the relationship between vestibular morphology and locomotion has uncovered a correlation between size of the semicircular canals and general mode of locomotion in several mammalian groups (e.g., Spoor et al., 1996, 2002; Spoor and Thewissen, chapter 16 in this volume), but surprisingly little attention has been given to nonmammalian amniotes. Furthermore, variables other than semicircular canal size are known to influence the dynamic characteristics of the canal system. But until now the adaptive significance of these variables has remained untested.

A direct connection between an organism's ability to perceive orientation and balance and its ability to locomote with facility has been known since the early nineteenth century when the physiologist Jean-Pierre-Marie Flourens (1828) studied the function of semicircular ducts in pigeons and rabbits. Despite this empirically demonstrable relationship, it was not until the turn of the twentieth century that a hypothesis stated a direct connection between a specific semicircular duct morphology and a specific mode of locomotion. Albert A. Gray (1906), as he began the first broad survey of vestibular morphology across vertebrates, noted the dramatic difference in semicircular duct morphology between most mammals and the three-toed sloth, *Bradypus tridactylus*, and, thus, hypothesized,

> It may be that this small size of the canals, associated with their irregular shape, may be in some way related to the sloth's clumsy and slow movements. The life which they lead, with the body inverted, as it almost continuously is, may also be connected in some way to the curious development of these organs. (p. 290)

This connection was not revisited in detail until Jones and Spells (1963) undertook a survey of semicircular duct dimensions across vertebrates and postulated a connection between the measures of size of the duct, the body mass of the organism, and the organism's locomotor capabilities (for a review see chapter 14 by Sipla and Spoor in this volume).

The implications of Jones and Spells's work led researchers to further examine this correlation (e.g., Spoor et al., 1994, 2002). However, the restrictions imposed by examining only bony tissue often limit these studies to the exclusive consideration of the metric referred to as "radius of curvature" by Jones and Spells (1963). The radius of curvature is the average radius of the approximately circular path followed by the canal. These studies do not take into consideration the shape of the canal path.

Canal shape is, to a large extent, phylogenetically conservative, and gross morphological similarities exist at the family or order level regardless of canal function or locomotor habit. Within those higher groups, however, there is considerable secondary variation in shape that appears to correlate with locomotor preference. In some cases, this secondary variation is so extreme as to appear as tight morphological convergence (e.g., the anterior canals of mosasaurs and aquatic crocodilians). Here, we consider two functional questions, both involving shape variation within the orders of amniotes: Do changes of canal shape correspond to differences between aquatic and terrestrial locomotion within closely related taxa? If this question is answered in the affirmative, the second question becomes: Can these differences be generalized across diverse groups?

Regarding the first question, variation from the "average" canal system within each group may represent the adaptation of individual taxa within that group to differing modes of locomotion. All of the previous studies that have sought to correlate semicircular canal morphology to locomotion have focused entirely on groups at these lower taxonomic levels (Spoor et al., 1994, 1996, 2002; Spoor and Zonneveld, 1998; J. S. Sipla, J. A. Georgi, C. A. Forster, unpublished). Using varying methods, and focusing on varying aspects, these studies have each demonstrated some correlation between aspects of canal morphology and locomotor behavior. Among these studies, Spoor et al. (2002) demonstrated a correlation between vestibular morphology and a return to aquatic locomotion in cetaceans. However, this connection has not been investigated for any group of secondarily aquatic nonmammalian amniotes.

Regarding the second question, although studies in the past have demonstrated correlations between semicircular canal size and locomotion, the taxonomic breadth to allow for generalizations of the patterns was not available. The diverse taxonomic groups incorporated into this study present the opportunity to evaluate such trends within nonmammalian amniotes.

This chapter investigates evolution of the shape of the semicircular canals in response to reinvasion of aquatic environments by turtles, varanoids (extant varanid lizards and mosasaurs), crocodilians, and birds. Despite a vast phylogenetic difference in canal shape between these groups, we show that there is an underlying consistent morphological response to the adoption of a secondarily aquatic lifestyle.

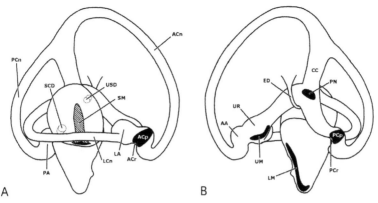

FIGURE 15.1. Generalized left membranous labyrinth in lateral (A) and medial view (B), based on *Iguana iguana*. Modified from Wever (1978). Abbreviations: AA, anterior ampulla; ACn, anterior semicircular canal; ACp, anterior cupula; ACr, anterior crista; CC, common crus; ED, endolymphatic duct; LA, lateral ampulla; LCn, lateral semicircular canal; PCn, posterior semicircular canal; PCp, posterior cupula; PCr, posterior crista; PN, papilla neglecta; SCD, sacculocochlear duct; SM, saccular macula; UM, utricular macula; UR, utricular recess; and USD, utriculosaccular duct.

GENERALIZED ANATOMY IN REPTILES AND BIRDS

SENSORY ENDORGANS

Of the eight sensory endorgans generally found within the amniote otic labyrinth, seven are associated with the sensation of balance. Though some caution should be exercised when starkly dividing these endorgans into vestibular and acoustic functions (Wever, 1978; Harada et al., 2001), the three cristae of the semicircular ducts, the three maculae, and the papilla neglecta all participate primarily in the sensation of balance and orientation. The neuronal integration of these different sensors allows the development of precise indication of the movements and position of the head in three-dimensional space.

Due to a high degree of conservation within the vestibular system, the anatomy of a membranous labyrinth generalized for nonmammalian amniotes (Fig. 15.1) does not differ significantly from that of the generalized gnathostome vertebrate briefly described in the previous chapter (Sipla and Spoor, chapter 14 in this volume). Traditional morphological description separates the entire structure into a

superior and an inferior division. The superior division of the vestibular system includes the three semicircular ducts and the utricle, with which the ducts are directly confluent. Thus, the superior division includes the three cristae and the utricular macula. It also includes the papilla neglecta (or macula neglecta), an endorgan found in the wall of the utricle. The inferior division encompasses the saccule, the cochlear duct, and the endolymphatic duct. Its two balance endorgans are the saccular macula and, within the cochlear duct, the lagenar macula.

The utricle is a roughly tubular vesicle with connections to each of the semicircular ducts and the saccule. The slightly expanded anterior portion of the utricle, the utricular recess, houses the macula. The semicircular canals of nonmammalian amniotes evince the standard morphology, with ampullae (each containing a single crista and its cupula) located at the junction of one duct end and the utricle. The anterior and lateral ampullae are adjacent to one another on the anterior end of the utricular recess, whereas the posterior ampulla is confluent with the posterior end of the utricle. The two vertical canals, the anterior and posterior, join to form a common crus, which structure rejoins the utricle near its posterior end; at this point of the utricular wall, situated near the common crus, is found the papilla neglecta. The utricle is confluent with the saccule via a small utriculosaccular duct.

The saccule is a bulbous vesicle with the saccular macula covering much of the medial and ventral walls. In addition to its connection with the utricle, the saccule is connected to the cochlear duct via the sacculocochlear duct. The endolymphatic duct arises from the medial wall of the saccule. The acoustic functions of the inner ear belonging to the cochlear duct and the basilar papilla are discussed elsewhere (Hetherington, chapter 12 in this volume; Nummela, chapter 13 in this volume). However, the cochlear duct also contains, usually on its ventral and anterior walls, a balance endorgan, the lagenar macula.

These membranous structures of the otic labyrinth are housed within three bones of the posterior braincase, the prootic anteriorly, the supraoccipital superiorly, and the opisthotic posteriorly. All three bones contribute more-or-less equally to the medial wall of the otic capsule and the bony cavity that houses the utricle and saccule. The bony semicircular canals, through which the semicircular ducts pass, each traverse two of the three bones (Fig. 15.2). The anterior halves of the anterior and lateral canals pass through the prootic, while the posterior halves of the lateral and posterior canals pass through the opisthotic. The remaining canal sections, including the common crus and associated elements of the anterior and posterior canals, are housed in the supraoccipital.

As a result of the perilymphatic space surrounding the membranous labyrinth, correlation of bony morphology with membranous morphology varies by structure. The large indistinct bony cavity for the utricle and saccule preserves little or no direct evidence of the specific morphologies of these two structures. In contrast, the semicircular canals reveal, with a reasonable measure of accuracy, the paths and planar orientations of each of the semicircular ducts. The bony canals, however, do not preserve

an accurate indication of the cross-sectional area of the semicircular ducts. It has been shown that no consistent relationship exists across different amniote taxa between bony canal and semicircular duct cross-sectional area (Gray, 1907, 1908; Ramprashad et al., 1984).

Due to this inconsistency in bony and soft tissue correlation, any study that seeks to understand the evolution of balance sensation is limited in scope to the shape, size, and planar orientation of the semicircular canals when considering fossil taxa. Thus, the rest of this chapter focuses on the evolution of the shape of the bony semicircular canals as their shape relates to differences in locomotion between terrestrial and aquatic members of diverse nonmammalian amniote groups.

SEMICIRCULAR CANAL ANATOMY

The term semicircular stems from of a generic mammalian view of canal morphology, for nearly all mammals evince a canal shape that closely approximates a circle. Within the rest of the amniotes, however, the shape of the canals can vary. Given this variation, it is difficult to label the morphology of any single canal system as a generalized example for all nonmammalian amniotes. Extant varanid lizards, however, have a canal shape that shares at least some characteristics with most other groups. Therefore, the canal system of varanid lizards is described here as a starting point.

SQUAMATES

In the varanid semicircular canal system (Fig. 15.3), the short common crus rises from the vertically expanded bony vestibular cavity medial to the rounded peak of that cavity. The branching of the posterior canal is oriented directly posterolaterally, and that of the anterior canal is oriented at a slight angle above horizontal as it heads anterolaterally. The anterior canal then traverses a very elongate course, the primary section of which is sublinear and elevated only a small distance above the obliquely oriented wall of the vestibular cavity. A sharp ventral turn at the anterior end brings the anterior

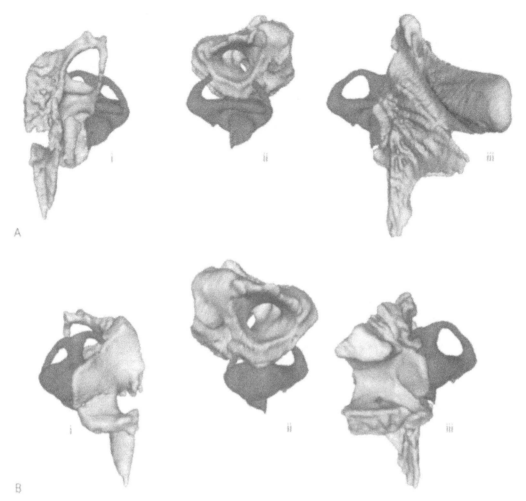

FIGURE 15.2. Three-dimensional reconstructions of the bony inner ear cavity of *Crocodylus porosus*, showing contributions of the prootic (i), supraoccipital (ii), and opisthotic (iii) to bony inner ear cavity in lateral (A) and medial (B) view. Bony labyrinth endocast is shown in dark gray in all reconstructions.

canal into communication with the ampulla, the bony contour of which merges with the vestibular wall to form a complete anterior canal circuit. The posterior canal, after branching from the common crus, runs a similar though less elongate path than the anterior canal. At its most posterior, where it gently bends ventrally, the bony posterior canal intersects, and is confluent with, the lateral canal, and its remaining course is ventral to the plane of the lateral canal. This communication is a feature only of the perilymphatic space and is not shared by the endolymph-filled membranous ducts. The lateral canal, not including the portion of its circuit composed by the vestibular wall, is the most rounded of the three canals. However, the vestibular wall coplanar with the lateral canal bulges toward the center of the path and produces a concave portion of the canal circuit, which gives the interior circuit of the lateral canal an overall crescentic appearance. In some taxa, such as *Varanus niloticus*, this vestibular bulge can become exaggerated and produce a markedly slender, but still curved, interior circuit. It is important to note that this bulge in the vestibular wall is the contour of the saccule and not the lateral semicircular duct, which passes around the saccule to join the utricle on the medial side vestibular cavity.

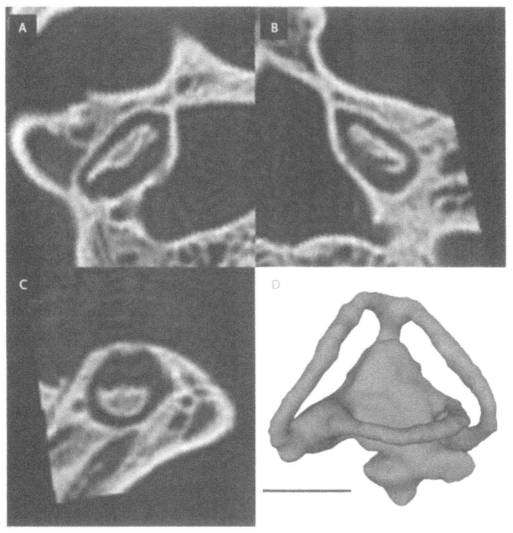

FIGURE 15.3. Computed tomography slices through anterior canal (A), posterior canal (B), and lateral canal (C), and three-dimensional reconstruction of endocast of the inner ear in superolateral view (D) of the squamate *Varanus niloticus*. Scale bar is 5 mm.

TURTLES

Variations from this generalized plan are dramatic. Within turtles, several different morphologies can be found. Turtles all share a vestibular cavity that is less bulbous than in varanids in relation to the position and size of the canals. Thus, the common crus, which is as short or far shorter than the varanid condition, arises from the very apex of the vestibular cavity.

For tortoises, such as those within the genus *Geochelone* (Fig. 15.4), all three canals are more robust and the anterior and posterior canals are more symmetrical than in the varanid. Each vertical canal branches from the common crus in an approximately horizontal plane and gently curves through an approximately 90° arc before reaching the ampulla. The ampulla, utricular region, and common crus are even thicker than the robust canals themselves. The lateral canal is very close in shape and thickness to the two vertical canals and also traces an even 90° arc as it passes from posterior utricle to the lateral ampulla.

Other turtles that exhibit primarily terrestrial behaviors share morphological features of their

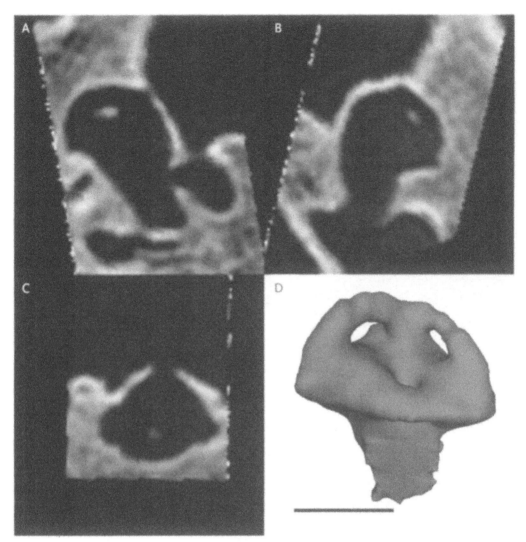

FIGURE 15.4. Computed tomography slices through the anterior canal (A), posterior canal (B), and lateral canal (C), and three-dimensional reconstruction of endocast of the inner ear in superolateral view (D) of the tortoise *Geochelone nigra*. Scale bar is 5 mm.

canal system with tortoises. Within the family Emydidae the terrestrial members, box turtles (genus *Terrapene*) and wood turtles (genus *Clemmys*), each have canals that echo the robust and symmetrical nature of the tortoise system.

CROCODILIANS

Most modern crocodilians have semicircular canals (Fig. 15.5) that are similar in shape: canals that are more rounded than the generalized varanid. The common crus is usually taller than in the varanid, and the branching of the two vertical canals is not equal. The anterior canal rises taller from the common crus than the posterior. In conjunction with the arced wall of the crus, the anterior canal forms a broad curve that continues its gentle contour toward the ampulla and then terminates in a much sharper curve. This results in the typical ovoid shape of the crocodilian anterior canal, with its apex at the ampullary end and its long axis running from the midpoint of the crus to a point just anterior to the ampulla. The posterior canal, though smaller in size than the anterior, shares the ovoid shape of the anterior canal. However, the shorter vertical excursion of the

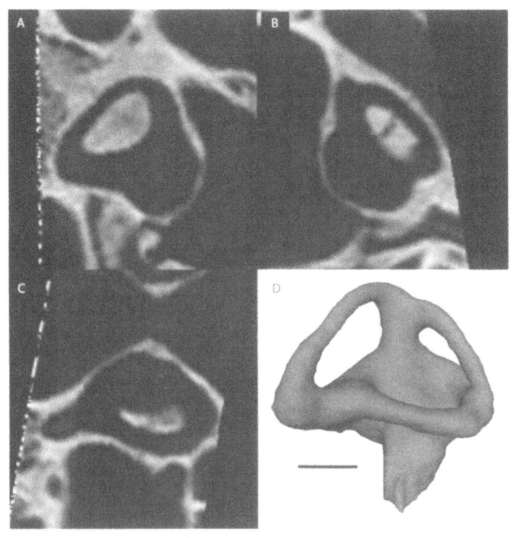

FIGURE 15.5. Computed tomography slices through the anterior canal (A), posterior canal (B), and lateral canal (C), and three-dimensional reconstruction of endocast of the inner ear in superolateral view (D) of *Crocodylus palustris*. Scale bar is 5 mm.

posterior canal from the common crus reduces the ovoid nature of the canal circuit and, at times, produces a far more evenly elliptical course. The lateral canal, though in some ways similar to the crescent-shaped varanid canals, is less sharply curved. The crocodilian lateral canal also has a much smoother interface between canal and vestibule wall, thus, rather than the sharp corners of the varanid lateral canal, it has more rounded elliptical ends and an overall more rounded circuit contour.

The extant crocodilians described above are capable of terrestrial locomotion to varying degrees. Nonetheless, they all exhibit a semi-aquatic habitat preference and locomotor mode. However, within the fossil lineage leading up to modern crocodilians there is a diverse range of habitats and locomotor behaviors (Parrish, 1987; Hua and Buffetaut, 1997) and an associated diversity of semicircular canal morphologies.

The basal crocodylomorph *Junggarsuchus sloani* exhibits a high degree of adaptation to a cursorial (i.e., fully terrestrial) locomotor mode (Clark et al., 2004). Its canal morphology differs from that of modern crocodilians only in the aspect ratio (defined as the height of a canal

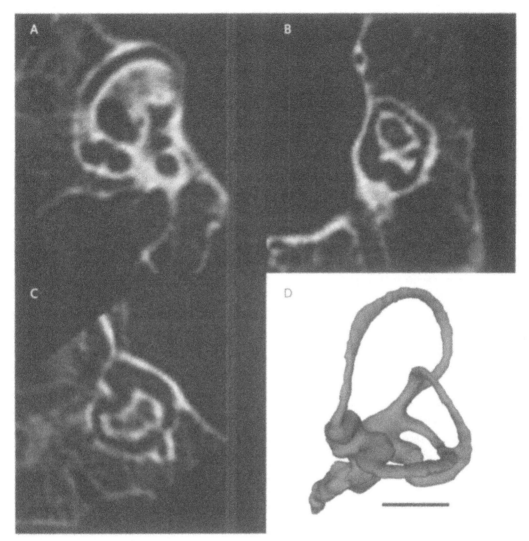

FIGURE 15.6. Computed tomography slices through the anterior canal (A), posterior canal (B), and lateral canal (C), and three-dimensional reconstruction of the inner ear in superolateral view (D) of the ostrich *Struthio camelus*. Scale bar is 5 mm.

divided by its width) of the vertical canals. Both canals maintain an approximately ovoid shape with the apex directed toward the ampulla; however, the greater height of the common crus and the further vertical progress of the canals after branching from the crus produce a canal path that is higher than it is wide. There is slight anterior-posterior elongation in the lateral canal relative to modern crocodilians.

BIRDS

Where the morphology of the anterior canal is concerned, it is the group Aves that deviates the most from the generalized plan (as represented by the terrestrial Ostrich, *Struthio*, Fig. 15.6). These deviations result from a dramatic increase in anterior canal circumference achieved primarily via a hyperelongation in the superioposterior direction. The posterior aspect to the elongation produces an unusual shape. The common crus, rather than passing directly superiorly, is deflected in a posterolateral direction, and the anterior canal branches from it at a sharp posteromedial angle. Thus, a significant portion of the anterior canal circuit travels posterior to the common crus before it curves anteriorly to join the ampulla in the typical position. In many species, the anterior canal shape is further convoluted along its anterior aspect, resulting from encroachment of the optic lobe of the brain into the developmental pathway of the canal during morphogenesis (Bissonnette and Fekete, 1996).

Despite the change in crus morphology, the avian posterior canal has more in common with the typical varanid morphology than most other taxa described here. The shape approximates the same ellipsoid, and the relationship to the lateral canal is the same, including the communication between the perilymphatic spaces. The large size of the avian canals in relation to the vestibular cavity means that there is no incursion of the vestibular wall into the approximately circular path of the lateral canal. In some taxa, however, the posterior portion of the lateral canal is hypertrophied to the extent that its utricular junction is located on the medial side of the posterior utricle.

The morphology described is typical for volant birds, but the overall shape of the system is approximated in secondarily flightless birds. Among terrestrial bird taxa such as *Struthio* and *Rhea*, the anterior canal remains highly elevated, but the degree of posterior expansion is reduced. Within the terrestrial taxa of this study, only *Struthio* demonstrates the posteriorly hypertrophied lateral canal.

SEMICIRCULAR CANAL FUNCTION

Gross anatomical differences in semicircular canal morphology have now been discussed at length. Why, however, we should expect some aspect or aspects of those differences to be related to a transition to an aquatic mode of life has only been briefly mentioned here (see also Sipla and Spoor, chapter 14 in this volume). The membranous organs within this system have been described as an organism's means of perceiving orientation and balance, but the connection to locomotion lies in how that information is used and what sort of perturbations stimulate the system. Specifically, we must understand both the movements that stimulate the semicircular ducts

and the movements imposed on the head by locomotion. The set of movements that these two profiles share is the answer to this puzzle.

The sensation of balance and orientation is traditionally divided into sensation of rotational movement, accredited to the semicircular ducts and their cristae, and sensation of linear movement, accredited to all the remaining nonacoustic endorgans. While it has been repeatedly shown that these distinctions are not so clear cut (e.g., Maxwell, 1919, 1920a, 1920b, 1921; Löwenstein and Compton, 1978; Brichta and Goldberg, 1998, 2000), such a basic description can still be applied regarding the primary function of the semicircular ducts. Numerical modeling and experimental evidence both suggest similar duct function (Van Buskirk et al., 1976), that is, ampullary cristae function as transducers of rotational accelerations and changes in rotational accelerations that occur on the scale of microseconds (Muller, 1994, 2000). Transduction of the remaining types of movement—linear and long-term rotational (i.e., large-scale changes of orientation with respect to the line of gravity)—is primarily accomplished by the macular endorgans. Once afferent nerves carry the signal to the vestibular nuclei, this information drives a number of motor reflexes that produce compensatory stabilization of the eyes (vestibulo-ocular reflex), head and neck (vestibulocolic reflex), spinal column (vestibulospinal reflex), and limbs (Löwenstein and Sand, 1940; Timerick et al., 1990; Avens et al., 2003).

The second piece of the puzzle, movements that affect the head, results from one of two factors. The first source of potential head movements is voluntary motion of the body. The second source is the interaction of the organism with the unpredictable or unstable aspects of its locomotor substrate or environment. The locomotor substrate is the surface or substance the organism pushes against in order to move; for example, the wing of a bird pushes against the air, while the foot of an elephant pushes against the ground, and the fin of a shark pushes against the water.

It is this second source that is of primary interest when considering the function of the semicircular ducts, for two reasons: (1) The voluntary movements produced by the body, if unimpeded, are predictable and will include simultaneous voluntary compensatory stabilization of the head and eyes. (2) These voluntary movements do not produce significant perturbations on the time scale that stimulates the ducts (Muller, 1990); that is, voluntary locomotor movements of most organisms produce rotations slower than the order of microseconds. It is, therefore, primarily interruptions and alterations of the body's predictable movements by unpredictable environmental and locomotor substrate factors that induce movements most suitable for detection by the semicircular ducts.

Three aspects of the locomotor mode of an organism can be expected to have some link to semicircular ducts and the bony canals housing them. The first aspect is the basic kinematic profile of that mode: are an organism's movements very quick and agile, or ponderous and deliberate? This aspect also includes the overall body mechanics typically employed by an animal over the course of its locomotor cycle, such as the lateral undulatory movements of swimming crocodilians or the parasagittal limb movements of terrestrial mammals. These factors influence how frequently interaction with the substrate occurs and the dominant orientations of movements that the substrate can interrupt or alter. The second aspect that connects locomotion and semicircular duct function is the substrate itself. The physical properties of the substrate influence, as does the first aspect, the kinds of perturbations that are introduced into the locomotor cycle. Each possible substrate presents a different suite of hindrances to smooth locomotion (e.g., wind gusts, churning waves, or loose rocks). The third aspect is environment, specifically the medium through which the organism is moving, regardless of the substrate used. While the other two aspects certainly have the potential to affect canal function, this third is of paramount importance in

discussing the transition to an aquatic environment. The density of the surrounding medium (water) in an aquatic environment makes it far more influential in terms of locomotor perturbations than does the medium of a terrestrial environment (air).

The effect of this influence on vestibular control of stability is demonstrated elegantly in several studies, including that by Avens et al. (2003). These authors demonstrated a reflexive compensatory control of flipper position in hatchling loggerhead sea turtles in response to whole-body rotations. Such rotations are a natural environmental perturbation the turtles encounter as they make their way from the beach and attempt to swim through churning surf.

It must be emphasized that the vestibular system is one of immense complexity, and here we are investigating only a few factors, while there are others that may also play a role. As is the case in any other liquid, the viscosity of endolymph varies with temperature. Thus, endothermic and ectothermic organisms may have differing endolymph mechanics, as may organisms inhabiting environments with significantly differing temperatures (see Sipla and Spoor, chapter 14 in this volume). Furthermore, many other aspects of vestibular morphology have effects on semicircular duct function (e.g., the relative size of the ampulla, the height of the common crus, cross-sectional area). The relationship between the canals and the other body structures with which they are associated (e.g., the eyes) also may play an important role in canal function (Ezure and Graf, 1984a, 1984b; Graf and Baker, 1985).

Thus, we must consider why we expect changes in canal shape to be indicative of changes in canal function. Most biomechanical models of endolymph flow within a semicircular duct system assume a perfectly circular duct path (e.g., Ten Kate et al., 1970; Van Buskirk et al., 1976; Muller, 1994). Despite the lack of shape consideration inherent to these models, they are still informative in this situation. Each model incorporates a factor that describes the proportion of the circuit that is allocated to the duct and the proportion allocated to the utricle. This ratio is something that can be dramatically affected by changes in duct geometry. Ducts with the same absolute utricular length can have different duct lengths, and therefore different duct length–to–utricle length ratios, if they have differently shaped duct circuits. Thus, diverse duct shapes will produce diverse endolymph responses, given the same initial rotation. Restated in an adaptive context, different duct shapes will result in different sensitivities to rotational accelerations.

Furthermore, in the second of a three-part series of papers, Muller and Verhagen (2002) addressed the relation of certain duct features to optimized duct function. Examining the length of the common crus in relation to the length of the anterior and posterior ducts, they concluded that to maintain optimal endolymph flow conditions, this feature would vary with duct circuit shapes. Again, the shape of the circuit has a significant effect on the response of the endolymph to rotation.

While canal shape does affect canal function, a full predictive model of semicircular canal shape across the transition to an aquatic mode of locomotion is not feasible, primarily due to the lack of a full shape-based model for so complex a biophysical system. Preliminary observations of a broad range of canals, however, indicate that such a unifying functional correlation might exist.

ANATOMY AND FUNCTION IN AQUATIC AMNIOTES

SQUAMATES

Within squamates, mosasaurs represent a fully aquatic radiation very closely related to varanid lizards, and the two are often compared (see the sections by Schwenk and Thewissen, and Thewissen and Hieronymus in chapter 1 in this volume for a brief discussion of reptile and bird diversity). In his monograph on American mosasaurs, Dale Russell (1967:59) stated: "The

otic labyrinth of mosasaurs is practically identical to that of *Varanus*." CT imagery of the mosasaur genera *Platecarpus* and *Tylosaurus* reveals that this is not the case (Fig. 15.7). In both instances, the anterior canals are practically identical to those of the aquatic fossil crocodilians (see below), and for *Tylosaurus*, this similarity is borne out for the other two canals as well. In *Platecarpus*, however, the posterior canal does preserve the typical shape and relationship with the lateral canal seen in varanids and birds. The lateral canal of *Platecarpus* is more similar to that of birds than varanids, meaning that the vestibular wall does not intrude into the approximately circular interior circuit of the canal.

TURTLES

Many members of the turtle family are fully or partially aquatic. Most of these exhibit a semicircular canal morphology similar to that described for the tortoises or box and wood turtles, with the notable difference that the system is elongated along an anteroposterior axis (Fig. 15.8). An example of this is found in the fully aquatic pig-nosed turtle, *Carettochelys insculpta* (Carettochelyidae). This stretching is most prominent in the shape of the anterior canal, though the posterior also appears to be slightly lengthened relative to its height. Where *C. insculpta* employs subaqueous flying as its aquatic mode of locomotion, the fully aquatic snapping turtles, *Macrochelys temminckii* and *Chelydra serpentina*, employ four-limb paddling. Despite this difference in locomotor behavior, the snapping turtles have semicircular canal shapes nearly identical to *C. insculpta*.

Sea turtles (Cheloniidae and Dermochelyidae) share a common mode of locomotion with *C. insculpta*. They exhibit, however, a divergent semicircular canal morphology. All the canals are closer in shape to the very thick and robust canals of the tortoises, but the vertical canals of some sea turtles (e.g., *Caretta caretta*, the loggerhead, and *Chelonia mydas*, the green sea turtle) do show minor elongation, particularly in the anterior canal. Others (e.g., *Dermochelys*

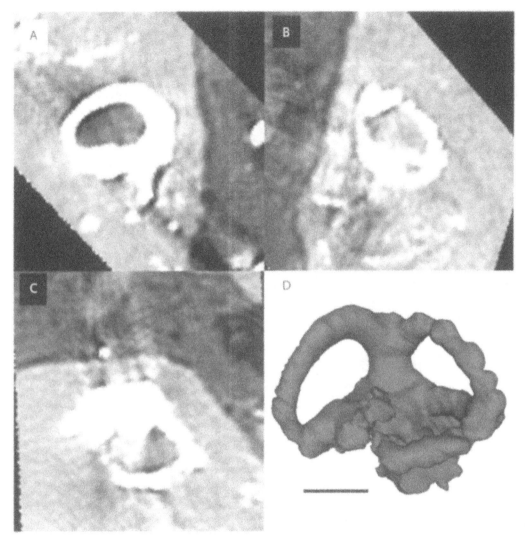

FIGURE 15.7. Computed tomography slices through the anterior canal (A), posterior canal (B), and lateral canal (C), and three-dimenstional reconstruction of the inner ear in superolateral view (D) of the squamate *Tylosaurus neopaeolicus*. Scale bar is 5 mm.

coriacea, the leatherback) do not show any elongation along the anteroposterior axis and differ from tortoises primarily in a rounder (i.e., less angular) canal circuit.

The pleurodires (family Cheliidae), *Chelus fimbriatus* (the mata mata), and its relative *Chelodina longicollis* (the common snake-necked turtle) are fully aquatic but do not employ subaqueous flying. These taxa walk on the beds of lakes and slow-moving rivers. The morphology of their semicircular canals deviates tremendously from that of other aquatic turtles, and instead they have anterior and lateral canals

that closely resemble the varanid condition; however, the anterior canal is even more highly elongated, and the lateral canal is hyperslender and has a more pronounced curvature to its crescentic shape resulting from an extreme convexity of the vestibular wall. The posterior canal of *C. fimbriatus*, however, more closely resembles its anterior counterpart than it resembles the posterior canal of varanids.

CROCODILIANS

One extant crocodilian is not as capable of terrestrial locomotion as all the others, the Indian

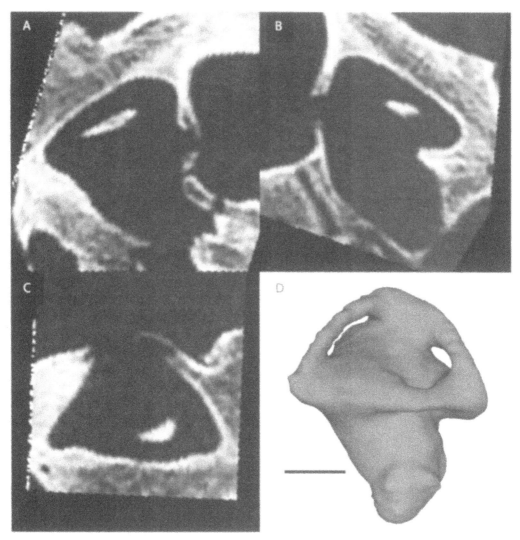

FIGURE 15.8. Computed tomography slices through the anterior canal (A), posterior canal (B), and lateral canal (C), and three-dimensional reconstruction of the inner ear in superolateral view (D) of the turtle *Macrochelys temminckii*. Scale bar is 5 mm.

gharial, *Gavialis gangeticus*. The gharial is somewhat capable of terrestrial locomotion, but, as an adult, it cannot employ the "high-walk" that all other extant crocodilians use for most terrestrial locomotion (Bustard and Singh, 1978). With a very long, slender snout and numerous teeth, *G. gangeticus* also shows a high degree of adaptation to piscivory, another indication that it is a highly aquatic creature.

As the locomotor capabilities of the gharial differ from other extant crocodilians, so too does the morphology of its semicircular canals (Fig. 15.9). The vestibular cavity is larger relative to the canals than in other crocodilians, and this gives the common crus a foreshortened appearance. Furthermore, the vertical canals do not continue significantly superiorly past their point of branching from the crus, and they follow a straighter path toward their respective ampullary regions. These factors combined give gharials vertical semicircular canals that appear far longer than they are tall, a change from the proportions seen in other more semiaquatic extant crocodilians. The lateral canal of the gharial shows far less difference with other extant crocodilians than the two vertical canals.

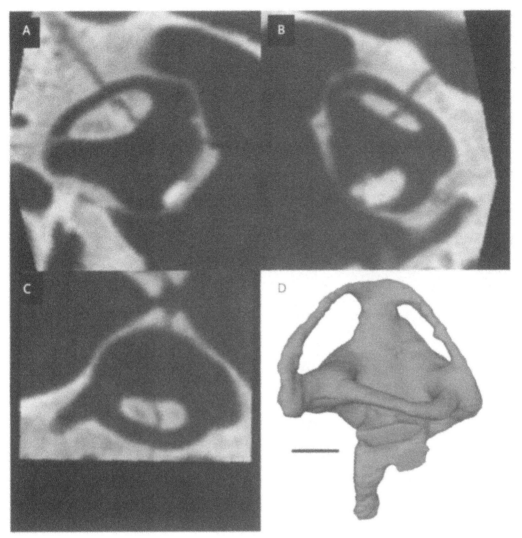

FIGURE 15.9. Computed tomography slices through the anterior canal (A), posterior canal (B), and lateral canal (C), and three-dimensional reconstruction of the inner ear in superolateral view (D) of the crocodilian *Gavialis gangeticus*. Scale bar is 5 mm.

As already seen, fossil ancestors of the crocodilians offer a much broader array of locomotor behaviors, including the terrestrial *Junggarsuchus sloani* (described above). In contrast, the fossil crocodylomorph *Metriorhynchus* (see Table 1.2 in this volume) is a member of a fully aquatic radiation of crocodilian ancestors. Its vertical canals exhibit an aspect ratio change opposite that seen in *J. sloani*. The common crus is shortened and there is little superior continuation of the canals beyond their branching from the crus. Thus, the vertical canals of *Metriorhynchus* are shorter relative to

the width of the canal than is seen in the modern crocodilians (except for *G. gangeticus*). The lateral canal exhibits no significant deviation from that of the typical modern crocodilian.

BIRDS

The subaqueous flying employed by penguins during aquatic locomotion closely approximates the flight stroke of volant birds (Clark and Bemis, 1979). Due to similarities between the characteristics of penguin swimming and typical avian flight, vast alterations of the penguin vestibular system are neither

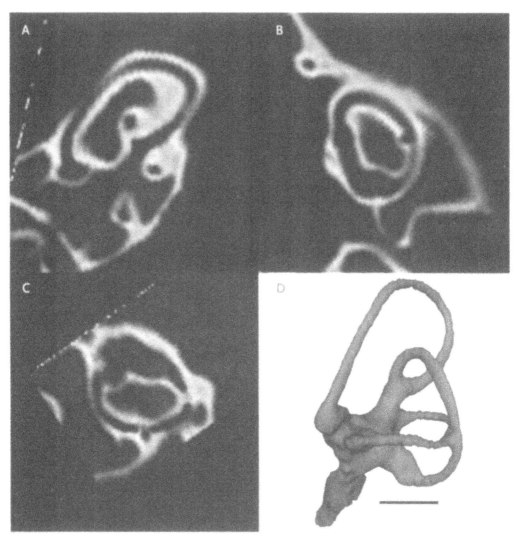

FIGURE 15.10. Computed tomography slices through the anterior canal (A), posterior canal (B), and lateral canal (C), and three-dimensional reconstruction of the inner ear in superolateral view (D) of the penguin *Aptenodytes forsteri*. Scale bar is 5 mm.

expected nor seen in the animals we examined (Fig. 15.10). Nevertheless, the anterior canal of the penguins trends in shape with other aquatic tetrapods in being dorsoventrally compressed and anteroposteriorly lengthened relative to those of the terrestrial taxa discussed above.

The anteroposterior lengthening of the already-hypertrophied anterior canal results in a slightly greater posterior deviation of the common crus than in most nonpenguin taxa. None of the penguins examined demonstrated the hypertrophy of the posterior region of the lateral canal described in *Struthio*.

EVOLUTIONARY PATTERNS ACROSS AQUATIC AMNIOTES

Based on the canal systems described above, it appears that aquatic forms typically demonstrate vertical canals that are more dorsoventrally compact and anteroposteriorly elongate than terrestrial forms; that is, they have a lower aspect ratio (Fig. 15.11). We tested these preliminary observations on our sample of 75 species of terrestrial, aquatic, and semiaquatic chelonians, crocodilians, varanoids, and birds with the quantified shape data produced through elliptical Fourier analysis (Fig. 15.12).

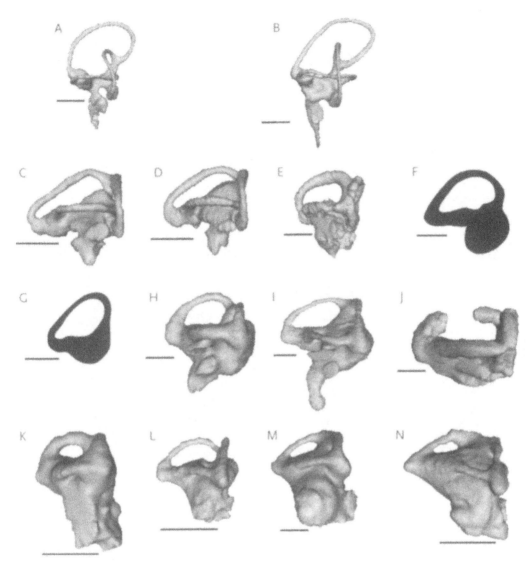

FIGURE 15.11. Three-dimensional reconstructions based on computed tomography (CT) scans with the anterior canal oriented in the plane of the page. Taxa in the *left column* are terrestrial, whereas those in the *second column* are semiaquatic. Taxa in the *third* and *fourth columns* are aquatic. The *first row* shows birds, the *second* squamates, the *third* crocodilians, and the *fourth* turtles. Birds, crocodilians, and turtles show a trend toward decreasing aspect ratio of the anterior semicircular canal from left to right, but this is not the case for squamates. (A) The ostrich *Struthio camelus*; (B) the penguin *Aptenodytes forsteri*; (C) the lizard *Varanus komodoensis*; (D) the lizard *Varanus salvator*; (E) the lizard *Tylosaurus neopaeolicus*; (F) the mosasaur *Platecarpus tympaniticus*; (G) the crocodilomorph *Junggarsuchus sloani*; (H) the crocodile *Crocodylus porosus*; (I) the crocodilomorph *Gavialis gangeticus*; (J) the crocodilomorph *Metriorhynchus* sp.; (K) the tortoise *Geochelone nigra*; (L) the turtle *Trachemys scripta*; (M) the turtle *Macrochelys temminckii*; and (N) the turtle *Carettochelys insculpta*. F and G are shown as silhouettes due to insufficient contrast in the original CT data for model production; the silhouettes were produced from planar-oriented images.

Figure 15.13A shows a canonical discriminant function plot of crocodilian ear shape data produced with SPSS 8.0 for Windows (SPSS Inc; Chicago, Illinois). The data set includes all the elliptical Fourier coefficients through the sixth harmonic (inclusive) for all three semicircular canals and compares single representatives from 20 extant crocodilian taxa and 6 fossil taxa divided into groups on the basis of locomotor behavior. The resulting two discriminant functions are both significant ($p < 0.05$), and the first corresponds to 80.6% of the variation

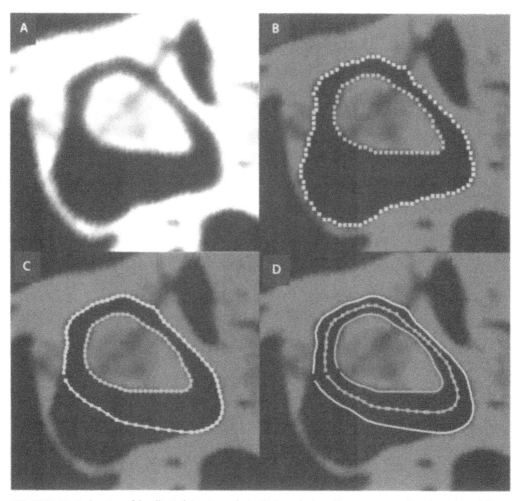

FIGURE 15.12. Explanation of the elliptical Fourier analysis (EFA) method used here (Ferson et al., 1985; Kuhl and Giardina, 1982; Rohlf and Archie, 1984). This method is not dependent on multiple homologous landmarks and quantifies the shape of a closed circuit path and is therefore more suitable for this study than landmark shape analysis (Bookstein, 1990). A Fourier decomposition is performed on parametric equations composed from a path's raw coordinates, and then EFA produces a series of coefficients describing elliptical functions (harmonics) that when concatenated, approximate the original outline. If the paths are controlled for starting point, size, and rotation, the coefficients of these harmonics can be treated as homologous data for statistical analysis. For EFA, each bony canal system was imaged via computed tomography (CT) (A). For each canal in the system, the volumetric CT data were reoriented parallel to the canal, and a median slice through it was captured (A) using VoxBlast 3D Visualization and Measurement Software version 3.1 (Vaytek, Inc., Fairfield, Iowa). Using SigmaScan 5.0 (1999; SPSS Inc.; Chicago, Illinois) the interior circuit and exterior circuit are traced automatically (B). Using Igor Pro 4.0 the circuits are controlled for start point (medial end of vestibular wall), binomial smoothing is applied to each circuit, and the vestibular wall from the interior circuit is copied to the exterior circuit (C). Using Igor Pro 4.0.4. (2001; WaveMetrics Inc.; Lake Oswego, Oregon) an average circuit path is calculated (D). These coordinates were processed using the Fourier module of NTsysPC (Rohlf, 2003; Exeter Software; Setauket, New York), which standardized the outlines for size based on the area enclosed.

between the groups. The first axis also shows a clear gradient, from left to right, of aquatic, semiaquatic, and terrestrial forms.

A similar analysis comparing 25 turtles and tortoises (Fig. 15.13B) does not show a comparable locomotor grade along either axis. While there is complete separation of the three groups, there is no one obvious interpretation of either axis. However, it may be that these data are influenced by the peculiar morphology of the sea turtle canals. A second discriminant analysis of all turtles and tortoises excluding the sea turtles (Fig. 15.13C) does show a functional grade from aquatic through semiaquatic to

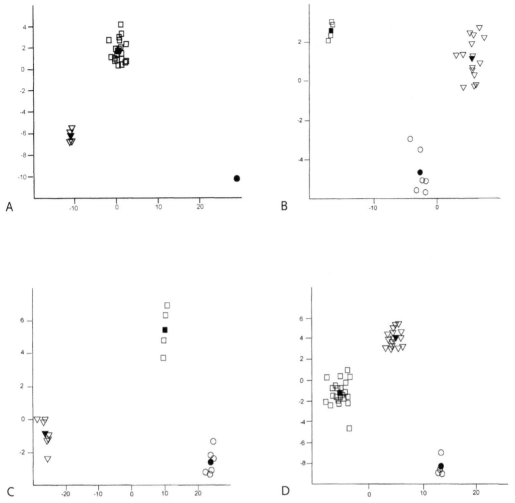

FIGURE 15.13. Discriminant function analysis of crocodilians (A), turtles and tortoises including sea turtles (B), turtles and tortoises excluding sea turtles (C), and crocodilians and birds (D). Aquatic taxa indicated as *triangles*, semiaquatic as *squares*, and terrestrial as *circles*. *Black symbols* represent the cluster centroids. In A and C, discriminant function one (*x* axis) is sufficient to separate the locomotor groups fully along a functional gradient from aquatic on the left to terrestrial on the right, but no such separation occurs in B. In D, the phylogenetic shape differences dominate the variation between groups. The semiaquatic group, on the left, has no avian specimens, while the aquatic and terrestrial groups, both on the right, are composed primarily of avian specimens. The *y* axis in D carries some of the functional information shown on the *x* axis in A and C.

terrestrial locomotor behaviors along the first axis. This axis accounts for 98.1% of the variation between the groups ($p < 0.05$).

Birds and varanids could not be reliably analyzed in this fashion. The birds in this study fall into only two of the locomotor groups (aquatic for penguins and terrestrial for the ratite birds). The varanid sample, while it contains representatives of each of the three groups, is too small by itself to produce statistically significant results.

A test of this correlation over a more inclusive taxonomic group, *Archosauria* (crocodilians and birds), produces the discriminant functions plotted in Figure 15.13D. Here we are presented with a demonstration of the sensitivity of this analysis to sampling bias. Again, both of the discriminant axes are significant, and the first axis represents 75.2% of the variation between the groups. However, the potential functional signal has been relegated to the second axis,

and on the first axis we are left with phylogenetic similarity. In this analysis, birds predomi-
nate in the terrestrial and aquatic groups (both
are composed of more than 70% avian specimens). No avian specimens are included in the
semiaquatic group. As a result, axis 1 separates
the groups by the absence or presence of the
highly derived and distinct avian canal shape.
The remainder of the variation, 24.8% ($p < 0.05$)
along axis 2, however, does describe function.

This result is not surprising. This chapter
began with a description of the dramatic variation in semicircular canal shape across various
orders of nonmammalian amniotes. It is to be
expected that this variation would play a strong
role in any analysis of semicircular canal shape.
The challenge is to incorporate methods that
minimize the effects of taxonomic variation in
canal shape while maximizing functional variation. The approach we use is to sample a large
breadth of taxa within each locomotor group,
and to sample each group as evenly as possible,
to prevent any one group from biasing the data
set. More phylogenetically restricted versions of
this analysis would benefit from independent
phylogenetic contrasts analysis (Garland et al.,
1999; Garland and Ives, 2000) to correct the
data matrix for nonindependence of canal
shape between species.

Lastly, we can test for the presence of a
potential functional signal throughout the
breadth of nonmammalian amniotes. Figure
15.14 shows a discriminant analysis of the same
coefficients as the previous analyses; this time
including 75 specimens encompassing birds,
fossil and extant crocodilians, turtles, tortoises,
varanid lizards, and mosasaurs. This plot is
close to the inverse of the one seen in Fig.
15.13D. The inclusion of the additional diverse
data adds sufficient strength to the functional
signal such that it now accounts for the majority of the variation between groups (73.0% of
the variation is on the first axis; $p < 0.05$).
There is a substantial trend from aquatic specimens on the right of the plot to terrestrial specimens on the left, with the intermediate semiaquatic specimens found in between. The

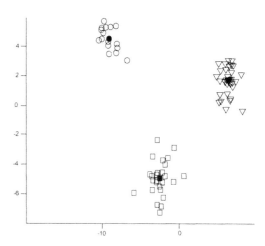

FIGURE 15.14. Discriminant function analysis of all specimens in this study. Across squamates, birds, crocodylo-
morphs, and turtles, functional differences in semicircular
canal shape are the primary source of variation between
aquatic, semiaquatic, and terrestrial groups; 73.0% of the
between-groups variation is accounted for by this functional signal. See Figure 15.13 for explanation of symbols.

second axis still appears to be dominated by the
effects of the absence or presence of avian morphologies within the groups.

These results show dual influences on semicircular canal shape in nonmammalian
amniotes. A primary factor in determining the
canal shape of these organisms is phylogeny,
and closely related organisms will share a basic
morphology. Within that phylogenetic framework, however, there is a second factor, locomotor behavior, which imposes a consistent
change in the canal shape. In this study, we find
one such change associated with transitions
between terrestrial and aquatic environments.
No matter the starting shape, avian, crocodilian, squamate, or turtle, aquatic taxa all share a
common shape change when compared to terrestrial taxa. The question remains, however,
what is that change?

One difficulty with elliptical Fourier analysis
is that intuitive interpretation of the meaning of
the coefficients is problematic. Furthermore,
statistical association of any single coefficient
with a given feature of shape is impossible
(Bookstein et al., 1982). Thus, we cannot examine the structure of the discriminant functions

and derive any meaningful picture of which factors of canal shape are driving the functional distinctions.

This leaves, instead, the original descriptive observations, and these can be summarized in a powerful observation: more aquatic organisms will have smaller aspect ratios of the anterior and posterior canals. Most aquatic forms exhibit either an increase in the width between the common crus and the anterior or posterior ampullae, or a dramatic reduction in the height between the canal and the vestibular wall when compared to their terrestrial relatives. Whichever the case, the aspect ratio of the canal, the ratio of height to width, is reduced in the aquatic forms. Due to the nature of Fourier analysis and also due to the inaccurate representation that the (investigated) bony features give of the (functional) soft tissues, however, it is impossible to say in each case, whether this change is brought about by changes in the canal or the vestibular wall. Furthermore, it is difficult to know how much these canal changes reflect the changes of the membranous ducts they surround.

There are a few notable exceptions to this observed generalization. In comparison to the modern varanids, the mosasaurs exhibit vertical canals with a similar (or even larger) aspect ratio (Fig. 15.11C–F). This, however, is partially an artifact of the expanded vestibular wall relative to the size of the canals in the varanids, whereas the mosasaurs (on average of a much larger body size than the varanids) have very large semicircular canals relative to the size of the vestibule. The other group examined that presents an exception to the pattern of aspect ratio change is the sea turtles. As demonstrated by Figure 15.13B and C, the canals of the sea turtles are sufficiently different from all other aquatic turtles as to completely mask the functional signal. One possibility here is that the very different locomotor pattern of these turtles (subaqueous flying) is driving the difference in shape more strongly than the aquatic environment. This may explain why the sea turtles have relatively vertically expanded anterior and posterior

canals, a feature also seen in the basic avian canal shape.

It may be that there are other aspects of this functional change that are more important. The change in aspect ratio also brings about a change in orientation of the major axis; canals taller than they are wide (e.g., *Junggarsuchus sloani*) have a more vertically oriented major axis, while those that are wider (e.g., *Tylosaurus neopaeolicus*) have a more horizontally oriented major axis. Thus, the functional relationship could be summarized in a second way: more aquatic organisms have smaller angle between the major axis of each vertical canal and the transverse plane. Again, this change cannot be interpreted functionally in a straightforward fashion, but it does constitute a dramatic morphological observation.

This second description still does not resolve the problems surrounding the sea turtles; however, it can be applied with some success to the varanids and mosasaurs. While the varanids and mosasaurs share very similar aspect ratios, the morphology of the mosasaur system shows a reorientation of the major axis of the anterior canal (Fig. 15.11E and F), and there is a reduction of the angle of that axis with the transverse plane.

This leaves a tantalizing problem. There is a clear correlation between canal shape and locomotor behavior. However, it cannot be determined precisely what the correlation means. We also cannot have a solid model for prediction and, thus, cannot know precisely where to begin looking for the meaning of the correlation.

Nonetheless, hypotheses based on the observations we have made may help drive future semicircular canal investigations. Others have proposed that inhabiting a three-dimensional environment such as an aerial, aquatic, or tunneled subterranean environment will place increased demands on the vestibular system (Graf and Vidal, 1996; McVean, 1999). Inhabiting such environments requires a constant, accurate gravitational frame of reference, but the macular endorgans, and not the semicircular canals, are responsible for sensory transduction of gravitoinertial motion. Where the semicircular canals are concerned, the demands of an aquatic environment require a capacity to compensate for the increased potential of external factors to rotate the organism in any direction, such as the interactions between hatchling loggerhead sea turtles and ocean surf, described by Avens et al. (2003). We hypothesize that such demands should result in a system of semicircular ducts with increased sensitivity.

A reduction in canal aspect ratio may represent an increase in duct sensitivity as the vestibular wall and, therefore, the utricle fall on the axis of the canal that is being elongated. Therefore, the elongation of the canal results in an increase of the proportion of the overall circuit that can be composed of utricle. The models of Van Buskirk et al. (1976), Ten Kate et al. (1970), and Muller (1994, adapted from Ten Kate et al., 1970) all indicate that an increase in the ratio of utricle to duct along the circuit will result in an increase in canal sensitivity. This is explained in simplified terms as a reduction in the frictional forces within the system; there is less resistance to flow in the relatively dilated utricle than there is in the relatively constricted duct. Thus, a utricle that makes up a larger proportion of the circuit leads to less frictional resistance to flow and causes the endolymph to flow more freely in response to a given acceleration.

There are other ways to increase the sensitivity of the semicircular ducts, such as increasing the cross-sectional area of the duct and increasing the length of the system. Regarding why aquatic nonmammalian amniotes attenuate the vertical canals rather than (or perhaps, in conjunction with) these other factors, we can only speculate.

This investigation is the first of its kind and is giving us only a glimpse of the perplexing questions that surround the connection between locomotion and semicircular canal morphology. Further work will elucidate this relationship more fully and could provide valuable analytical tools for researchers investigating locomotion in evolutionary contexts. Other taxa within nonmammalian amniote groups also provide interesting models of aquatic adaptation. Analysis of euryapsids (ichthyosaurs and plesiosaurs) or the hydrophiids (sea snakes) or even *Amblyrhynchus cristatus* (the marine iguana) would add further weight to the breadth of this analysis. As well, more generalized models of semicircular duct function and the development of better analytical methodologies may reveal more precisely which characteristics of the semicircular canal system are related to aquatic behaviors and to full independence from life on land.

LITERATURE CITED

Avens, L., J. H. Wang, S. Johnsen, P. Dukes, and K. J. Lohmann. 2003. Responses of hatchling sea turtles to rotational displacements. Journal of Experimental Marine Biology and Ecology 288:111–124.

Bissonnette, J. P., and D. M. Fekete. 1996. Standard atlas of the gross anatomy of the developing inner ear of the chicken. Journal of Comparative Neurology 368:620–630.

Bookstein, F. L. 1990. Introduction to methods for landmark data; pp. 215–225 in F. J. Rohlf and F. L. Bookstein (eds.), Proceedings of the Michigan Morphometrics Workshop. Special Publication 2. University of Michigan Museum of Zoology, Ann Arbor.

Bookstein, F. L., R. E. Strauss, J. M. Humphries, B. Chernoff, R. L. Elder, and G. R. Smith. 1982. A comment upon the uses of Fourier methods in systematics. Systematic Zoology 31:85–92.

Brichta, A. M., and J. M. Goldberg. 1998. The papilla neglecta of turtles: a detector of head rotations with unique sensory coding properties. Journal of Neuroscience 18:4314–4324.

Brichta, A. M., and J. M. Goldberg. 2000. Morphological identification of physiologically characterized afferents innervating the turtle posterior crista. Journal of Neurophysiology 83:1202–1223.

Bustard, H. R., and L. A. K. Singh. 1978. Studies on the Indian gharial *Gavialis gangeticus* (Gmelin) (Reptilia, Crocodilia). Change in terrestrial locomotory pattern with age. Journal of the Bombay Natural History Society 74:534–536.

Clark, D. B., and W. Bemis. 1979. Kinematics of swimming of penguins at the Detroit Zoo. Journal of Zoology (London) 188:411–428.

Clark, J. M., X. Xu, C. A. Forster, and Y. Wang. 2004. A Middle Jurassic "sphenosuchian" from China and the origin of the crocodylian skull. Nature 430:1021–1024.

Ezure, K., and W. Graf. 1984a. A quantitative analysis of the spatial organization of the vestibulo-ocular reflexes in lateral- and frontal-eyed animals. I. Orientation of semicircular canals and extraocular muscles. Neuroscience 12:85–93.

Ezure, K., and W. Graf. 1984b. A quantitative analysis of the spatial organization of the vestibulo-ocular reflexes in lateral- and frontal-eyed animals. II. Neuronal networks underlying vestibulo-oculomotor coordination. Neuroscience 12:95–109.

Ferson, F., F. J. Rohlf, and R. K. Koehn. 1985. Measuring shape variation of two-dimensional outlines. Systematic Zoology 34:59–68.

Flourens, J.-P.-M. 1828. Les canaux semi-circulaires de l'oreille, dans les mammaferes. Read at the Académie Royale des Sciences, Paris.

Garland, T., and A. R. Ives. 2000. Using the past to predict the present: confidence intervals for regression equations in phylogenetic comparative methods. American Naturalist 155:346–364.

Garland, T., P. E. Midford, and A. R. Ives. 1999. An introduction to phylogenetically based statistical methods, with a new method for confidence intervals on ancestral values. American Zoologist 39:374–388.

Graf, W., and R. Baker. 1985. The vestibuloocular reflex of the adult flatfish. I. Oculomotor organization. Journal of Neurophysiology 54:887–899.

Graf, W., and P.-P. Vidal. 1996. Semicircular canal size and upright stance are not interrelated. Journal of Human Evolution 30:175–181.

Gray, A. A. 1906. Observations on the labyrinths of certain animals. Proceedings of the Royal Society of London B 78:284–296.

Gray, A. A. 1907. The Labyrinths of Animals, Vol. 1. Churchill Press, London.

Gray, A. A. 1908. The Labyrinths of Animals, Vol. 2. Churchill Press, London.

Harada, Y., S. Kasuga, and S. Tamura. 2001. Comparison and evolution of the lagena in various animal species. Acta Oto-Laryngologica 121:355–363.

Hua, S., and E. Buffetaut. 1997. Introduction to part V: Crocodylia; pp. 357–374 in J. M. Callaway and E. L. Nicholls (eds.), Ancient Marine Reptiles. Academic Press, San Diego.

Jones, G. M., and K. E. Spells. 1963. A theoretical and comparative study of the functional dependence of the semicircular canal upon its physical dimensions. Proceedings of the Royal Society of London B 157:403–419.

Kuhl, F. P., and C. R. Giardina. 1982. Elliptical Fourier features of a closed contour. Computer Graphics and Image Processing 18:236–258.

Löwenstein, O., and G. J. Compton. 1978. A comparative study of the responses of isolated first-order semicircular canal afferents to angular and linear acceleration, analyzed in the time and frequency domains. Proceedings of the Royal Society of London B 202:313–338.

Löwenstein, O., and A. Sand. 1940. The mechanism of the semicircular canal. A study of the responses of single-fibre preparations to angular accelerations and to rotations at constant speed. Proceedings of the Royal Society of London B 129:256–275.

Maxwell, S. S. 1919. Labyrinth and equilibrium. I. A comparison of the effect of removal of the otolith organs and of the semicircular canals. Journal of General Physiology 2:123–132.

Maxwell, S. S. 1920a. Labyrinth and equilibrium. III. The mechanism of the static functions of the labyrinth. Journal of General Physiology 3:157–162.

Maxwell, S. S. 1920b. Labyrinth and equilibrium. II. The mechanism of the dynamic functions of the labyrinth. Journal of General Physiology 2:349–355.

Maxwell, S. S. 1921. The equilibrium functions of the internal ear. Science 53:423–429.

McVean, A. 1999. Are the semicircular canals of the European mole, Talpa europaea, adapted to a subterranean habitat? Comparative Biochemistry and Physiology A 123:173–178.

Muller, M. 1990. Relationships between semicircular duct radii with some implications for time constants. Netherlands Journal of Zoology 40:173–202.

Muller, M. 1994. Semicircular duct dimensions and sensitivity of the vertebrate vestibular system. Journal of Theoretical Biology 167:239–256.

Muller, M. 2000. Biomechanical aspects of the evolution of semicircular duct systems. Netherlands Journal of Zoology 50:279–288.

Muller, M., and J. H. G. Verhagen. 2002. Optimization of the mechanical performance of a two-duct semicircular canal system. 2. Excitation of endolymph movements. Journal of Theoretical Biology 216:425–442.

Parrish, J. M. 1987. The origin of crocodilian locomotion. Paleobiology 13:396–414.

Ramprashad, F., J. P. Landolt, K. E. Money, and J. Laufer. 1984. Dimensional analysis and dynamic response characterization of mammalian peripheral vestibular structures. American Journal of Anatomy 169:295–313.

Rohlf, F. J. 2003. NTSYSpc 2.11. Exeter Software, Setauket, NY.

Rohlf, F. J., and J. W. Archie. 1984. A Comparison of Fourier methods for the description of wing shape in mosquitoes (Diptera: Culicidae). Systematic Zoology 33:302–317.

Russell, D. A. 1967. Systematics and morphology of American mosasaurs. Bulletin of the Peabody Museum of Natural History 23:1–241.

Spoor, F., B. Wood, and F. Zonneveld. 1994. Implications of early hominid labyrinthine morphology for evolution of human bipedal locomotion. Nature 369:645–648.

Spoor, F., B. Wood, and F. Zonneveld. 1996. Evidence for a link between human semicircular canal size and bipedal behavior. Journal of Human Evolution 30:183–187.

Spoor, F., and F. Zonneveld. 1998. Comparative review of the human bony labyrinth. Yearbook of Physical Anthropology 41:211–251.

Spoor, F., S. Bajpai, S. T. Hussain, K. Kumar, and J. G. M. Thewissen. 2002. Vestibular evidence for the evolution of aquatic behavior in early cetaceans. Nature 417:163–166.

Ten Kate, J. H, H. H. Barneveld, and J. W. Kuiper. 1970. The dimensions and sensitivities of semicircular canals. Journal of Experimental Biology 53:501–514.

Timerick, S. J. B., D. H. Paul, and B. L. Roberts. 1990. Dynamic characteristics of the vestibular-driven compensatory fin movements of the dogfish. Brain Research 516:318–321.

Van Buskirk, W. C., R. G. Watts, and Y. K. Liu. 1976. The fluid mechanics of the semicircular canals. Journal of Fluid Mechanics 78:87–98.

Wever, E. G. 1978. The Reptile Ear. Princeton University Press, Princeton, NJ.

16

Comparative and Functional Anatomy of Balance in Aquatic Mammals

Fred Spoor and J. G. M. Thewissen

The first comprehensive comparative study to associate morphological diversity of the mammalian inner ear with particular locomotor behaviors is the seminal two-volume monograph published by Albert Gray in 1907 and 1908. That the otolith and semicircular canal systems of the inner ear are not part of the organ of hearing, but form a separate organ of balance, had been discovered not long before (Breuer, 1874; Brown, 1874; Mach, 1875). Subsequent studies focused on the functional relationship between the semicircular canals and locomotion. These studies used empirical evidence by comparing animals with diverse behaviors, as well as biophysical models of the canal system associating morphological properties with physiological characteristics such as sensitivity and response speed (reviewed in Spoor, 2003 and Spoor et al., 2007). In this chapter, we consider the comparative and functional anatomy of the organ of balance, or vestibular system, in aquatic and semiaquatic (amphibious) mammals, on the basis of published observations, as well as newly collected data on the bony labyrinth of the inner ear.

GENERALIZED ANATOMY OF THE ORGAN OF BALANCE IN MAMMALS

In mammals other than monotremes, the part of the inner ear associated with hearing is

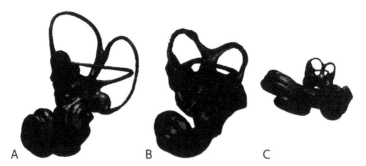

FIGURE 16.1. Inferolateral view of the bony labyrinth of the primate *Galago sene-galensis* (A), the artiodactyl *Hippopotamus amphibius* (B), and the dolphin *Tursiops truncatus* (C). Three-dimensional reconstructions made from computed tomography and size-corrected for body mass according to the average canal regression for noncetacean mammals given in Table 16.2.

macroscopically distinctly different from that in other tetrapods, in that it forms a spiral-shaped cochlea. In contrast, the phylogenetically older organ of balance is basically similar in bauplan in all gnathostome vertebrates (Sipla and Spoor, chapter 14 in this volume; Georgi and Sipla, chapter 15 in this volume). Thus, the mammalian inner ear has three membranous semicircular ducts, anterior, posterior, and lateral, and these serve the perception of angular motion (Fig. 16.1). In addition, it has two otolith organs, one in the utricle, the other in the saccule, which perceive linear motion and gravity. Housed in a bony otic capsule, the ducts are enclosed by bony semicircular canals, and the utricle and saccule are located in the vestibule (see chapter 14 by Sipla and Spoor in this volume for the terminology used).

This review mainly discusses the semicircular canal system, because its comparative and functional diversity among mammals is far better researched and understood than the otolith system. Differences between aquatic and terrestrial mammal species are largely known from studies of the bony semicircular canals. Valid functional interpretation of these osteological data is possible because each canal reliably reflects the functionally important length of the enclosed membranous duct

(along its arc). A second functionally important aspect of the canal system, the lumen size (cross section) of each duct, may be more difficult to assess on the basis of the bony canal, because the duct occupies the canal lumen to a variable degree (Gray, 1907, 1908; Ramprashad et al., 1984). Nevertheless, the relationship tends to be consistent within mammalian families, and some inferences can be made, in particular where the duct consistently fills most of the canal (i.e., when there is very little perilymphatic space surrounding the ducts). This issue is addressed in more detail below, in the review of individual taxa. Some functionally important aspects of the semicircular canal system, such as the viscosity of the endolymph, and the structure of the ampullar apparatus (including the cupula), are not discussed at all, because meaningful comparative data are not available.

First, we review our current knowledge of the organ of balance of aquatic and semiaquatic species, considering cetaceans, artiodactyls, sirenians, rodents, carnivores, and monotremes. In the subsequent discussion section the comparative patterns are placed in a functional context, relating anatomical diversity to locomotor behaviors and the demands of an aquatic environment.

MATERIALS AND METHODS

Apart from a review of the literature, we provide new data and analyses of the arc size of the semicircular canals, and of the lumen size of the ducts. Information regarding the mammal species investigated prior to this study can be found in work by Spoor et al. (2002) and their supplementary information. Specimens added since this study include several aquatic and semiaquatic species, as well as specific terrestrial species, artiodactyls and carnivores in particular, that provide a comparative context and are closely related to the aquatic and semiaquatic taxa of interest (Table 16.1).

Skulls or isolated petrosals were investigated with computed tomography (CT), using three different scanners. These are the following, with pixel size, slice thickness, and slice increment of the images in parentheses: General Electric High Speed scanner at the Bromley Hospital, London, courtesy of Peter Morris (0.2, 1.0, 0.5 mm), Norland XCT Research M scanner at the Northeastern Ohio Universities College of Medicine (0.07, 0.50, 0.25 mm), X-tech BIR at the Center for Quantitative Imaging, Penn State University, courtesy of Alan Walker (voxel sizes in the range 0.03 to 0.04 mm). Additional images were obtained from the Digimorph archive (www.digimorph.org) (voxel sizes in the range 0.08 to 0.50 mm). Specimens were typically scanned coronally.

Measurements were taken from planar reformatted images showing the full extent of each canal, using Voxblast 3.1 (VayTek, Inc., Fairfield, Iowa). The arc size of the semicircular canals is quantified by the radius of curvature (R), defined as half the average of the arc height and width, measured to the center of the canal lumen (Fig. 1a in Spoor and Zonneveld, 1995). In nearly all taxa considered here the ducts fill most of the canal lumen, and the arc shape is semicircular. Hence, the radius of curvature of the canal provides a good estimate of the functionally important parameter, the central length of the enclosed duct along its arc.

The duct lumen size is quantified by its radius (r). In noncetacean species reliable measurements were taken directly from soft-tissue inner ear preparations (Table 16.1). Reliable duct lumen radii are not available for cetaceans. However, cetaceans have little perilymphatic space around the ducts (Gray, 1908; personal observation of unpublished mysticete labyrinth prepared by A. Gray), although the amount may increase with overall size (Yamada and Yoshizaki, 1959). Since the specific question investigated concerns the degree of reduction in lumen size of the cetacean ducts, we took a conservative approach when estimating the duct lumen size from measurements of the bony canal lumen (casts and CT imaging, described in Spoor et al., 2002). The presented values take duct size as 85% of canal size, although the actual cetacean duct sizes are likely smaller because mammalian duct radii of more than 80% of the canal radii have not been observed (Curthoys et al., 1977; Ramprashad et al., 1980, 1984; Igarashi et al., 1981). For estimated duct radii (Jones and Spells, 1963; Muller, 1990) of 43 mammal species prepared by Albert Gray (1907, 1908), this maximum is 72%.

The relationships of both the canal radii of curvature and the radius of the duct lumen to body mass are assessed using bivariate double logarithmic plots, and by calculating Spearman rank correlation coefficients (r_{rank}), which were not applied to the smallest samples of just three or four species. The overall interspecific scaling pattern is characterized by the reduced major axis (RMA), and the residuals represent the deviation of individual species from the general trend, and the biological adaptations that are of particular concern here. Body mass data were taken from a range of sources, indicated in the tables. In addition to the bivariate analyses, the relative arc sizes of the three canals are considered by expressing the radius of each canal as a percentage of the sum of all three radii.

TABLE 16.1

Arc Size of the Semicircular Canals and Body Mass (in Grams) of Cetaceans and Artiodactyls

	N	LABEL	BM	BM SOURCE	SC-R	%ASC	%PSC	%LSC	SC DATA SOURCE
Artiodactyla									
Antilocapra americana	1	+	51,500	1	2.7	32	33	35	6
Aepyceros melampus	1	+	50,000	1	2.5	36	36	28	6
Alcelaphus buselaphus	1	+	134,250	1	3.1	35	35	30	6
Bison bison	1	+	480,000	1	3.0	38	35	27	6
Bos taurus	3	+	290,000	1	3.1	37	34	29	6, 9, 10
Capra hircus	1	+	46,750	1	2.6	35	36	29	6
Cephalophus silvicultor	1	+	62,500	2	2.9	34	36	30	6
Damaliscus lunatus	1	+	121,500	1	3.1	39	33	28	6
Gazella bennetti	1	+	23,000	3	2.5	34	37	29	6
Gazella granti	1	+	46,750	1	2.5	34	37	29	6
Gazella thompsonii	1	+	18,333	1	2.5	34	37	29	6
Kobus ellipsiprymnus	1	+	153,333	1	3.0	35	37	28	6
Madoqua guentheri	1	+	5,270	1	2.2	33	35	31	6
Oryx beisa	1	+	200,250	1	3.4	35	33	33	6
Ovis aries	1	+	33,750	1	2.3	37	36	27	6
Redunca redunca	1	+	40,000	1	3.2	37	35	27	6
Syncerus caffer	1	+	592,667	1	3.8	35	37	28	6
Taurotragus oryx	1	+	325,750	1	3.4	35	33	33	6
Tragelaphus euryceros	1	+	253,000	1	3.3	34	33	33	6
Tragelaphus scriptus	2	+	40,000	1	2.9	37	36	27	6
Tragelaphus strepsiceros	1	+	194,500	1	3.5	34	37	30	6
Camelus dromedarius	1	+	415,000	1	4.1	36	35	29	6
Axis porcinus	1	+	77,833	1	2.6	36	35	29	6
Capreolus capreolus	1	+	23,900	1	2.1	36	35	29	6
Cervus elaphus	1	+	85,000	1	2.3	35	34	31	6
Cervus nippon	1	+	23,250	1	2.4	36	35	28	6
Dama dama	1	+	56,450	1	2.7	37	34	29	6
Muntiacus reevesi	1	+	14,800	1	2.2	34	34	32	6
Odocoileus virginianus	1	+	72,800	1	2.3	34	36	30	6
Giraffa camelopardalis	2	+	895,929	1	3.8	36	34	30	6, 10
Hexaprotodon liberiensis	1	h	235,000	1	3.3	38	33	28	7

Label	Species	n	Size	SC-R	%ASC	%PSC	%LSC	Sources
H	*Hippopotamus amphibius*	2	1,140,500	3.5	37	34	29	6, 10
+	*Phacochoerus africanus*	1	58,750	2.3	38	33	29	6
+	*Potamochoerus porcus*	1	54,000	2.4	38	34	28	6
+	*Sus scrofa*	1	47,283	2.0	37	33	29	6
—		1	3,096	1.6	33	34	32	6
Mysticeti								
1	*Balaena mysticetus*	1	110,000,000	[3.1]	—	—	—	11
2	*Eubalaena glacialis*	2	56,525,000	3.2	34	31	35	12, 13
3	*Caperea marginata*	1	4,500,000	2.5	34	28	38	8
4	*Eschrichtius robustus*	1	26,750,000	2.5	33	32	35	8
5	*Balaenoptera acutorostrata*	1	6,801,244	2.3	34	29	37	8
6	*Balaenoptera musculus*	1	139,000,000	2.8	37	29	34	14
7	*Balaenoptera physalus*	1	62,500,000	3.1	35	29	36	13
8	*Megaptera novaeangliae*	2	48,350,000	2.4	33	32	35	8, 13
Odontoceti								
9	*Physeter catodon*	1	27,500,000	1.9	30	26	43	13
10	*Kogia sp.*	1	268,167	0.9	34	30	37	14
11	*Berardius bairdii*	1	7,810,000	1.6	37	29	34	13
12	*Mesoplodon densirostris*	1	1,050,000	1.7	32	29	38	8
12	*Ziphius cavirostris*	1	3,342,250	1.5	33	27	40	14
14	*Platanista gangetica*	4	88,779	1.6	39	26	36	8
15	*Inia geoffrensis*	5	128,250	1.1	38	26	36	8
16	*Pontoporia blainvillei*	1	50,000	0.9	37	25	38	8
17	*Delphinapterus leucas*	1	804,000	1.1	32	24	43	8
18	*Monodon monoceros*	1	1,200,000	1.1	32	28	40	8
19	*Delphinus sp.*	2	77,592	1.0	32	24	44	14
20	*Lagenorhynchus obliquidens*	1	151,667	0.9	30	30	40	13
21	*Orcinus orca*	1	4,909,750	1.2	37	22	41	8
22	*Tursiops truncatus*	3	189,875	1.1	34	28	38	6
23	*Neophocaena phocaenoides*	1	29,750	0.9	34	26	40	14
24	*Phocoena phocoena*	1	62,250	0.7	37	23	40	12

DATA SOURCES: 1, Silva and Downing (1995); 2, www.ultimateungulate.com; 3, www.wildlifeofpakistan.com; 4, Reeves and Brownell (1989); 5, Best and Da Silva (1989); 6, Norland XCT Research M scanner; 7, General Electric High Speed scanner; 8, as in Supplementary Information to Spoor et al. (2002); 9, O. Gray cast, in care of FS; 10, O. Gray cast, Royal College of Surgeons, London; 11, Muller (1999), based on lateral canal value of 3.2 mm, and relative canal sizes of other mysticetes; 12, Gray (1908); 13, cast, National Science Museum, Tokyo; 14, Yamada and Yoshizaki (1959).

NOTE: The mean radius of curvature of the three canals (SC-R) is given in millimeters, and the relative size in percentage (sum of the three canal radii is 100), for the anterior (%ASC), posterior (%PSC), and lateral (%LSC) canal. The number of specimens studied (n), and the label for each species used in Fig. 16.2 are also given.

COMPARATIVE ANATOMY OF THE ORGAN OF BALANCE IN AQUATIC MAMMALS

CETACEA AND ARTIODACTYLA

HISTORICAL PERSPECTIVE

Early studies of the cetacean inner ear have been reviewed by Hyrtl (1845), Boenninghaus (1903), and Reysenbach de Haan (1957). Central in these historical accounts is the initial observation of Petrus Camper in the second half of the eighteenth century that the semicircular canals are missing, an error ascribed to the limitations of dissection techniques at the time, as the canals are unusually small in size. Both the lumen and the arc of the cetacean semicircular canals are indeed the smallest among mammals relative to body size, as was shown when Hyrtl (1845) made extensive comparisons of mammalian bony labyrinths, including those of *Delphinus*, *Monodon*, *Phocoena*, *Physeter*, *Balaena*, and *Balaenoptera*. Subsequently, it was noticed that cetaceans also stand out by the relative size of the three canals. Their lateral canal was found to be the largest, and the posterior canal the smallest, whereas in terrestrial mammals the lateral canal is generally the smallest (Denker, 1902; Gray, 1908; Yamada and Yoshizaki, 1959). Two exceptions to the unique cetacean canal morphology have been reported. Yamada and Yoshizaki (1959) noticed that in *Eubalaena glacialis* the canal system is not reduced in size and is rather "terrestrial" looking. Furthermore, Anderson (1878) and Purves and Pilleri (1973) describe the canals of *Platanista gangetica* as larger relative to the cochlea than in other cetaceans.

In the 1990s the original debate regarding whether cetaceans have fully developed semicircular canals came full circle. Ketten and Wartzok (1990) and Ketten (1992, 1993, 1997) indicated that in odontocetes the canals are not only substantially reduced in size, but that they are incomplete, tapering to fine threads that do not form complete channels. Apart from the earliest report by Camper, such a vestigial state has not been observed in other studies. In rela-

tion to the sirenian inner ear, Ketten et al. (1992) noted that their CT scan resolution is inadequate to determine whether the canals are incomplete. Given that cetacean canals are smaller than those of sirenians, this may be a plausible explanation for the reported absence of complete odontocete canals. Most recently, Spoor et al. (2002) used CT as well and did find complete canals in all cetaceans examined. The study provides the first quantitative analysis of cetacean semicircular canal size, which considers the relationship with body mass, makes comparisons with a large sample of other mammal species, and follows the evolution of the canals using fossils. Here we expand on this work, assessing each of the three canals individually, and using an increased comparative sample. The artiodactyls are considered in particular, both because they are the closest living relatives of the cetaceans, and to assess the morphology of the hippopotamids, which is the family most adapted to an aquatic environment. Moreover, the hippopotamids are also of interest because they may possibly form the direct sister group of the cetaceans (Nikaido et al., 1999; see also the section by Pihlström in chapter 1 of this volume).

ARC SIZE OF THE SEMICIRCULAR CANALS

The sample analyzed comprises 24 extant cetacean species and 174 other mammal species, representing all major orders. Table 16.1 lists absolute and relative arc sizes of the semicircular canals for the cetaceans and 36 artiodactyl species. Correlation and RMA regression statistics for the relationship between canal arc size and body mass are given in Table 16.2.

The canal arc size of cetaceans scales with body mass with similar, strong negative allometry, as it does in other mammals (Fig. 16.2; Table 16.2). The RMA slope values obtained for the three canals of the cetaceans are somewhat higher than for the combined sample of non-cetacean mammals, and closer to the artiodactyl values. However, when comparing these slopes

TABLE 16.2

Correlation and Regression Statistics of the Relationship Between the Radius of Curvature (R) and Duct Radius (r)
of the Semicircular Canals (SR) and Body Mass

	SC	ASC	PSC	LSC
Radius of curvature (R)				
Mammals, excluding				
cetaceans				
rrank	0.885 xxx	0.854 xxx	0.893 xxx	0.850 xxx
RMA slope	0.142 +/− 0.008	0.138 +/− 0.009	0.152 +/− 0.009	0.143 +/− 0.010
Int.	−0.218	−0.171	−0.256	−0.261
Cetacea				
rrank	0.895 xxx	0.831 xxx	0.871 xxx	0.887 xxx
RMA slope	0.175 +/− 0.034	0.177 +/− 0.041	0.200 +/− 0.044	0.160 +/− 0.031
Int.	−0.913	−0.919	−1.154	−0.762
Odontoceti				
rrank	0.753 xx	0.612 xx	0.715 xx	0.738 xxx
RMA slope	0.139 +/− 0.054	0.143 +/− 0.062	0.156 +/− 0.065	0.138 +/− 0.051
Int.	−0.727	−0.739	−0.921	−0.653
Mysticeti				
rrank	0.643 ns	0.607 ns	0.643 ns	0.619 ns
Artiodactyla				
rrank	0.796 xxx	0.808 xxx	0.738 xxx	0.772 xxx
RMA slope	0.160 +/− 0.027	0.166 +/− 0.027	0.160 +/− 0.031	0.168 +/− 0.033
Int.	−0.340	−0.343	−0.321	−0.432
Terrestrial Carnivora				
rrank	0.925 xxx	0.908 xxx	0.873 xxx	0.890 xxx
RMA slope	0.141 +/− 0.018	0.148 +/− 0.024	0.135 +/− 0.021	0.150 +/− 0.024
Int.	−0.237	−0.244	−0.203	−0.314
Duct lumen radius (r)				
Mammals				
Excl. Cetacea				
rrank	0.839 xxx			
RMA slope	0.097 +/− 0.021			
Int.	−1.230			
Cetacea				
SD–r = 0.85×SC–r				
rrank	0.951 xxx			
RMA slope	0.132 +/− 0.020			
Int.	−1.593			

NOTE: Data are presented for each of the canals separately (ASC, anterior semicircular canal; PSC, posterior SC; LSC, lateral SC) and for the mean value of all three canals (SC). Listed are the Spearman rank correlation coefficient (r_{rank}) with its level of significance, as well as the slope with 95% confidence interval and intercept (int.) of the reduced major regression axis (RMA). x, $p < 0.05$; xx, $p < 0.01$; xxx, $p < 0.001$; ns, not significant ($p > 0.05$).

it is important to realize that values vary with the taxonomic level of the sample. For example, the bovid sample has slopes of 0.13, 0.11, and 0.14 for the anterior, posterior, and lateral canal, respectively, which is less than the equivalent slopes of the entire artiodactyl sample. Likewise, when odontocetes and mysticetes are considered separately, the slope of the former is lower than for the entire sample and very similar to the general mammalian values. For the mysticete sample, the correlation between canal arc and body mass is not statistically significant,

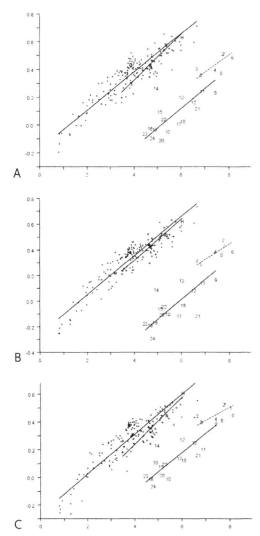

A

B

C

FIGURE 16.2. Double logarithmic plots of the radius of curvature of the anterior (A), posterior (B), and lateral (C) semicircular canals (in millimeters) against body mass (in grams). The codes for artiodactyls and cetaceans are presented in Table 16.1. Reduced major axis (RMA) regressions are given for the overall noncetacean mammal sample, the artiodactyls, and the odontocetes. The RMA calculated for the mysticetes is also given *(dashed line)*, although its correlation is not statistically significant.

owing to the low number of species. Nevertheless, the tentative RMA slopes for the mysticetes are rather similar to those for the odontocete ones (Fig. 16.2).

The plots of Figure 16.2 demonstrate that the cetaceans have much smaller semicircular canal arcs for their body mass than other mammals, with respective interspecific ranges of variation that do not overlap. Moreover, on average, odontocetes have even smaller canals than mysticetes. In broad terms these differences mean that the canals of the blue whale, *Balaenoptera musculus,* are just smaller than those of an average human, and those of the bottlenose dolphin, *Tursiops truncatus,* are smaller than those of the brown rat, *Rattus norvegicus.* Among the odontocetes, *Platanista gangetica* stands out by its substantially larger canals, although it is also out of the range of variation of noncetacean mammals (Fig. 16.2, point 14). This confirms quantitatively the observations by Anderson (1878) and Purves and Pilleri (1973). On the other hand, despite suggestions otherwise (Yamada and Yoshizaki, 1959), *Eubalaena glacialis* does not differ from other mysticetes in the size of either its semicircular canals or its cochlea (Fig. 16.2, point 2) (Spoor et al., 2002). The two hippopotamid species, *Hexaprotodon liberiensis* and *Hippopotamus amphibius,* fall very close to the RMA regressions for all artiodactyls, demonstrating that their amphibious lifestyle is not associated with semicircular canals of unusual size (Fig. 16.2, points H and h).

Using the RMA regressions, the average difference (grade shift) in canal arc size between groups can be quantified (Table 16.3). The posterior canal of cetaceans shows the largest difference from that of other mammals (4 times smaller), the lateral canal shows the smallest difference (2.5 times smaller), and the anterior canal is intermediate (3 times smaller). When compared to the artiodactyls, these differences are usually slightly less, and the differences are greater for odontocetes than for mysticetes. That the three canals are not uniformly smaller in cetaceans causes the pattern of relative canal sizes observed previously, with the lateral canal being the largest, and the posterior one the smallest (Table 16.1). It is worth noting that this is not the case for all cetacean specimens investigated, but it is unclear whether this represents interspecific or intraspecific variation, and no obvious underlying taxonomic pattern is apparent. Moreover,

TABLE 16.3
Radius of Curvature Reduction of the Semicircular Canals in Cetaceans

	SC	ASC	PSC	LSC
Canal radius (*R*)				
Cetaceans, nc mammals	32	32	25	40
Cetaceans, artiodactyls	33	31	26	42
Odontocetes, nc mammals	30	29	23	38
Odontocetes, artiodactyls	31	30	24	41
Mysticetes, mammals	38	38	32	44
Mysticetes, artiodactyls	37	35	32	43
Duct lumen radius (*r*)				
Cetaceans, nc mammals	68			
Cetaceans, artiodactyls	76			

NOTE: The observed radii are given as a percentage of the value predicted by the RMA equation calculated for noncetacean (nc) mammals or artiodactyls (Table 16.2). Data are presented for each of the canals separately (ASC, anterior semicircular canal; PSC, posterior SC; LSC, lateral SC) and for the mean value of all three canals (SC).

among artiodactyls the specimen of *Antilocapra americana* examined shows a lateral canal that is the largest of the three as well.

When corrected for body mass, the overall size of the cochlea in cetaceans is similar to that in other mammals (Spoor et al., 2002). This confirms that the unprecedented size reduction specifically concerns the semicircular canals rather than the entire cetacean inner ear. This also means that the size of the cochlea can be used for reference when considering the size of the semicircular canals, as an alternative when body mass is not available. This is demonstrated by calculating an index between the mean of the three canal radii of curvature and cochlea size (Table 16.4). The differences in value for cetaceans and other mammals, and for odontocetes and mysticetes are all highly statistically significant ($p < 0.001$).

Spoor et al. (2002) studied arc size in the fossil cetaceans that document the transition from land to water in that order. They found that the semicircular canals reach modern proportions in the middle Eocene, and that only the earliest cetaceans (pakicetids) have canals with the proportions of land mammals. As documented by the fossil record, there are no intermediate forms known: pakicetid cetaceans of around 50 million years ago had canals that resembled those of land mammals, whereas 43- to 46-million-year-old remingtonocetids and protocetids had canals that fall within the size range of modern odontocetes.

LUMEN SIZE OF THE SEMICIRCULAR DUCTS

In a preliminary analysis, the estimated lumen size of the semicircular ducts of 12 cetacean species is compared with actual values obtained for 27 noncetacean mammals (Fig. 16.3; Table 16.5). The results show that the duct lumen size of cetaceans is about three-quarters of that of the three artiodactyls investigated, and smaller still compared with most other mammals (Table 16.3). Although not as spectacular as the three-times reduction of the arc size, this finding is nevertheless noteworthy because lumen size directly affects a canal's response speed and also has a greater impact on its sensitivity than arc size (McVean, 1999; Rabbitt et al., 2004). The duct lumen size of hippopotamids is not known.

OTOLITH ORGAN

Little is known about the comparative anatomy of the mammalian otolith organ, and whether any interspecific variation relates functionally to aspects of locomotor behavior. Hyrtl (1845) noticed that cetaceans have a substantial

TABLE 16.4

Index of the Mean Radius of Curvature of the Three Semicircular Canals (SC) to Cochlea Size (CO)

	N	MEAN	MIN	MAX	SD
Noncetacean mammals	55	51	35	77	7.8
Cetaceans	23	16	11	24	3.2
Mysticetes	7	20	17	24	2.1
Odontocetes	16	15	11	20	2.5

NOTE: Means across species are based on species values calculated using 100SC/CO. Cochlea size is quantified by the mean of the "slant height" and the diameters of the first and second turn. The slant height is the distance from the apex to the superior-most edge of the round window (Gray, 1907, 1908), and the diameters were measured inferosuperiorly or mediolaterally, depending on the orientation of the cochlea (see the supplementary information to Spoor et al. 2002, for species list).

otolith mass, and Gray (1908) reported that *Phocoena phocoena* has two relatively large otoliths that can be recognized with the naked eye. Solntseva (2001) found that the otolith membrane of cetaceans is significantly thicker than in terrestrial and semiaquatic mammals. This would suggest that the otolith organ of cetaceans is altogether well developed, but none of the studies provides quantitative information or takes into account any scaling effects. The otolith organ of hippopotamids has not been studied.

INNERVATION

Given the strongly reduced size of the canal system, it is worth considering whether this is reflected in its afferent nerve supply. Gray (1908) observed that the ampullae of the porpoise *Phocoena* are small, but that each is supplied with a nerve. According to Ketten (1997) the ampullae of odontocetes are nevertheless nearly acellular. In striking contrast, a histological study of the odontocetes *Stenella attenuata*, *Delphinus delphis*, *Tursiops truncatus*, and *Delphinapterus leucas*, as well as *Balaenoptera acutorostrata*, found well-developed ampullar crests, which occupy a significant part of the ampullar space, and are entirely covered by receptor epithelium (Solntseva, 2001). This thus suggests a fully functional ampullar apparatus with ample hair cells and afferent nerve supply.

Other studies comment on the size of the vestibular component of the cetacean vestibulocochlear nerve only, without distinguishing between the ampullar and otolith parts. Jansen and Jansen (1969) report that the vestibular nerve makes up only about 10% of the cross section of cranial nerve VIII, but that its size, presumably in absolute terms, nevertheless compares favorably with that of terrestrial mammals. According to Ketten (1997) this value is less than 5% for odontocetes, compared with around 30% in most mammals. Vestibular ganglion cell counts, reviewed by Wartzok and Ketten (1999), indicate that *Delphinus delphis* (4100) has half the number that *Cavia porcellus* has (8200), a third that of *Felis catus* (12,400), and less than a fourth that of *Homo sapiens* (156,000). Values for *Lipotes*, *Neophocaena*, *Phocoena*, *Sousa*, and *Tursiops* are less still (3200 to 3600). Wartzok and Ketten (1999) also give ratios between vestibular and auditory ganglion cell count and the number of optic nerve fibers. However, these values are difficult to interpret because of unknown scaling effects, and the possibility that cetaceans have an increased number of auditory ganglion cells and a decreased number of optic nerve fibers.

SIRENIA

Hyrtl (1845) included the sirenians in with the cetaceans and described the semicircular canal

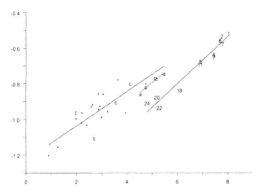

FIGURE 16.3. Double logarithmic plot of the radius of the semicircular duct lumen (SD-r in millimeters), against body mass (in grams). The reduced major axis regression lines are for the overall noncetacean sample, the artiodactyls (*dashed;* correlation not statistically significant), and the cetaceans. Number codes for cetaceans are presented in Table 16.1. Abbreviations: a, artiodactyls; P, *Phoca groenlandica;* c, other carnivores.

size of *Dugong dugon* as small for its body size, but still more terrestrial-like than in "true" *(eigentlichen)* cetaceans. Gray (1908) confirmed this observation and added that the perilymphatic space can be traced fully along the ducts, something shown in his stereophotographs of the *D. dugon* labyrinth as well (Fig. 16.4). Ketten et al. (1992) used CT to examine the inner ear of *Trichechus manatus* and described its canals as poorly developed, but they indicated that their scan resolution is inadequate to trace the course of the canal lumen through the dense periotic. Absolute and relative arc sizes of the semicircular canals of *D. dugon* and *T. inunguis* are listed in Table 16.6, as are those of *Elephas maximus* and *Loxodonta africana,* the closest living relatives of sirenians. Figure 16.5 confirms Hyrtl's (1845) observation that the sirenian canals are small, but within the range of noncetacean mammals. For the mean of the three canal radii, they take a similar position as the elephantids. However, the relative size of their canals differs substantially (Table 16.6). Both elephant species are characterized by a lateral canal that is by far the smallest of the three. On the other hand, in *T. inunguis* the lateral canal is the largest of the three, and in *D. dugon* it is just larger than the posterior canal. The main dif-

ference between these two sirenians is that the anterior canal of *D. dugon* is larger than in *T. inunguis.*

The vestibule of the sirenians is not reduced in overall size (Gray, 1908; Ketten et al., 1992; *T. inunguis* examined here). Gray (1908) found an otolith in the utricle of *D. dugon* that can be recognized with the naked eye.

CARNIVORA

Apart from the many fully terrestrial taxa, the order Carnivora includes arctoid species that are adapted to aquatic behavior to a variable degree, from pinnipeds to lutrines, *Mustela vison* and *Ursus maritimus* (see the section by Pihlström in chapter 1 in this volume for a brief overview of carnivore diversity). Early comparative studies of the inner ear showed that the phocids stand out by displaying a significantly wider canal lumen than other carnivores (Hyrtl, 1845; Gray, 1907). In contrast, the lumen of the enclosed membranous duct is approximately similar, after correcting for body size, and the perilymphatic space is therefore particularly wide in phocids, whereas it is narrow in otariids and fissipeds (Gray, 1907; Ramprashad et al., 1984) (Fig. 16.6). In fact, together with humans, phocids have the widest perilymphatic space recorded for mammals, with a duct lumen occupying only about 10% of the canal lumen (Gray, 1907; Ramprashad et al., 1984). The exact duct lumen has been measured only for *Phoca groenlandica* (Ramprashad et al., 1984). Scaled against body mass it falls within the range of fissipeds, close to *Panthera tigris* (Fig. 16.3). The membranous labyrinth of *Odobenus rosmarus* is not known, but casts of its bony labyrinth give some insight. The posterior and lateral canals are distinctly thin and round in cross section, as in otariids (Fig. 16.6B), whereas the ampullar half of the arc of the anterior canal is wider and flattened in the plane of the arc, a shape that gradually tapers toward the common crus. This suggests that there is little perilymphatic space in the canals of *O. rosmarus,* except for the ampullar half of the anterior canal.

TABLE 16.5

The Radius of the Membranous Semicircular Duct in Millimeters (SD-r) for a Sample of Mammals,
Compared with the Estimated Values for Cetaceans

	N	LABEL	SD-r	SOURCE
Afrosorida				
Erinaceus europaeus	1	o	0.12	1
Talpa europaea	10	o	0.11	2
Chiroptera				
Pteropus giganteus	1	o	0.14	1
Myotis lucifugus	4	o	0.06	3
Primates				
Homo sapiens	6	o	0.16	4, 5
Macaca mulatta	2	o	0.11	6
Saimiri sciureus	5	o	0.14	7
Aotus trivirgatus	1	o	0.12	7
Xenarthra				
Choloepus sp.	2	o	0.17	7
Lagomorpha				
Oryctolagus cuniculus	5	o	0.11	1, 6, 7
Rodentia				
Rattus norvegicus	18	o	0.09	6, 8, 9
Mus musculus	1	o	0.07	6
Mesocricetus auratus	1	o	0.10	6
Cavia porcellus	18	o	0.10	1, 4, 6
Meriones unguiculatus	4	o	0.10	7
Chinchilla laniger	4	o	0.12	7
Cryptomys hottentotus	4	o	0.11	10
Cryptomys mechowi	4	o	0.12	10
Spalax ehrenbergi	8	o	0.11	9
Carnivora				
Panthera tigris	1	c	0.17	1
Felis catus	1	c	0.13	4, 6
Canis familiaris	2	c	0.16	1, 6
Mustela putorius	1	c	0.08	6
Phoca groenlandica	1	P	0.17	7
Artiodactyla				
Bos taurus	1	a	0.18	11
Ovis aries	1	a	0.14	1
Sus scrofa	2	a	0.15	1, 11
Cetacea/Mysticeti				
Balaena mysticetus	1	1	0.31	12
Eubalaena glacialis	3	2	0.30	12, 13, 14
Eschrichtius robustus	1	4	0.23	15
Balaenoptera acutorostrata	1	5	0.21	15
Balaenoptera physalus	1	7	0.27	14
Megaptera novaeangliae	2	8	0.28	14, 15

TABLE 16.5 *(continued)*

	N	LABEL	SD-r	SOURCE
Cetacea/Odontoceti				
Physeter catodon	2	9	0.23	14, 16
Berardius bairdi	1	11	0.20	14
Monodon monoceros	1	18	0.15	16
Lagenorhynchus obliquidens	1	20	0.13	14
Tursiops truncatus	2	22	0.12	17
Phocoena phocoena	1	24	0.12	13

DATA SOURCES: 1, Jones and Spells (1963) from O. Gray specimen; 2, McVean (1999); 3, Ramprashad et al. (1980); 4, Curthoys et al, (1977b); 5, Curthoys and Oman (1987); 6, Jones and Spells (1963), from CS Hallpike histological sections; 7, Ramprashad et al. (1984); 8, Curthoys and Oman (1986); 9, Lindenlaub et al. (1995); 10, Lindenlaub and Burda (1994); 11, O. Gray specimen in care of first author; 12, Muller (1999); 13, Gray (1908); 14, cast, National Science Museum, Tokyo; 15, Siemens Somatom AR.SP; 16, Hyrtl (1845); 17, Norland XCT Research M scanner.

NOTE: The estimated values for cetaceans are calculated as 85% of the bony canal lumen (85% SC-r). The number of specimens (*n*) is given, as well as the labels used in Figure 16.3.

Arctoids highly adapted to aquatic behavior of which the canal arc sizes were examined include three phocids *(Halichoerus grypus, Phoca groenlandica, P. vitulina)*, four otariids *(Arctocephalus australis, A. pusillus, Otaria byronia, Zalophus californianus)*, and *O. rosmarus*. In addition, we assessed four semiaquatic species with proficient terrestrial locomotor capability, *Enhydra lutris, Lutra lutra, Lontra canadensis,* and *Mustela vison*, as well as 23 fully terrestrial carnivores to provide a comparative context. Table 16.7 lists absolute and relative arc sizes of the semicircular canals of these species. A complication when analyzing this sample is the unusually large perilymphatic space in the phocids, which causes the arc size of the canals to provide a poor estimate of the functionally important arc size of the enclosed ducts. The ducts are positioned consistently along the outer canal wall, rather than in the center (Fig. 16.6A). Hence, the canal arc size of phocids underestimates the duct arc size, whereas this is not the case for otariids and fissipeds with ducts that occupy most of the canal lumen. Direct measurements or scaled photographs of soft tissue specimens (Gray, 1907; Ramprashad et al., 1984) make it possible to estimate the arc size to the center of the ducts for all three phocid species investigated here. It is these more compatible phocid values that are considered, but the values to the center of the bony canal are given in Table 16.7 as well.

Canal arc size is plotted against body mass in Figure 16.7, and the associated correlation and RMA regression statistics for fully terrestrial species are given in Table 16.2. Moreover, it was statistically tested whether the values of the semiaquatic species differ significantly from those predicted by that regression (Table 16.8). First, all terrestrial arctoids fall on or close to the terrestrial carnivore regression (Fig. 16.7, point x), and the latter thus is a good reference for comparison of the semiaquatic arctoids. *Odobenus rosmarus* falls close to, and is not significantly different from, the terrestrial regression for each of the three canals. For the anterior and posterior canals, the phocids and otariids form separate clusters, falling above and below the regression, respectively. Indeed, the phocids have larger anterior and posterior canals, and the otariids smaller ones than predicted by the terrestrial trend, although the differences are not all statistically significant when considering each species individually (Table 16.8). The phocids also have a larger lateral canal than the fissipeds. On the other hand, the lateral canal of the otariids is not different, with the exception of *Arctocephalus pusillus*, where it is distinctly smaller. The marked difference with the closely related species *A. australis* is particularly striking, and additional

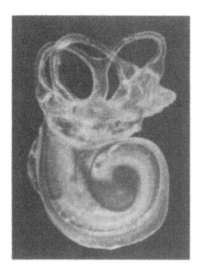

FIGURE 16.4. Inferolateral view of the right inner ear of *Dugong dugon*, showing both the bony and membranous labyrinth. Modified after Gray (1908).

individuals of *A. pusillus* should be examined to assess whether the small lateral canal measurements given by Gray (1907) are correct and representative of this species. The different patterns in canal sizes among the pinnipeds are well summarized by their relative sizes (Table 16.7).

Gray (1907) found two large otoliths in *Phoca vitulina*, which he characterized as the largest in his entire sample of mammals, and one large otolith in the utricle of *A. pusillus*. In contrast, no otoliths could be seen in the specimen of *Halichoerus grypus*, possibly as the results of acid treatment of the specimen (Gray, 1907). The presence of an unusually large otolithic mass in phocids is supported by the observation of Solntseva (2001) that their otolith membrane is thicker than in otariids, *Odobenus*, and terrestrial mammals.

Finally, the semiaquatic mustelid *Mustela vison* is characterized by semicircular canal sizes that fall on the terrestrial regression (Fig. 16.7; Table 16.8). The anterior and posterior canal sizes of the lutrines all fall below the terrestrial regression, and the difference is statistically significant, although not for each of the species individually (Fig. 16.7; Table 16.8). The lutrine lateral canal size is not different from the terrestrial regression. Similar to fully terres-

trial carnivores, *Lutra lutra* has little perilymphatic space surrounding the ducts and lacks visible otoliths (Gray, 1907).

RODENTIA

A number of rodent species show variable degrees of aquatic behavior. Table 16.6 lists absolute and relative arc sizes of the semicircular canals of four of these, *Castor canadensis*, *Ondatra zibethicus*, *Hydrochaeris hydrochaeris*, and *Myocastor coypus*, as well as a small sample of other rodents. Figure 16.5 plots the mean arc size of the three semicircular canals, as the results for each of the three are similar. The largest canals are shown by three fast and agile species, *Chinchilla laniger*, *Dipus sagitta*, and *Sciurus vulgaris*, a pattern that is consistent with findings for primates and other mammals (Spoor et al., 2007). The semiaquatic species, on the other hand, are not substantially different from the other rodents and mammals of similar body mass. No RMA regressions were calculated, because the rodent sample is too limited to provide meaningful statistical results, especially as the larger species in the sample are all semiaquatic (Fig. 16.5).

Comparative information regarding soft-tissue aspects of the vestibular system of amphibious rodents is limited to the qualitative observations of Gray (1907, 1908) for *H. hydrochaeris*. As in other rodents, this species does not have otoliths visible to the naked eye. However, it does stand out among the rodents examined by Gray in that it has relatively thick canals, with fairly well marked perilymphatic space surrounding the ducts. However, this may well be a size-related difference, because all small mammals typically lack a marked perilymphatic space, and the condition in larger rodents, other than *H. hydrochaeris*, is not known.

MONOTREMATA

The semicircular canals of the semiaquatic species *Ornithorhynchus anatinus* are well within the variation shown by noncetacean mammals (Fig. 10.5 and Table 10.6 in this volume). The perilymphatic space is fairly well marked (Gray, 1908). Any comparisons of the canal system

TABLE 16.6
Arc Size of the Semicircular Canals and Body Mass (BM, in Grams) of a Variety of Mammals

	N	BM	SOURCE	SC-R	%ASC	%PSC	%LSC	SOURCE
Rodentia, Bathyergidae (r)								
Cryptomys hottentotus	4	95	1	1.1	38	29	33	1
Cryptomys mechowi	4	400	1	1.3	38	28	34	1
Rodentia, Castoridae (C)								
Castor canadensis	1	18,667	2	2.6	36	35	29	14
Rodentia, Caviidae (r)								
Cavia porcellus	20	1,000	3	1.7	36	31	33	10
Rodentia, Chinchillidae (r)								
Chinchilla laniger	4	450	4	1.9	35	34	31	4
Rodentia, Dipodidae (r)								
Dipus sagitta	1	66	2	1.3	38	34	28	11
Rodentia, Hydrochaeridae (H)								
Hydrochoeris hydrochaeris	1	39,540	2	2.8	35	33	32	11
Rodentia, Muridae								
Meriones unguiculatus (r)	4	100	4	1.2	33	35	32	4
Mus musculus (r)	2	19	2	0.7	40	34	26	11, 13
Ondatra zibethicus (O)	1	995	2	1.7	32	35	33	15
Rattus norvegicus (r)	9	270	5	1.4	40	30	30	5, 11
Rodentia, Myocastoridae (M)								
Myocastor coypus	1	7,855	2	1.8	36	32	32	15
Rodentia, Sciuridae (r)								
Sciurus vulgaris	2	322	2	1.7	40	30	30	12, 14
Rodentia, Spalacidae (r)								
Spalax ehrenbergi	8	170	5	1.3	37	30	33	1
Sirenia, Dugongidae (D)								
Dugong dugon	1	425,000	6	3.2	37	31	32	12
Sirenia, Trichechidae (T)								
Trichechus inunguis	1	425,000	7	2.8	30	32	37	16
Proboscidea, Elephantidae								
Elephas maximus (E)	1	3,450,000	8	3.6	37	38	25	17
Loxodonta africana (L)	1	4,540,000	9	4.8	36	38	25	16
Monotremata (A)								
Ornithorhynchus anatinus	2	1752	2	1.6	37	34	29	12, 13

DATA SOURCES: 1, Lindenlaub and Burda (1994); 2, Silva and Downing (1995); 3, Nowak (1991); 4, Ramprashad et al. (1984); 5, Lindenlaub et al. (1995); 6, Blanshard (2001); 7, Emmons (1990); 8, Sukumar (1989); 9, Norwood (2002); 10 Curthoys et al. (1977a, 1977b); 11, Gray (1907); 12, Gray (1908); 13, www.digimorph.org; 14, X-tech BIR scanner; 15, Norland XCT Research M scanner; 16, General Electric High-Speed scanner; 17, O. Gray cast in care of first author.

NOTE: The mean radius of curvature (R) of the three semicircular canals (SC) is given in millimeters, and the relative size, in percentage, for anterior (A), posterior (P), and lateral (L) semicircular canal (sum of the three canal radii is 100). The number of specimens (n) is given, as well as the symbols used in Figure 16.4.

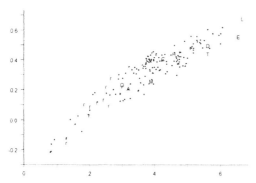

FIGURE 16.5. Double logarithmic plots of the mean radius of curvature of the three semicircular canals (in millimeters), against body mass (in grams). The rodent (r), sirenian (M, D), elephant species (E, L), and *Ornithorhynchus anatinus* (O) are labeled following the code given in Table 16.6.

beyond the observation that it is not strongly reduced as in cetaceans are difficult because of the limited information available for the other extant monotremes, *Tachyglossus aculeatus* and *Zaglossus bruijnii* (Hyrtl, 1845; Denker, 1901). The otolith organ of the former has been described (Jorgensen and Locket, 1995), but the equivalent aspects are unknown in *O. anatinus*.

FUNCTIONAL INTERPRETATION AND EVOLUTIONARY HYPOTHESIS

This review of the vestibular system in aquatic and semiaquatic mammals suggests that only the cetaceans show functionally relevant morphology well outside the range of variation of all other mammals. Their semicircular canal system is characterized by a strong reduction in arc and lumen size, and similarly reduced afferent innervation. However, there is no unambiguous evidence that it is less than fully functional. Among noncetacean mammals, only the pinnipeds are found to differ substantially in the radius of the semicircular canals from their terrestrial relatives. Pinnipeds do not show a uniform pattern, with phocids having a larger arced canal system than other carnivores, whereas otariids have somewhat smaller arced anterior and posterior canals. Sirenians, which are with cetaceans the only obligately aquatic

mammals, are difficult to interpret because of the lack of a closely related, extant nonaquatic outgroup. None of the sirenian inner ear morphology investigated thus far falls outside the variation shown by noncetacean mammals. The otolith system of cetaceans, pinnipeds, and sirenians is insuffciently studied to draw definitive conclusions, but there is some evidence that all three groups have relatively large otoliths (otolithic mass of otoconia). If correct, this could imply an increased sensitivity for linear acceleration (see work by Rabbitt et al. [2004] for a review of the biomechanics of the otolith system).

We now consider the reduced cetacean canal system in more detail and briefly assess how our interpretation of its functional morphology compares with the findings for other aquatic and semiaquatic mammals.

CETACEA

The functional implications of the unparalled reduction of the cetacean canal system are discussed at some length by Boenninghaus (1903). The study makes reference to the empirical observation of Panse (1899) that animals with "clumsy" and "awkward" movements have canals with a wide lumen, as well as to the biomechanical notion that a narrow lumen renders the cetacean canals less sensitive. However, rather than applying these observations to the locomotion and canal morphology of cetaceans, Boenninghaus (1903) suggests an explanation based on spatial constraints. He proposes that the cetacean inner ear is dominated by auditory function, and that limited space is available for vestibule and semicircular canals. Alternatively, Gray (1908:23) associates canal function with "the extent and delicacy of the movements of the head upon the trunk" and proposes that the reduced canal size in cetaceans is caused by the ankylosis of cervical vertebrae. Ketten (1991) concludes that if the canal system is vestigial, only linear acceleration is detected, which would permit rapid rotations without nauseating side effects. Regarding the origin of the cetacean condition, Ketten

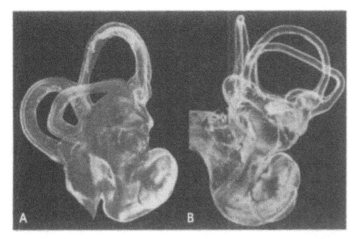

FIGURE 16.6. Inferolateral view of the right inner ear of the phocid seal *Phoca vitulina* (A), and inferoposterior view of the right inner ear of otariid fur seal *Arctocephalus pusillus* (B), modified from Gray (1908), showing both the bony and membranous labyrinth. The perilymphatic space around the semicircular ducts is wide in the phocid and nearly absent in the otariid.

expands upon Gray's (1908) hypothesis: fusion of the cervical vertebrae may have resulted in limited head movements and substantially fewer inputs to the vestibular system, leading to loss of related receptors (Ketten, 1992, 1997). Wartzok and Ketten (1999:141) add that this "does not mean that cetaceans do not receive acceleration and gravity cues but rather that the neural 'budget' for these cues is less."

Most recently, Spoor et al. (2002) conclude that the small semicircular canal size of cetaceans is unlikely to represent a vestigial condition. Rather, the observed pattern of allometric scaling, with its changing intercept but similar slope (scaling coefficient) to terrestrial mammals, points at structured functional adjustment of the system, instead of degeneration and redundancy. The hypothesis explaining the small canal arc size put forward by Spoor et al. (2002) reinterprets and combines two key factors, each noted previously as having possible relevance (Gray, 1908; Ketten, 1991). The first one is that extant cetaceans, freed from the restrictions of gravitational pull and the need for substrate contact, are particularly agile and acrobatic when compared with terrestrial animals of similar body size and mass (for example, compare *Orcinus* with *Loxodonta*, or *Tursiops* with the larger bovid species). The second factor is that cetaceans have integrated their head and trunk to streamline the body, and in most species the strongly shortened and frequently fused cervical vertebrae allow very little neck motility. Ketten

(1997) concluded that the relatively immobile neck resulted in fewer head movements and reduced input into the vestibular system, but this is not the case. Rather the opposite, the movements of the body cause increased head movements and vestibular stimuli because a motile neck plays a vital role in isolating the head from the locomoting body, minimizing angular accelerations and velocities, both passively by inertia, and actively via compensatory neck movements generated by the vestibulocollic reflex. This mechanism not only contributes to the stabilization of gaze, but also allows the semicircular canals of agile species to be sensitive, without the risk of overstimulation. As part of a feedback system the canals supply the vestibulocollic reflex, which stabilizes the head, thus keeping the input signal of the canals within limits. Sensitive canals can resolve small changes in angular head motion, and this is thought to be important for body coordination of animals showing fast and highly maneuverable locomotion (Spoor et al., 2002; Spoor, 2003). Indeed, among noncetacean mammals, fast and acrobatic species do show larger-arced, and thus more sensitive, canals than related species that are characterized by less agile locomotor behaviors (Spoor et al., 2007). In cetaceans, on the other hand, the combination of ineffective head stabilization and acrobatic locomotion implies that the semicircular canal system is likely to experience substantially higher levels of angular motion (resulting from

TABLE 16.7
Arc Size of the Semicircular Canals and Body Mass (BM, in Grams) of Carnivores

	N	BM	SOURCE	SC-R	%ASC	%PSC	%LSC	SOURCE
Carnivora, Canidae (+)								
Canis familiaris	50	13,850	10	2.2	37	32	31	10
Canis latrans	1	12,650	1	2.3	35	33	32	14
Nyctereutes procyonoides	42	4,500	1	1.7	38	34	28	10
Otocyon megalotis	1	21,500	1	2.1	35	35	30	14
Vulpes vulpes	27	5,000	1	2.0	37	32	31	10
Carnivora, Felidae (+)								
Acinonyx jubatus	1	50,167	1	2.4	34	35	32	15
Felis catus	28	3,601	1	1.8	35	34	31	11, 12
Felis concolor	1	56,425	1	2.5	35	35	30	11
Felis sp. (serval, caracal, or aurata)	1	11,008	1	2.2	35	35	31	14
Lynx rufus	1	9,750	1	2.3	34	35	30	16
Panthera leo	1	148,750	1	3.3	34	34	32	14
Panthera tigris	2	150,100	1	3.0	36	32	32	11, 18
Carnivora, Herpestidae (+)								
Ichneumia albicauda	1	2,625	1	2.1	35	35	30	14
Carnivora, Hyaenidae (+)								
Crocuta crocuta	2	58,600	1	2.9	37	35	28	14, 15
Proteles cristatus	1	8,550	1	2.1	42	30	29	11
Carnivora, Mustelidae								
Enhydra lutris (E)	2	27,050	1	2.1	32	35	33	16
Lontra canadensis (C)	1	9,250	1	1.8	32	34	34	14
Lutra lutra (L)	1	8,175	1	1.9	35	31	34	11
Mustela nivalis (x)	1	87	1	1.1	33	36	31	11
Mustela vison (M)	1	1,478	1	1.6	35	35	30	14
Spilogale putorius (x)	1	857	1	1.5	37	36	27	14
Taxidea taxus (x)	1	7,190	1	2.1	32	33	35	16
Carnivora, Procyonidae (x)								
Procyon cancrivorus	1	6,994	1	2.2	32	37	31	11
Procyon lotor	1	9,000	1	2.1	34	35	31	14
Carnivora, Ursidae (x)								
Ursus americanus	1	136,000	2	3.1	36	33	30	16
Carnivora, Viverridae (+)								
Genetta genetta	1	1,500	1	1.6	37	34	29	14
Herpestes ichneumon	1	2,981	1	1.6	31	37	32	11
Carnivora, Odobenidae (O)								
Odobenus rosmarus	1	1,013,500	3	3.8	33	34	33	18
Carnivora, Otariidae								
Arctocephalus australis (A)	1	91,000	4	2.4	32	31	37	17
Arctocephalus pusillus (P)	1	57,000	4	1.9	34	34	32	11
Otaria byronia (B)	1	150,000	5	2.8	33	32	36	17
Zalophus californianus (Z)	1	100,000	6	2.6	31	31	38	17

TABLE 16.7 (continued)

	N	BM	SOURCE		SC-R	%ASC	%PSC	%LSC	SOURCE
Carnivora, Phocidae (H)									
Halichoerus grypus	1	194,000	7	Duct	4.3	39	32	29	11
				Canal	3.7	42	31	27	
Phoca groenlandica	1	127,500	8	Duct	4.1	33	37	30	13
				Canal	3.7	33	38	30	
Phoca vitulina	1	81,250	9	Duct	4.0	35	34	31	11
				Canal	3.4	39	31	31	

DATA SOURCES: 1, Silva and Downing (1995); 2, Ohio Dept. of Natural Resources (www.dnr.ohio.gov/); 3, Fay (1981); 4, Arnould (2002); 5, Cappozzo (2002); 6, Heath (2002); 7, Bonner (1981); 8, Ronald and Healey (1981); 9, Markussen et al. (1989); 10, Takahashi (1971); 11, Gray (1907); 12, Curthoys et al. (1977a); 13, Ramprashad et al. (1984); 14, Norland XCT Research M scanner; 15, www.digimorph.org; 16, X-tech BIR scanner; Siemens Somatom AR.SP scanner; 17, General Electric High Speed scanner; 18, O. Gray, in care of first author.

NOTE: The mean radius of curvature (R) of the three semicircular canals (SC) is given in millimeters, and the relative size in percentage of the anterior (A), posterior (P), and lateral (L) canal (sum of the three canal radii is 100). The number of specimens (n) is given, as well as the symbols used in Fig. 16.7.

movements of the entire body) than in terrestrial mammals of similar body size. Spoor et al. (2002) suggest that the small arc size of cetaceans serves to reduce their mechanical sensitivity to match the high levels of uncompensated angular motion, thus avoiding overstimulation of the canal system and resultant disorienting signals. The loss of canal sensitivity, in response to streamlining of the body, is arguably less critical in an aquatic environment than in, for example, an arboreal setting, where less accurate sensory clues easily impair locomotor control.

Our "reduced-sensitivity" hypothesis is consistent with the current observation that the odontocetes show a somewhat stronger reduction of their canal arcs than mysticetes (Table 16.3), given that the former are the more agile group, as predators actively pursuing large and mobile prey species. Moreover, that the anterior and posterior canals are more reduced than the lateral one could well reflect caudal dorsoventral oscillation as the main mode of cetacean locomotion, producing the strongest body motion in the pitch plane, sensed by the two vertical canals. It is worth noting that the well-understood role of the canal system in gaze stabilization, through the vestibulo-ocular reflex, is likely minimal in extant cetaceans, as their eyes are positioned laterally, and visual clues are not central in navigation.

Possible corroboration of the specific link between neck motility and canal size is provided by the near-blind Ganges river dolphin, *Platanista gangetica*. Compared with other extant cetaceans this species has a relatively long, more motile neck with unfused cervical vertebrae, apparently used in a specialized method of navigation, which is characterized by side-swimming and active neck movements (Pilleri, 1974; Reeves and Brownell, 1989). At the same time, *Platanista* also has larger-arced canals than other extant odontocetes (Fig. 16.2, point 14), and the combination suggests a more effective vestibulocollic interaction than in other cetaceans. That the link between canal size and neck motility is not a simple one is demonstrated by *Delphinapterus leucas*, which shows a substantial degree of head movement (Brodie, 1989; O'Corry-Crowe, 2002), but its canals are among the most reduced in the sample (Fig. 16.2, point 17). However, it is not clear whether its head movements play any role in reflex head stabilization during fast and agile swimming, or whether these are employed only during specific activities when vestibular input is low. Likewise, it was noticed that the fossil whale *Remingtonocetus* has small semicircular canals, whereas it has neck vertebrae that are long, and presumably a mobile neck (Spoor et al., 2002). It was hypothesized that its large,

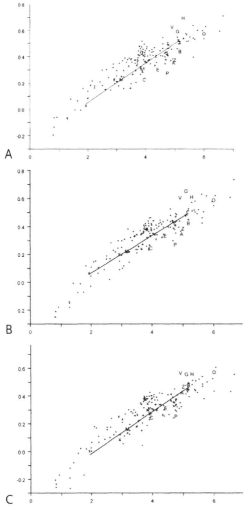

FIGURE 16.7. Double logarithmic plots of the radius of curvature of the anterior (A), posterior (B), and lateral (C) semicircular canals (in millimeters), against body mass (in grams). The reduced major axis regression line pertains to the terrestrial carnivores. Labels refer to carnivore species as listed in Table 16.7.

of interest because they mark the upper- and lower-corner frequencies of the range of frequencies over which the canal system acts as an integrating angular accelerometer, with a response that is proportional to instantaneous head angular velocity (reviewed in Wilson and Melvill Jones, 1979; Rabbitt et al., 2004).

Before assessing these properties, however, it is important to emphasize the limitations of focusing on mechanical response behavior of the canals. The latter should be seen primarily as a description of the input into the hair cells of the ampullary apparatus, rather than as a characterization of the afferent signals transmitted by the canals to the brain. This is because the hair cells and afferent neural components are known to perform a substantial amount of signal processing, and major differences have been found, even between individual afferent neurons (Goldberg, 2000; Holstein et al., 2004; Rabbitt et al., 2004, 2005; Highstein et al., 2005). As yet, there is no clear understanding of these issues in a comparative context, and of how any interspecific differences relate to vestibular function at a whole organism level. Hence, with these "known unknowns" the biomechanical aspects of the canal system should be interpreted with caution. Nevertheless, with strong empirical evidence for major differences in functionally relevant morphology, as is shown by the cetacean semicircular canals, it seems justified to assume that these represent a meaningful biological signal, rather than random modification with an afferent output entirely neutralized or even reversed by neural processing. The view that semicircular canal radius is highly responsive to locomotor function is further supported by interpreting these radii in the context of an established cetacean phylogeny (Fig. 16.8). It is not apparent that a phylogenetic signal is preserved in semicircular canal sizes. Finally, a practical point is also that if one wants to apply any comparative results to the fossil record, it is only the features reflected by the bony labyrinth that can be considered.

Mechanical sensitivity of the canal system is defined here as the average cupula displacement

remarkably long snouted head favored hydrodynamic integration of the head and trunk to reduce drag, minimizing neck movements at the cost of effective reflex stabilization of the head (Spoor et al., 2002).

Having empirically established the degree of reduction in arc radius (R) and lumen radius (r) of cetacean semicircular ducts (Table 16.3), it is possible to combine these results with biomechanical models of the canal system to assess the impact on the mechanical sensitivity and the short and long time constants. The latter are

TABLE 16.8

Difference Between Observed Radii of Curvature of Semicircular Canals in Semiaquatic Carnivores, and Those Predicted by the Reduced Major axis Regression for Terrestrial species

	ASC			PSC			LSC		
	y pred	95%	y obs	y pred	95%	y obs	y pred	95%	y obs
Carnivora, Odobenidae									
Odobenus rosmarus	4.4	3.5-5.6	3.8 =	4.0	3.3-5.0	3.9 =	3.9	3.1-4.9	3.8 =
Carnivora, Phocidae	3.3	2.9-3.8	4.4 >	3.1	2.7-3.5	4.3 >	2.9	2.5-3.3	3.7 >
Halichoerus grypus	3.5	2.8-4.3	5.0 >	3.2	2.7-3.9	4.1 >	3.0	2.4-3.8	3.7 =
Phoca groenlandica	3.3	2.6-4.1	4.0 =	3.1	2.5-3.7	4.6 >	2.8	2.3-3.5	3.7 >
Phoca vitulina	3.1	2.5-3.8	4.3 >	2.9	2.4-3.5	4.1 >	2.7	2.1-3.3	3.7 >
Carnivora, Otariidae	3.1	2.8-3.6	2.3 <	3.0	2.7-3.3	2.3 <	2.7	2.4-3.1	2.6 =
Arctocephalus australis	3.1	2.5-3.8	2.3 <	2.9	2.4-3.5	2.3 <	2.7	2.2-3.3	2.7 =
Arctocephalus pusillus	2.9	2.3-3.6	1.9 <	2.7	2.3-3.3	1.9 <	2.5	2.0-3.1	1.8 <
Otaria byronia	3.3	2.7-4.2	2.8 =	3.1	2.6-3.8	2.7 =	2.9	2.3-3.6	3.0 =
Zalophus californianus	3.1	2.5-3.9	2.4 <	3.0	2.5-3.6	2.5 <	2.7	2.2-3.4	3.0 =
Carnivora, Lutrinae	2.4	2.1-2.7	1.9 <	2.3	2.1-2.5	1.9 <	2.1	1.8-2.3	1.9 =
Enhydra lutris	2.6	2.1-3.2	2.0 <	2.5	2.1-3.0	2.2 =	2.3	1.8-2.8	2.1 =
Lontra canadensis	2.2	1.8-2.7	1.7 <	2.1	1.8-2.6	1.8 <	1.9	1.6-2.4	1.8 =
Lutra lutra	2.2	1.8-2.7	2.0 =	2.1	1.8-2.5	1.8 =	1.9	1.5-2.3	1.9 =
Carnivora, Mustelidae									
Mustela vison	1.7	1.4-2.1	1.7 =	1.7	1.4-2.0	1.7 =	1.5	1.2-1.8	1.5 =

NOTE: Data are presented for each of the canals separately (ASC, anterior semicircular canal; PSD, posterior SC; LSC, lateral SC). The 95% confidence interval of the y value predicted (y pred) by the body mass of the investigated species was calculated, and compared with the actual y value (y obs), with an indication (<, =, >) of the degree of similarity of these (Sokal and Rohlf, 1995).

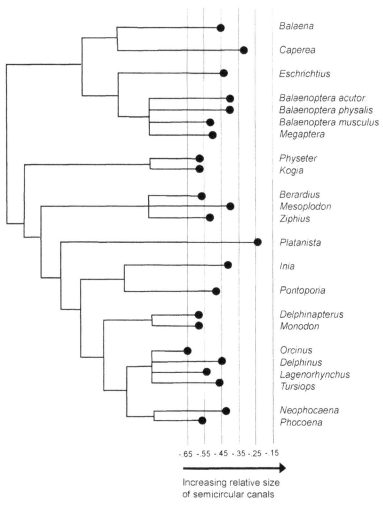

FIGURE 16.8. Cladogram of modern cetaceans, with the end points of the branches indicating the size reduction of the semicircular canals expressed as the residuals from the regression for the average canal radius of artiodactyls (which includes their modern sister group, compare to Fig. 16.2). This figure shows that semicircular canal size varies greatly within cetacean families and appears to lack a strong phylogenetic signal.

per unit head angular velocity. Given the circular duct shape in cetaceans, artiodactyls, and most other mammals, a simplified formula can be used that indicates sensitivity changes in proportion to R and r^4 (Oman et al., 1987; Lindenlaub et al., 1995; McVean, 1999; Rabbitt et al., 2004). The reduction in R and r of cetacean semicircular ducts is 33% and 76%, respectively, relative to artiodactyls, and 32% and 68% relative to all noncetacean mammals (Table 16.3). Given these differences, and assuming that all else is similar, this implies that the mechanical sensitivity of the cetacean

duct is 11% of that of artiodactyls, and 7% of that of noncetacean mammals overall.

The fast (short) time constant, which indicates the response speed, or the time for the cupula to attain maximum displacement, is proportional to r^2 and independent of R (Jones and Spells, 1963; Muller, 1994; Rabbitt et al., 2004). This implies that the short time constant of cetaceans is 58% of that of artiodactyls, and 47% of that of noncetacean mammals overall. Finally, the slow (long) time constant represents the restoration or recovery time of the cupula to passively return to the initial state, after acceleration stops. Most recently, Rabbitt et al. (2004) modeled this time constant taking into account the area and thickness of the cupula. No relevant comparative data of cupula dimensions are available, but it might be reasonable to hypothesize that cupula area scales as r^2, and cupula thickness as r. With these assumptions the slow time constant scales as R/r (modified formula 4.24 [Rabbitt et al., 2004]), and in cetaceans is 43% of that of artiodactyls, and 47% of that of noncetacean mammals overall. Taking the less-likely scenario that the cupula dimensions would be unchanged in cetaceans, only the duct R and r are reduced, the slow time constant is proportional to R/r^4, and in cetaceans is 97% of that in artiodactyls, and 148% of that of noncetacean mammals overall (i.e., remains unchanged, and half as long, respectively).

In broad terms the comparative findings mean that the mechanical sensitivity of cetacean canals is about a tenth of that expected on average for other mammals of the same size, and that the fast time constant is approximately half the expected value. The latter implies a doubling of the upper-corner frequency of the frequency range over which a velocity-sensitive response can be obtained. Changes to the lower-corner frequency of the bandwidth are less clear because the slow time constant is strongly influenced by the unknown factor of cupula morphology. The substantial reduction in sensitivity is consistent with the explanation of Spoor et al. (2002) for the small canal dimensions of cetaceans. Nevertheless, it can be argued that optimizing sensitivity may not be the driving factor, because there are mechanisms to adjust this property that seem biologically easier to achieve, for example at the level of the sensory epithelium by modifying the numbers of synaptic contacts and hair cells. In contrast, there may not be simple nonmechanical ways to obtain flat velocity-sensitive canal responses over a specific frequency range. From this perspective, it could be adjustments of the time constants that drive adaptive size changes of the canal system (Highstein et al., 2005). The shift to higher bandwidth frequencies is consistent with the notion that

cetaceans are more acrobatic than other mammals of a similar size. Moreover, this explanation is simpler than that based on sensitivity adjustment because it does not require the additional factor of the loss of neck motility and effective head stabilization.

The problem with the adaptive expansion to higher bandwidth frequencies as an explanation is that it applies only to the reduction in duct lumen r; it is this property, not the arc radius R, that influences the fast time constant, and thus the upper-corner frequency. Another issue is that any explanation for size changes of the canal system should preferably apply to all mammals, and other vertebrate groups as well, so that no special pleading is required for the morphology of cetaceans. At present no comparative study has assessed how r compares in vertebrates with contrasting locomotor agility. However, such data are available for R and indicate that fast and acrobatic species of birds and terrestrial mammals have larger-arced semicircular canals than those with slower and less maneuverable types of locomotion (Sipla et al., 2003; Spoor et al., 2007). Hence, cetaceans are the only known exception: they combine small-arced canals with agile locomotion. If changes to the velocity-sensitive bandwidth are the driving factor behind this pattern, it would follow that acrobatic species other than cetaceans aim to expand the bandwidth to lower frequencies, as it is only the lower-corner frequency that is influenced by R. It is difficult to envisage why this would be beneficial, and why cetaceans alone show the opposite trend. As it stands, the explanation of Spoor et al. (2002), based on adjustment of sensitivity levels, and the importance of head stabilization, appears most realistic in that it is consistent with both biomechanical models, and all available comparative evidence for mammals and birds.

A comparative issue worth briefly considering here is the notion that fishes have larger-arced canals than most other vertebrates (Jones and Spell, 1963). They share with cetaceans the distinct aquatic environment, as well as the absence of a motile neck, but have larger rather than smaller canals than other vertebrates. It could thus be argued that this discounts any functional explanation of the cetacean canal system offered so far. Several studies have attempted to explain the large canals of fishes by making broad comparisons with tetrapods (Jones and Spell, 1963; Ten Kate and Kuiper, 1970; Howland and Masci, 1973; Bernacsek and Carroll, 1981; Young, 1981; Muller, 1999; Maisey, 2001). However, only some of these studies have considered not just the general allometric trend, but also the wide-ranging diversity in canal morphology among fish species (Howland and Masci, 1973; Young, 1981; Maisey, 2001). Importantly, no study has as yet systematically and quantitatively analyzed this diversity in relation to different modes of swimming. Thus, it could well be that among fish species the same locomotion-related pattern of canal morphologies is present as among mammal species, and that the overall allometric grade shift is based on underlying physiological and neurological baseline factors. It is these interspecific patterns that need to be investigated before major differences between remotely related classes of vertebrates can be interpreted. A complicating factor in such future analyses will be that fishes are poikilotherm, and the viscosity of endolymph is not a constant (Ten Kate et al., 1969).

One important aspect of the link between cetacean canal morphology and aquatic behavior not yet discussed is the exact nature of head motion in cetaceans, and how it compares with that of terrestrial mammals of similar size. This information is essential to verify assertions that cetaceans are indeed more agile and experience higher levels of angular accelerations and velocities in the absence of head stabilization. No full quantitative kinematic descriptions of head motion in cetaceans have yet been published, although some initial results for captive *Tursiops truncatus* have been reported in an abstract (Hullar and Armand, 2004). Rotational accelerations and velocities in the yaw plane (bilateral movement) are said to commonly exceed 1250 deg/s^2, and 400 deg/s, respectively, whereas those in other planes are lower, with velocities rarely exceeding 50 deg/s during regular swimming. The dominant frequency of head motion

is reported as approximately the same in all planes and unique to the animal, with values between 1.5 and 2 Hz regardless of the activity. The values are difficult to interpret, because no information is available for the head motion of similar-sized terrestrial or semiaquatic artiodactyls, such as *Hexaprotodon*, *Oryx*, or smaller breeds of *Bos* (see Table 16.1 for comparative body masses). Nevertheless, the highest recorded velocities in the yaw plane are surprising given the fact that aquatic locomotion is based on dorsoventral undulation, which is expected to produce body movement predominantly in the pitch plane. Meaningful interpretations of these intriguing results will have to await a full description of the routines performed by the test animals, and of their kinematic analysis.

CARNIVORA

The semicircular canals of several aquatic and semiaquatic carnivores are different in arc size from those of related terrestrial species. However, none show the dramatic size reduction seen in cetaceans, an expected result in the context of the reduced-sensitivity hypothesis because none combine great aquatic locomotor agility with lack of neck motility (Spoor et al., 2002). On the other hand, the "frequency-range" hypothesis seems inconsistent with the absence of cetaceanlike canal reduction in phocids in particular. These are arguably as agile as cetaceans in their marine environment and nearly as dedicated to swimming, and they do not show any competent, limb-supported terrestrial locomotion that could restrain changes to the vestibular system. However, where phocids, and other pinnipeds, differ from cetaceans is their long and motile neck. The head with large forward-looking eyes leads changes in direction while swimming, and the rest of the body follows in a flowing, flexible fashion. Head stabilization is thus as effective as in any terrestrial carnivore, and this is expressed in their canal morphology, as predicted by the reduced-sensitivity hypothesis.

The pinnipeds clearly do show differences in canal size from terrestrial carnivores. Perhaps the most surprising result of this review is that the pattern of these differences shows opposite trends in phocids and otariids. That phocids have larger canals than terrestrial carnivores is expected, as they are particularly agile. This thus follows the normal pattern seen among noncetacean mammals and birds. On the other hand, the smaller anterior and posterior canals of otariids are more difficult to understand. Otariids differ from phocids by using an entirely different mode of propulsion, a birdlike forelimb flight stroke, as opposed to bilateral hindlimb undulation, and this will affect the main plane of body motion (pitch and yaw, respectively). In relation to this forelimb-driven mode of locomotion, otariids have a longer neck with a center of gravity that is located further forward than in phocids. The anterior and posterior canals are the ones that perceive pitch motion, but it is not clear how the reduced mechanical sensitivity that follows from their smaller arc size would relate to the otariid locomotion pattern or body plan. A more comprehensive survey of the pinniped inner ear, and kinematic studies of pinniped swimming modes may offer the best way forward to gain a better understanding of this issue.

Another aspect of pinniped semicircular canal morphology to be considered is the possible significance of the particularly wide perilymphatic space surrounding the semicircular ducts of phocids. Although some studies have proposed that the perilymph plays a significant role in semicircular canal mechanics (Rejtoe, 1939; Anliker and Van Buskirk, 1971), others have convincingly demonstrated that this is very unlikely, mainly because the dense trabecular pattern in the perilymphatic space prohibits free fluid movement (Dohlman and Kühn, 1973; McCabe and Ryu, 1973). It seems most likely that the wide perilymphatic space in phocids reflects distinct otic capsule development affecting the bony lumen. Making comparisons with the other species that stands out by its wide perilymphatic space, humans, could possibly shed light on the underlying developmental processes leading to this morphology.

The lutrines have anterior and posterior canals that tend to be smaller arced than in terrestrial arctoids, the same pattern as shown by otariids. Of the three lutrine species examined, *Enhydra* is undoubtedly the most adapted to aquatic behavior, but this is not expressed in a clearly distinct canal morphology. Active swimming modes involve hindlimb paddling, aided by dorsoventral body undulation when diving (Estes and Bodkin, 2002), and this would suggest a predominant pitch plane of locomotor motion. As is the case for otariids, this pitch plane is perceived by the smaller, less-sensitive anterior and posterior canals, but it is not clear if there is any functional association. It may well be that it is a specific pattern of head movements during particular activities, such as feeding or hunting, rather than the rhythmic propulsive motion associated with locomotion, that is important in this context.

Mustela vison has semicircular canals that are similar to those of fully terrestrial carnivore species. This is expected because it combines swimming with a full terrestrial locomotor repertoire.

SIRENIA, RODENTIA, AND MONOTREMATA

Sirenians are obligatorily aquatic and do show reduced neck motility. However, they are slow and cautious in their swimming, so that fast and effective head stabilization is not a factor of importance. Their canal size is at the lower end of the noncetacean mammalian variation, as are the canals of terrestrial species that are slow and cautious in their locomotion.

The semiaquatic rodents have canal morphologies that are not substantially different from those of related species that are fully terrestrial. They combine a full terrestrial locomotor repertoire with swimming, much of which is on the surface and not more agile in character than their terrestrial gaits.

Among monotremes the vestibular morphology of *Ornithorhynchus anatinus* cannot be assessed in detail because of the lack of comparative context. Nevertheless, because of its competent terrestrial locomotion, it can be speculated that it will not differ substantially from other monotremes.

ACKNOWLEDGMENTS

We thank the following people for help and discussion: Brian Day, Michael Gresty, Nathan Jeffery, Jeanette Killius, Kornelius Kupczik, Meave Leakey, Peter Morris, Sirpa Nummela, Andreas Pommert, Richard Rabbitt, D. H. Rumsfeld, Tim Ryan, Alan Walker, and the late Jane Moore. Supported by NSF grants EAR-0207370 and BCS-0003920.

LITERATURE CITED

Anderson, J. 1878. Anatomical and zoological researches: comprising an account of the zoological results of the two expeditions to western Yunnan in 1868 and 1875; and a monograph of the two cetacean genera, *Platanista and Orcella*. B. Quaritch, London.

Anliker, M., and W. Van Buskirk. 1971. The role of perilymph in the response of the semicircular canals to angular acceleration. Acta Otolaryngologica 72:93–100.

Arnould, J. P. Y. 2002. Southern fur seals; pp. 1146–1152 in W. P. Perrin, B. Wursig, and J. G. M. Thewissen (eds.), Encyclopedia of Marine Mammals. Academic Press, San Diego.

Bernacsek, G. M., and R. L. Carroll. 1981. Semicircular canal size in fossil fishes and amphibians. Canadian Journal of Earth Sciences 18:150–156.

Best, R. C. and V. M. F. da Silva. 1989. *Inia geoffrensis*; pp. 1–23 in S. H. Ridgway and R. Harrison (eds.), Handbook of Marine Mammals, Vol. 4. Academic Press, London.

Blanshard, W. H. 2001. Dugong strandings. Veterinary Conservation Biology: Wildlife Health and Management in Australasia, Workshop Proceedings, 1 July 2001, Taronga Zoo, Sidney, Australia.

Boenninghaus, G. 1903. Das Ohr des Zahnwales, zugleich ein Beitrag zur Theorie der Schalleitung. Eine biologische Studie. Fischer Verlag, Jena.

Bonner, W. N. 1981. Southern fur seals *Arctocephalus* (Geoffroy Saint-Hillaire and Cuvier, 1826); pp. 111–144 in S. H. Ridgway and R. Harrison (eds.), Handbook of Marine Mammals, Vol. 2. Academic Press, London.

Breuer, J. 1874. Über die Funktion der Bogengänge des Ohrlabyrinthes. Wiener Medizinisches Jahrbuch 4:72.

Brodie, P. F. 1989. The white whale *Delphinapterus leucas* (Pallas, 1776); pp. 119–144 in S. H. Ridgway and R. Harrison (eds.), Handbook of Marine Mammals, Vol. 4. Academic Press, London.

Brown, A. C. 1874. On the sense of rotation and the anatomy and physiology of the semicircular canals of the internal ear. Journal of Anatomy and Physiology 8:327–331.

Cappozzo, H. L. 2002. South American sea lion, *Otaria flavescens*; pp. 1143–1146 in W. P. Perrin, B. Würsig, and J. G. M. Thewissen (eds.), Encyclopedia of Marine Mammals. Academic Press, San Diego.

Curthoys, I. S., and C. M. Oman. 1986. Dimensions of the horizontal semicircular duct, ampulla and utricle in rat and guinea pig. Acta Otolaryngologica 101:1–10.

Curthoys, I. S., and C. M. Oman. 1987. Dimensions of the horizontal semicircular duct, ampulla and utricle in human. Acta Otolaryngologica 103:254–261.

Curthoys, I. S., R. H. I. Blanks, and C. H. Markham. 1977a. Semicircular canal radii of curvature (R) in cat, Guinea pig and man. Journal of Morphology 151:1–16.

Curthoys, I. S., C. H. Markham, and E. J. Curthoys. 1977b. Semicircular duct and ampulla dimensions in cat, Guinea pig and man. Journal of Morphology 151:17–34.

Denker, A. 1901. Zur Anatomie des Gehoerorgans der Monotremata. Denkschrifte der Medizinische-Naturwissenschaftliche Gesellschaft (Jena) 6:635–662.

Denker, A. 1902. Zur Anatomie des Gehoerorgans der Cetacea. Anatomische Hefte 19:421–447.

Dohlman, G. F., and L. A. Kühn. 1973. The role of the perilymph in semicircular canal stimulation. Acta Otolaryngologica 75:396–404.

Emmons, L. 1990. Neotropical Rainforest Mammals. University of Chicago Press, Chicago.

Estes J. A., J. L. Bodkin. 2002. Otters; pp. 842–858 in W. P. Perrin, B. Würsig, and J. G. M. Thewissen (eds.), Encyclopedia of Marine Mammals. Academic Press, San Diego.

Fay, F. H. 1981. Walrus, *Odobenus rosmarus*; pp. 1–23 in S. H. Ridgway and R. Harrison, eds.), Handbook of Marine Mammals, Vol. 1. Academic Press, London.

Goldberg, J. M. 2000. Afferent diversity and the organization of central vestibular pathways. Experimental Brain Research 130:277–97.

Gray, A. A. 1907. The Labyrinth of Animals, Vol. 1. Churchill, London.

Gray, A. A. 1908. The Labyrinth of Animals, Vol. 2. Churchill, London.

Heath, C. B. 2002. California, Galapagos, and Japanese sea lions, *Zalophus californianus*, *Z. wollebaeki*,

Z. japonicus; pp. 180–186 in W. P. Perrin, B. Wursig, and J. G. M. Thewissen (eds.), Encyclopedia of Marine Mammals. Academic Press, San Diego.

Highstein, S. M., R. D. Rabbitt, G. R. Holstein, and R. Boyle. 2005. Determinants of spatial and temporal coding by semicircular canal afferents. Journal of Neurophysiology 93:2359–2370.

Holstein, G., R. C. Rabbitt, R. D. Boyle, G. P. Martinelli, V. L. Friedrich, and S. M. Highstein. 2004. Convergence of excitatory and inhibitory hair cell transmitters shape responses of primary vestibular afferent neurons. Proceedings of the National Academy of Sciences USA 101:15767–15771.

Howland H. C., and J. Masci. 1973. The phylogenetic allometry of the semicircular canals of small fishes. Zeitschrift für Morphologie der Tiere 75:283–296.

Hullar, T. E., and M. Armand. 2004. Head motion of freely swimming dolphins. Journal of Vestibular Research 14:144–145.

Hyrtl, J. 1845. Vergleichend-anatomische Untersuchungen über das innere Gehörorgan des Menschen und der Säugethiere, 3. F. Ehrlich, Prag, Poland.

Igarashi, M., T. O-Uchi, and B. R. Alford. 1981. Volumetric and dimensional measurements of vestibular structures in the squirrel monkey. Acta Otolaryngologica 91:437–444.

Jones G. M., and K. E. Spells. 1963. A theoretical and comparative study of the functional dependence of the semicircular canal upon its physical dimensions. Proceedings of the Royal Society of London B 157:403–419.

Jorgensen J. M., and N. A. Locket. 1995. The inner ear of the echidna *Tachyglossus aculeatus*: the vestibular sensory organs. Proceedings of the Royal Society, Biological Sciences 260:183–189.

Ketten, D. R. 1992a. The cetacean ear: form, frequency, and evolution; pp. 53–57 in J. Thomas, R. A. Kastelein, and A. Y Supin (eds.), Marine Mammal Sensory Systems. Plenum Press, New York.

Ketten, D. R. 1992b. The marine mammal ear: specializations for aquatic audition and echolocation; pp. 717–754 in D. Webster, R. Fay, and A. Popper (eds.), The Evolutionary Biology of Hearing. Springer-Verlag, New York.

Ketten D. R. 1997. Structure and function in whale ears. Bioacoustics 8:103–135.

Ketten D. R., and D. Wartzok. 1990. Three-dimensional reconstructions of the dolphin ear; pp. 81–105 in J. Thomas and R. Kastelein (eds.), Sensory Abilities of Cetaceans. Plenum Press, New York.

Ketten D. R., D. K. Odell, and D. P. Domning. 1992. Structure, function, and adaptation of the manatee ear; pp. 77–95 in J. A. Thomas, R. A. Kastelein and A. Y. Supin, eds.), Marine Mammal Sensory Systems. Plenum Press, New York.

Lindenlaub, T., and H. Burda. 1994. Functional allometry of the semicircular ducts in subterranean mole-rats *Cryptomys* (Bathyergidae, Rodentia). Anatomical Record 240:286–289.

Lindenlaub T., H. Burda, and E. Nevo. 1995. Convergent evolution of the vestibular organ in the subterranean mole-rats, *Cryptomys* and *Spalax*, as compared with the aboveground rat, *Rattus*. Journal of Morphology 224:303–311.

Mach, E. 1875. Grundlinien der Lehre von den Bewegungsempfindungen. Engelmann, Leipzig, Germany.

Maisey, J. G. 2001. Remarks on the inner ear of elasmobranchs and its interpretation from skeletal labyrinth morphology. Journal of Morphology 250:236–264.

Markussen, N. H., A. Bjørge, and N. A. Øritsland. 1989. Growth in harbour seals *(Phoca vitulina)* on the Norwegian coast. Journal of Zoology 219:433–440.

McCabe, B. F., and J. H. Ryu. 1973. Does perilymph modify cupula deflection? Acta Otolaryngologica 75:405–407.

McVean, M. A. 1999. Are the semicircular canals of the European mole, *Talpa europaea*, adapted to a subterranean habitat? Comparative Biochemistry and Physiology A 123:173–178.

Muller, M. 1990. Relationship between semicircular duct radii with some implications for time constants. Netherlands Journal of Zoology 40:173–202.

Muller, M. 1994. Semicircular duct dimensions and sensitivity of the vertebrate vestibular system. Journal of Theoretical Biology 167:239–256.

Muller, M. 1999. Size limitations in semicircular duct systems. Journal of Theoretical Biology 198:405–437.

Nikaido, M., A. P. Rooney, and N. Okada. 1999. Phylogenetic relationships among cetartiodactyls based on insertions of short and long interpersed elements: hippopotamuses are the closest extant relatives of whales. Proceedings of the National Academy of Sciences USA 96:10261–10266.

Norwood, L. 2002. *Loxodonta africana*. Available at http://animaldiversity.ummz.umich.edu/site/accounts/information/Loxodonta_africana.html.

Nowak, R. M. 1991. Walker's mammals of the world, Vol. 1. 5ed. Johns Hopkins University Press, Baltimore.

O'Corry-Crowe, G. M. 2002. Beluga whale, *Delphinapterus leucas*; pp. 94–99 in W. P. Perrin, B. Würsig, and J. G. M. Thewissen (eds.), Encyclopedia of Marine Mammals. Academic Press, San Diego.

Pilleri, G. 1974. Side-swimming, vision and sense of touch in *Platanista indi* (Cetacea, Platanistidae). Experientia 30:100–104.

Purves, P. E., and G. Pilleri. 1973. Observations on the ear, nose, throat and eye of *Platanista indi*. Investigations on Cetacea 5:13–57.

Rabbitt, R. D., E. R. Damiano, and J. W. Grant. 2004. Biomechanics of the vestibular semicircular canals and otolith organs; pp. 153–201 in S. M. Highstein, A. Popper, and R. Fay (eds.), The Vestibular System. Springer-Verlag, New York.

Rabbitt, R. D., R. Boyle, G. Holstein, and S. Highstein. 2005. Hair-cell versus afferent adaptation in the semicircular canals. Journal of Neurophysiology 93:424–436.

Ramprashad, F., J. P. Landolt, K. E. Money, and J. Laufer. 1980. Neuromorphometric features and dimensional analysis of the vestibular end organ in the little brown bat *(Myotis lucifugus)*. Journal of Comparative Neurology 192:883–902.

Ramprashad, F., J. P. Landolt, K. E. Money, and J. Laufer. 1984. Dimensional analysis and dynamic response characterization of mammalian peripheral vestibular structures. American Journal of Anatomy 169:295–313.

Reeves, R. R., and R. L. Brownell. 1989. Susu, *Platanista gangetica* (Roxburgh, 1801) and *P. minor* Owen, 1853; pp. 69–99 in S. H. Ridgway and R. Harrison (eds.), Handbook of Marine Mammals, Vol. 4. Academic Press, London.

Rejtoe, A. 1939. Die Rolle der Perilymphe in der Entstehung des kalorischen Nystagmus. Acta Otolaryngologica 27:270–280.

Reysenbach de Haan, F. W. 1956. De ceti auditi. Over de gehoorzin bij de walvissen. Ph.D. Thesis, Utrecht University, Netherlands.

Ronald, K., and P. J. Healey. 1981. Harp seal, *Phoca groenlandica* Erxleben, 1777; pp. 55–87 in S. H. Ridgway and R. Harrison (eds.), Handbook of Marine Mammals, Vol. 2. Academic Press, London.

Silva, M. and J. A. Downing. 1995. Handbook of Mammalian Body Masses. CRC, Boca Raton.

Sipla, J. S., J. A. Georgi, and C. A. Forster. 2003. The semicircular canal dimensions of birds and crocodilians: implications for the origin of flight. Journal of Vertebrate Paleontology Supplement 23:97.

Sokal, R. R., and F. J. Rohlf. 1995. Biometry. Freeman and Co., San Francisco.

Solntseva, G. N. 2001. Comparative analysis of vestibular system development in various groups of mammals living under different environmental conditions. Russian Journal of Developmental Biology 32:171–174.

Spoor, C. F., and F. W. Zonneveld. 1995. Morphometry of the primate bony labyrinth: a new method based on high-resolution computed tomography. Journal of Anatomy 186:271–286.

Spoor, F. 2003. The semicircular canal system and locomotor behaviour, with special reference to hominin evolution. Courier Forschungs-Institut Senckenberg 243:93–104.

Spoor, F., S. Bajpai, S. T. Hussain, K. Kumar, and J. G. M. Thewissen. 2002. Vestibular evidence for the evolution of aquatic behavior in early cetaceans. Nature 417:163–166.

Spoor, F., Th. Garland, G. Krovitz, T. M. Ryan, M. T. Silcox, and A. Walker. 2007. The Primate Semicircular Canal System and Locomotion. Proceedings of the National Academy of Sciences 104:10808–10812.

Sukumar, R. 1989. The Asian elephant: ecology and management. Cambridge University Press, Cambridge.

Takahashi, H. 1971. Comparative anatomical study of the bony labyrinth of the inner ear in Canidae. Kaibogaku Zasshi Tokyo 46:85–98.

Ten Kate, J. H., and J. W. Kuiper. 1970. The viscosity of the pike's endolymph. Journal of Experimental Biology 53:495–500.

Wartzok, D., and D. R. Ketten. 1999. Marine mammal sensory systems; pp. 117–175 in J. E. Reynolds III and S. A. Rommel (eds.), Biology of Marine Mammals. Smithsonian Institution Press, Washington, DC.

Yamada, M., and F. Yoshizaki. 1959. Osseous labyrinth of Cetacea. Scientific Reports of the Whale Research Institute (Tokyo) 14:291–304.

Young, J. Z. 1981. The Life of Vertebrates, 3rd Ed. Clarendon Press, Oxford.

Mechanoreception

17

The Physics and Physiology
of Mechanoreception

Guido Dehnhardt and Björn Mauck

Mechanosensory Information: Physics and Basic
   Stimuli
   Physical Parameters
   Active Touch Information
   Medium Disturbances
Mechanoreceptors

Mechanoreception can be considered as a fundamental sensory modality developed in all animal phyla. In tetrapods it comprises the auditory system (Hetherington, chapter 12 in this volume; Nummela, chapter 13 in this volume), the lateral line system of amphibians (not covered in this section, but see Bleckmann, 1994; Coombs et al., 1989), and somatosensory systems. Generally, the somatosensory system of secondarily aquatic tetrapods can be divided into the spinal system innervating the body surface, deep tissues, and extremities, and the trigeminal system innervating facial areas such as the muzzle of mammals or the beak of birds. In the next chapter, we focus on mechanoreception associated with the trigeminal system, as in secondarily aquatic reptiles, birds, and mammals facial areas serve specialized tactile functions primarily used for actively seeking environmental information. This view is supported by electrophysiological studies on the bottlenose dolphin (Ridgway and Carder, 1990) and the northern fur seal (Ladygina et al., 1985), showing that in both species, areas of the head are most significant for the reception of tactile information. Passive tactile sensitivity of the entire body surface, other cutaneous senses such as nociception (mediating pain caused by mechanical stimulation, heat, or chemical agents) and thermoreception, as well as deep receptors such as muscle spindles and joint receptors are not or only indirectly considered. In this chapter, we introduce the physics, receptors, and basic physiology of mechanosensory systems.

MECHANOSENSORY INFORMATION: PHYSICS AND BASIC STIMULI

Mechanosensory systems provide organisms with information about their nearby environment. Information about inanimate objects as well as about biological stimulus sources such as conspecifics, predators, and prey may be derived either directly from a physical contact of the tactile organ to the object of interest (active touch information or haptics) or indirectly from interpreting mechanosensory information about how the respective object has interfered with its surrounding medium. The latter represents the perception of medium motion that provides information about moving objects—to a certain extent independent of distance and time. For aquatic species, relevant water movements or hydrodynamic information may consist of midwater pressure waves or water surface waves (Bleckmann, 1994).

PHYSICAL PARAMETERS

Physical parameters influence mechanosensory information mainly while it is transmitted through the surrounding medium. Although belonging to acoustics and thus not to mechanoreception as considered in this chapter, the best-known example may be sound velocity. Being roughly the quotient of elasticity modules and density (both depend on temperature), sound propagation velocity in water exceeds that in air by a factor of 4.35 (pressure waves propagate $\sim$1435 m s^{-1} in water as opposed to $\sim$330 m s^{-1} in air).

There are implications for the perception of the hydrodynamic component of midwater pressure waves generated by, for instance, swimming fish. The high density of water (approximately 1000 kg m^{-3}) as opposed to that of air (1.202 kg m^{-3}) results in a comparatively high buoyancy of submerged material, allowing body appendages serving as mechanoreceptive units for this kind of hydrodynamic event to be larger than they could be in air. Due to the high surface tension of water (0.07275 N m^{-1}), the water surface can also provide hydrodynamic information in the form of surface waves caused by, for instance, objects fallen into the water.

As a result of the low molar mass of water (18.015), its heat capacity (approximately 4.2 kJ K^{-1} kg^{-1}) is higher than that of most other liquids and exceeds that of the same volume of air

(approximately 1 kJ K^{-1} kg^{-1}) by a factor of almost 3500. The heat conductivity of water is still higher than that of air by a factor of approximately 25. These differences lead to considerably higher heat-transfer rates from an animal's body surfaces in water as compared to air and may result in excessive cooling of sensory organs. Aquatic animals can be assumed and are found to have adapted to this well-known impairing effect on tactile sensitivity.

ACTIVE TOUCH INFORMATION

During active touch or the haptic exploration of an object, mechanosensory stimuli may convey information about the material, size, shape, and texture of an object, as well as the way it moves (lateral or rotational direction and/or frequency of pulsation). When touching an object, the respective tactile organ receives information about the object's different properties through different mechanical stimulations such as differences in pressure and/or stretch (e.g., resulting from size differences) or different frequencies of vibration (e.g., resulting from a moving object or an object's surface texture). Since these object properties usually do not alter with the surrounding medium, the respective information shall not be affected by the medium as well (Renouf, 1991). However, depending on the physical properties of the surrounding medium, it may thoroughly affect stimulus reception of the sensory system. For the vibrissae (see below) of harbor seals, G. Dehnhardt and M. Schmidt (unpublished), for example, have shown that in accordance with the above considerations, haptic size discrimination capabilities remained essentially unaltered in air and in water. In contrast, texture difference thresholds were 4.5 times lower underwater than in air. This factor reflects the difference in transmission speed of sound between the two media, and there is good reason to assume that the vibratory properties of seal vibrissae, determined by structural specializations of the hair shaft, might be adapted to the physical demands of water. However, improved texture discrimination can also be explained by the water operating as an intermediate layer between the tactile hair and a textured surface, thus reducing potential masking effects and/or affecting the mechanical coupling between these two. A similar effect is known from studies on active touch in humans, where a thin intermediate sheet of paper moved with the fingertip across a textured surface has been found to enhance the perception of roughness (Gordon and Cooper, 1975; Lederman, 1978, 1982).

The sensory information received in an active touch process and the resulting neural code are not only a function of the physical properties of the stimulus touched but decisively determined by the movement profile applied with the tactile organ to that stimulus (Darian-Smith et al., 1982; Klatzky et al., 1987; Lederman and Klatzky, 1987). This implies that besides mechanoreception, a haptic process always involves contributions of proprioception or kinaesthesis (Gibson, 1962; Loomis and Lederman, 1986), thus providing a good example of behavioral skills based on a synergism of different sensory modalities. Movement patterns of tactile organs can be interpreted as a function of medium density.

MEDIUM DISTURBANCES

Most medium disturbances in both water and air occur as abiotic turbulence modifying a more or less directed medium flow. Examples are differences in air pressure causing wind modified by, for example, local thermal events, rigid or moving obstacles causing turbulence in air flow, differences in potential causing steady water currents (rivers, tides) modified by turbulences behind underwater obstacles, or raindrops generating ringlike surface ripples at the water-air interface. Thus it must be assumed that most biologically generated medium disturbances in both media are masked to some extent by medium movement noise. Nevertheless, it might be at least possible to differentiate between biogenic and abiotic stimulus sources from thorough analysis of received information.

In some cases it might even be possible to reconstruct the type of biological stimulus source that was responsible for the perceived medium disturbances.

As mentioned above, hydrodynamic stimuli can be subdivided into water-surface and midwater events. Water-surface waves are caused by, for instance, terrestrial insects that have fallen into the water, aquatic and semiaquatic invertebrates and vertebrates contacting the water surface from below, or terrestrial animals disrupting the water surface while drinking. Thus water-surface disturbances might provide a silent predator with information about profitable prey. Propagation of water-surface waves is mainly determined by the asymmetry of forces at the water-air interface. Although surface waves are generally influenced by water depth, wave shape and transmission speed of biologically relevant water surface stimuli are rather independent of this factor, as water depth is normally large compared to displacement amplitudes of surface waves. In a propagating surface wave, particles move in circles or in ellipses. (Equations describing the displacement components and dispersion relationship of surface waves have been provided by Lighthill [1980].) Attenuation of surface waves during propagation depends on stimulus frequency and distance the stimulus has traveled (for calculations of attenuation see work by Bleckmann [1988]). Waves caused by insects fallen into the water can last for more than a minute and show a low displacement amplitude and an irregular time course, whereas those generated by, for example, fish or frogs are usually of much shorter duration (1 s) and more regular in time course (Bleckmann, 1994).

In hydrodynamic midwater events such as dipole stimuli (generated by, for instance, a vibrating sphere) used in most studies on hydrodynamic receptor systems, several physical components can be described. The displacement component of the hydrodynamic stimulus is generated by the moving object displacing water particles from their original location, which return as soon as the object moves back to its starting position. Furthermore, object movement causes a local compression (increase in pressure, i.e., sound) by pushing water particles more closely together, and a corresponding drop in pressure occurs when the object returns to its original place. Both components of midwater stimuli—displacement and pressure fluctuation—are interconnected and propagate away from the stimulus source as a chain reaction of particle collisions (for equations describing the propagation and attenuation of hydrodynamic midwater events see work by Kalmijn [1988] or Bleckmann [1994]). In brief, the pressure component (which is the adequate stimulus for perception of sound and thus can be called acoustic pressure) can be subdivided into the acoustic near-field pressure and the acoustic far-field pressure. Both start propagating at the sound source, but due to strong attenuation of the near-field pressure with distance—proportional to 1 divided by the square of the distance from the source and to the wave number—the far-field pressure dominates at some distance from the source. The far-field pressure is proportional to 1 divided by the distance from the source and to the square of the wave number, and thus outranges near-field pressure with increasing distance. Thus, it is the acoustic far-field pressure component that mainly represents a propagating sound wave at some distance and that can be perceived acoustically. However, in the narrow sense of mechanoreception discussed here, the pressure component of midwater stimuli is a rather inadequate stimulus, as it would require a pressure-sensitive device (pressure-displacement transducer), which in most animals is related to the acoustic sense.

The displacement component of a wave stimulus (which would be an adequate stimulus for hydrodynamic receptor systems and can be also expressed as velocity or acceleration) can be subdivided into a radial and a tangential component. The actual water velocity at any given point is the vector sum of velocities of both components. The equations for both components contain terms to calculate the so-called local flow and intermediate flow. Local flow is proportional to 1 divided by the cubed distance to the source, while intermediate flow is proportional to 1 divided by the squared distance to the source; thus local flow attenuates more rapidly with distance. Furthermore, the equation for the radial component contains a term that is proportional to only 1 divided by the distance to the source and to the squared wave number, and thus describes the particle movements (displacement, velocity, or acceleration) caused by the propagating sound source at some distance. As the radial displacement (or velocity or acceleration) component—and likewise the pressure component—is proportional to the cosine of the angle of radiation and thus becomes zero in the direction perpendicular to the direction of the moving sound source ($\cos 90° = 0$), only the tangential component (which is proportional to the sinus of the angle of radiation) represents the respective hydrodynamic event, for example, when a test animal is situated in front of a vertically oscillating sphere ($\sin 90° = 1$; compare Dehnhardt et al., 1998).

As a function of frequency, a hydrodynamic receptor system stimulated with dipole water movements may respond almost proportionally to particle displacement, water velocity, or water acceleration. For the detection of high-frequency components of water movements (e.g., 60 to 120 Hz) acceleration sensitivity is suggested to be most useful (Bleckmann et al., 1991).

As particle velocity attenuates rapidly with distance from the flow-generating source, the accepted view for a long time was that hydrodynamic stimulus detection plays a role only in the immediate vicinity of the receiving animal (Bleckmann, 1994). Although this is true for dipole water movements caused by a stationary vibrating sphere, the wake behind a swimming fish shows a vortex structure with particle velocities above threshold of most hydrodynamic receptors even several minutes after the fish has passed by (Hanke et al., 2000). This means that a swimming fish leaves a hydrodynamic trail of considerable length that a piscivorous predator equipped with a hydrodynamic sense might use for long-range prey detection. The structure of fish-generated hydrodynamic trails depends on the body shape and the swimming style of the species (Hanke and Bleckmann, 2004). This means that a predator may read more out of a fish-generated hydrodynamic trail than simply the presence of a wake generator. For example, the decay of vorticity in the vortex street provides information about the time passed since the fish swam by. The swimming direction of a fish is suggested to be derivable from the velocity gradient as well as the direction of gross water flow in its wake (Hanke et al., 2000). The theoretical option to read this kind of information from a hydrodynamic trail is concluded from flow visualizations using particle image velocimetry. The concept of hydrodynamic trail following is presented in Figure 17.1, showing a schematic view of a seal encountering a hydrodynamic trail consisting of a three-dimensional ladderlike chain of vortices left by a swimming fish (compare Blickhan et al., 1992; Nauen and Lauder, 2002).

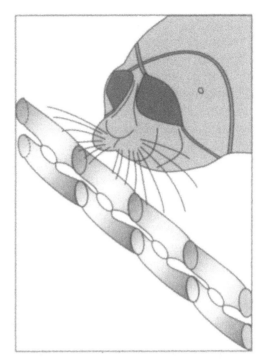

FIGURE 17.1. Schematic view of a seal encountering a hydrodynamic trail. Summarizing empirically determined hydrodynamic forces created by swimming fish led to models of fish wakes consisting of a three-dimensional ladderlike chain of counterrotating vortices with a central jet flow meandering through these rings (e.g., Blickhan et al, 1992; Nauen and Lauder 2002). The three-dimensional arrangement of the vibrissal system should allow a seal to simultaneously probe the trail at various points in order to obtain manifold information about the hydrodynamic trail and its generator (e.g., age and direction of the trail, fish species).

MECHANORECEPTORS

Terrestrial and secondarily aquatic tetrapods lost the highly sensitive lateral line system of fish and aquatic amphibians. Instead their skin is sensitive to mechanical stimulation due to the development of new types of cutaneous receptors (for details on mechanoreceptors in vertebrates see work by Andres [1966] and Andres and von Düring [1990]). As is known for the visual sense, where different retinal receptors (rods and cones) respond to different wavelengths of the light spectrum (Reuter and Peichl, chapter 10 in this volume), different types of tactile information are mediated by a variety of mechanoreceptors showing characteristic morphological properties. Especially with regard to functional aspects of mechanoreceptors, our knowledge is quite limited in reptiles and birds as compared to mammals. In general, mechanoreceptors are physiologically subdivided into slowly adapting (SA) receptors (Iggo and Gottschaldt [1974] describe the differentiation of SA-I and SA-II receptors) and rapidly adapting (RA) receptors. Stimulated with a ramplike stimulus that arrives at a plateau, for example, a blunt needle indented into the skin up to a certain depth, SA receptors usually respond to the ramp (during needle indentation) as well as to the plateau of the stimulus (needle arriving at a certain depth). Such a receptor is well suited to code the intensity of pressure stimuli. Although of considerable morphological and thus perhaps functional diversity (Baumann et al., 2003), a common SA receptor found in reptiles, birds, and mammals is the dermal Merkel cell neurite complex (Andres and von Düring, 1973;

Düring and Miller, 1979). SA receptors coding for stretch are the Ruffini endings found primarily in the hairy skin of mammals and the beak skin of birds (e.g., geese [Gottschaldt et al., 1982]). A receptor type described only in aquatic birds, and especially in the bill-tip organ (see Dehnhardt and Mauck, chapter 18 in this volume), are the Grandry corpuscles (Gottschaldt, 1985; Necker, 2000). Although generally larger, Grandry corpuscles morphologically resemble Merkel cells. However, electrophysiological results indicate that they are RA receptors, which respond for example, only during the ramp of needle indentation (velocity detectors). Very rapidly adapting receptors, responding only to the acceleration phase of the ramp, for example, at the beginning of needle indentation, are lamellated receptors like the Pacinian corpuscles in mammals or the Herbst corpuscles in birds. Lamellated receptors in crocodilians such as caimans share structural characteristics of both Pacinian and Herbst corpuscles (Düring and Miller, 1979). These sensory corpuscles respond best to vibrations of rather high frequency (up to 1000 Hz), with threshold amplitudes less than 0.1 μm (Necker, 2000).

LITERATURE CITED

Andres, K. H. 1966. Über die Feinstruktur der Rezeptoren an Sinushaaren. Zeitschrift für Zellforschung 75:339–365.

Andres, K. H., and M. von Düring. 1973. Morphology of cutaneous receptors; pp. 3–28 in A. Iggo (ed.) Handbook of Sensory Physiology: Somatosensory Systems, Vol. 2. Springer-Verlag, New York.

Andres, K. H., and M. von Düring. 1990. Comparative and functional aspects of the histological organization of cutaneous receptors in vertebrates; pp. 1–17 in W. Zenker and W. Neuhuber (eds.), The Primary Afferent Neuron: A Survey of Recent Morpho-functional Aspects. Plenum Press, New York.

Baumann, K. I., I. Moll, and Z. Halata. 2003. The Merkel Cell: Structure, Development, Function, and Cancerogenesis. Springer-Verlag, Berlin.

Bleckmann, H. 1988. Prey identification and prey localization in surface-feeding fish and fishing spiders; pp. 619–641 in J. Atema, R. R. Fay, A. N. Popper, and W. N. Tavolga (eds.). Sensory Biology of Aquatic Animals. Springer-Verlag, New York.

Bleckmann, H. 1994. Reception of Hydrodynamic Stimuli in Aquatic and Semiaquatic Animals. Fischer Verlag, Stuttgart, Jena, New York.

Bleckmann, H., T. Breithaupt, R. Blickhahn, and J. Tautz. 1991. The time course and frequency content of hydrodynamic events caused by moving fish, frogs and crustaceans. Journal of Comparative Physiology A 168:749–757.

Blickhan, R., C. Krick, D. Zehren, and W. Nachtigall. 1992. Generation of a vortex chain in the wake of a subundulatory swimmer. Naturwissenschaften 79:220–221.

Coombs, S., P. Görner, and H. Münz. 1989. The Mechanosensory Lateral Line: Neurobiology and Evolution. Springer-Verlag, New York.

Darian-Smith, I., M. Sugitani, J. Heywood, K. Karita, and A. Goodwin. 1982. Touching textured surfaces: cells in somatosensory cortex respond both to finger movement and to surface features. Science 218:906–909.

Dehnhardt, G., B. Mauck, and H. Bleckmann. 1998. Seal whiskers detect water movements. Nature 394:235–236.

Düring, M. von, and M. R. Miller. 1979. Sensory nerve endings of the skin and deeper structures of reptiles; pp. 407–441 in C. Gans (ed.), Biology of the Reptilia, Vol. 9, Neurobiology A. Academic Press, London.

Gibson, J. J. 1962. Observations on active touch. Psychological Review 69:477–491.

Gordon, I. E., and C. Cooper. 1975. Improving one's touch. Nature 256:203–204.

Gottschaldt, K.-M. 1985. Structure and function of avian somatosensory receptors; pp. 375–461 in A. S. King and J. McLelland (eds.), Form and Function in Birds, Vol. 3. Academic Press, London.

Gottschaldt, K.-M., H. Fruhstorfer, W. Schmidt, and I. Kraeft. 1982. Thermosensitivity and its possible fine-structural basis in mechanoreceptors in the beak skin of geese. Journal of Comparative Neurology 205:219–245.

Hanke, W., and H. Bleckmann. 2004. The hydrodynamic trails of Lepomis gibbosus (Centrarchidae), Colomesus psittacus (Tetraodontidae) and Thysochromis ansorgii (Cichlidae) investigated with scanning particle image velocimetry. The Journal of Experimental Biology 207: 1585–1596.

Hanke, W., C. Brücker, and H. Bleckmann. 2000. The ageing of the low-frequency water disturbances caused by swimming goldfish and its possible relevance to prey detection. Journal of Experimental Biology 203:1193–1200.

Iggo, A., and K.-M. Gottschaldt. 1974. Cutaneous mechanoreceptors in simple and in complex sensory structures. Abhandlungen der Rheinisch-Westfälischen Akademie der Wissenschaften 53:153–174.

Kalmijn, A. J. 1988. Hydrodynamic and acoustic field detection; pp. 83–130 in J. Atema, R. R. Fay, A. N. Popper, and W. N. Tavolga (eds.), Sensory Biology of Aquatic Animals. Springer-Verlag, New York.

Klatzky, R. L., S. J. Lederman, and C. L. Reed. 1987. There's more to touch than meets the eye: the salience of object attributes for haptics with and without vision. Journal of Experimental Psychology, General 116:356–369.

Ladygina, T. F., V. V. Popov, and A. Y. Supin. 1985. Somatotopic projections in the cerebral cortex of the fur seal (Callorhinus ursinus). Academy of Sciences Moscow 17:344–351.

Lederman, S. J. 1978. "Improving one's touch" . . . and more. Perception and Psychophysics 24:154–160.

Lederman, S. J. 1982. The perception of texture by touch; pp. 130–167 in W. Schiff and E. Foulke (eds.), Tactual Perception: A Sourcebook. Cambridge University Press, New York.

Lederman, S. J., and R. L. Klatzky. 1987. Hand movements: a window into haptic object recognition. Cognitive Psychology 19:342–368.

Lighthill, J. M. 1980. Waves in Fluids. Cambridge University Press, Cambridge.

Loomis, J. M., and S. J. Lederman. 1986. Tactual perception; pp. 1–41 in K. R. Boff, L. Kaufman, and J. R. Thomas (eds.), Handbook of Perception and Human Performance, Vol. 2. Wiley, New York.

Nauen, J. C., and G. V. Lauder. 2002. Hydrodynamics of caudal fin locomotion by chub mackerel, Scomber japonicus (Scombridae). The Journal of Experimental Biology 205:1709–1724.

Necker, R. 2000. The somatosensory system; pp. 57–69 in C. G. Whittow (ed.), Sturkies Avian Physiology. Academic Press, New York.

Renouf, D. 1991. Sensory reception and processing in Phocidae and Otariidae; pp. 345–394 in D. Renouf (ed.), The Behaviour of Pinnipeds. Chapman and Hall, London.

Ridgway, S. H., and D. A. Carder. 1990. Tactile sensitivity, somatosensory responses, skin vibrations, and the skin surface ridges of the bottlenose dolphin, Tursiops truncatus; pp. 163–179 in J. Thomas and R. Kastelein (eds.), Sensory Abilities of Cetaceans. Plenum Press, New York.

18

Mechanoreception in Secondarily Aquatic Vertebrates

Guido Dehnhardt and Björn Mauck

Although for a long time mechanoreception has been considered as a passive and thus merely receptive sense (Krueger, 1982), this chapter provides evidence that especially secondarily aquatic tetrapods may actively use this sensory modality for the exploration of their environment. Reptiles, birds, and mammals are sensitive to tactile stimulation almost over the entire body surface; however, each class has developed special tactile sense organs the animals can use for actively seeking mechanosensory information. As secondarily aquatic tetrapods often live in environments where visibility is poor, mechanosensory information that might supplement or even substitute for vision could be of special importance.

STRUCTURE AND FUNCTION OF MECHANOSENSORY SYSTEMS

REPTILES

Reptiles, the first tetrapods that completed the transition from an aquatic to a terrestrial lifestyle, faced the trade-off of keeping cutaneous sensitivity while simultaneously developing protective horny scales (up to the armored crocodilians). All orders of reptiles solved the problem by developing touch papillae (Fig. 18.1), spotlike elevations scattered primarily over the horny scales of the head (von Düring, 1975; von Düring and Miller, 1979). Touch papillae are characteristic tactile organs in this animal class, and in most species they show a high and complex supply with mechanoreceptors.

As in terrestrial snakes, the integument of aquatic species such as file snakes (Acrochordidae) (see the section by Schwenk and Thewissen

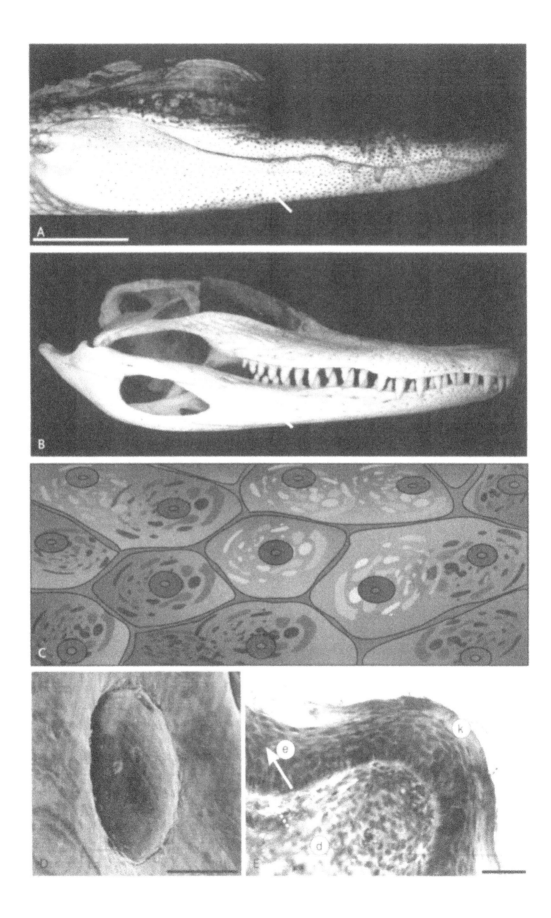

in chapter 1 in this volume for a brief overview of reptile diversity) and sea snakes (e.g., *Lapemis curtus*, Elapidae) possesses modified touch papillae, so-called scale sensillae (Povel and Kooij, 1997). In *Acrochordus granulatus* and *A. javanicus* each scale on the head is provided with a scale sensilla, but labial scales contain up to seven of them. Scale sensillae consist of a central tuft of fine hairlike projections and one or more platelike organs. From the tips of the hairlike protrusions small bristles branch off, increasing the surface of the tactile organ and thus guaranteeing good coupling to the medium by friction. The innervation of scale sensillae is not well described in aquatic snakes; however, preliminary histological results indicate that it resembles that of touch papillae (see below) of other reptiles (Povel and Kooij, 1997). Although their structure already suggests a special mechanosensory function, a recent study provided the first physiological evidence that at least the piscivorous sea snake *Lapemis curtus* can sense low-amplitude water motions produced by prey fish. Using hydrodynamic dipole stimuli (vibrating sphere) ranging from 50 to 200 Hz, Westhoff et al. (2005) recorded evoked potentials from the midbrain of *L. curtus*. In terms of water displacement the lowest-threshold amplitude of the most sensitive snake was 1.8 μm at 100 Hz. However, the nearly constant velocity threshold values in the range 50 to 100 Hz indicate that the snake responded to the velocity component of the hydrodynamic stimulus (Fig. 18.2). Although no exclusion tests have been conducted, it is likely that this sensitivity to hydrodynamic stimuli is a function of the scale sensillae. As the lateral line system of some fish species is more sensitive by up to one or two orders of magnitude, Westhoff et al. (2005) conclude from their

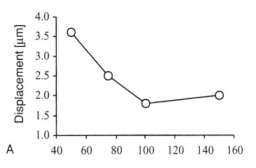

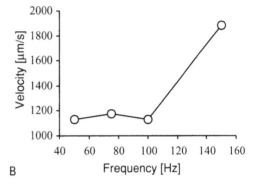

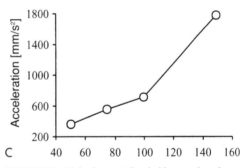

FIGURE 18.2. Hydrodynamic threshold curves based on evoked potentials in the sea snake *Lapemis curtus* (Elapidae, see chapter 1) plotted as a function of particle displacement (A), velocity (B), and acceleration (C) against frequency of the stimulus, a vibrating sphere. Adapted from Westhoff et al. (2005).

results that sea snakes do not possess a specialized hydrodynamic sense, but that their sensitivity is sufficient for the detection of fish-generated water movements. However, as

FIGURE 18.1. (Opposite page.) (A) Touch papillae, also called dome pressure receptors (DPRs) in alligators appear as black dots all over the jaw, and (B) their corresponding foramina on the skull. On the rostral part of the mandible one foraminen corresponds to approximately three DPRs. (C) Schematic drawing of horny scales with touch papillae as they appear on the jaws of Caiman crocodiles (adapted from von Düring, 1975), and (D) scanning electron micrograph of an individual DPR in a juvenile alligator, with (E) a sagittal section of a DRP under nissl staining. Abbreviations: d, dermis; e, epidermis; and k, keratin layer. Scale bars are 100 μm. A, B, D, and E courtesy of Daphne Soares.

will be described for the hydrodynamic sense in pinnipeds, thresholds of the sea snakes are low enough to enable the animals to perform complex behaviors based on hydrodynamic information.

Hydrodynamic reception has also been shown in crocodilians *(Alligator mississippiensis)*, but unlike the response to hydrodynamic midwater stimuli, crocodiles use the air-water interface as a source of sensory information. In a "sit and wait" strategy, these armored reptiles respond to very slight disturbances of the water surface and are suggested to use this hydrodynamic information for prey location (Soares, 2002). The tactile organs serving this function are the touch papillae, which measure about 0.3 to 0.5 mm in diameter and seem to be especially well developed in crocodilians (von Düring and Miller, 1979). In alligators touch papillae are round, domelike structures lacking pores or protruding hairs, and they cover the entire face (Fig. 18.1). They show an interesting innervation pattern with a concentration of different mechanoreceptors in a very small area. In the stratum spinosum, directly below the very thin and thus pliable keratin layer (Fig. 18.1D), there are numerous (60 to 100) discoid receptors, corpuscular mechanoreceptors of unknown response characteristic (von Düring and Miller, 1979). Below the stratum spinosum, dermal Merkel cells can be found that are closely arranged in columns. These Merkel cell columns are associated with lamellated receptors as well as free nerve endings. Due to this complex innervation pattern, von Düring and Miller (1979) have suggested that touch papillae are well suited to provide crocodilians with detailed information about the environment. In behavioral experiments, Soares (2002) could demonstrate that in complete darkness, half-submerged alligators precisely oriented themselves to surface waves caused by single water droplets (Fig. 18.3). This occurred only when the animals' faces were located on the air-water interface, but not when the alligators were completely submerged or their entire

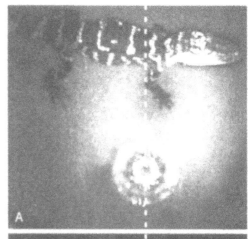
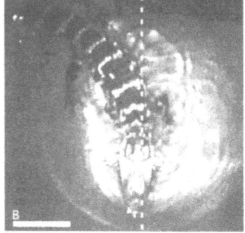

FIGURE 18.3. Infrared images showing a juvenile alligator responding to water droplets. (A) Position of the alligator at the moment the water droplet reaches the water surface, and (B) the animal orientates toward the center (indicated by the *dashed white line*) of the water surface waves caused by the droplet. Scale bar is 10 cm. Courtesy of Daphne Soares, modified.

head was out of the water. In exclusion tests, where the touch papillae were covered with a thin plastic elastomer, the orientation toward the surface wave was abolished completely. Soares (2002) suggested the functional term dome pressure receptors, because (1) they appear as small pigmented domes at the skin surface and (2) electrophysiological results indicated that the touch papillae are sensitive to pressure differences. The appearance of osteological markings of dome pressure receptors on

FIGURE 18.4. Bill-tip organ of the lower bill of a male western sandpiper (Charadriidae, see Table 1.3), with the bill skin removed. Similar to the skull of alligators, foramina of the bill tip correspond to the 88 sensory units counted within the rostral 2 mm of the tip of the lower bill. Scale bar is 1 mm. Courtesy of Silke Nebel, reproduced with permission from BRILL.

fossil crocodilian skulls in the form of foramina suggests that these mechanosensory receptors are a conserved feature of semiaquatic crocodilians.

BIRDS

In aquatic birds the beak is of special significance for the reception of mechanosensory information. The avian beak is primarily involved in feeding and in this context is suggested to serve as a prehensile tactile organ (obviously best developed in nonaquatic parrots) comparable to the precision grip of humans measuring objects between their thumb and index finger (Necker, 2000). While for such haptic capabilities almost the entire beak is involved, many birds have a special tactile organ under the keratin layer of the last few millimeters of the tip of the beak. This so-called bill-tip organ (Fig. 18.4) consists of many specialized dermal papillae, each of which contains numerous Grandry corpuscles and vibration-sensitive Herbst corpuscles (Gentle and Breward, 1986; Necker, 2000; Nebel et al., 2005). Shorebird species (Charadriiformes) (see the section by Thewissen and Hieronymus in chapter 1 in this volume for a brief overview of bird diversity) that probe for buried prey in wet sediments have considerably larger numbers of these dermal papillae than visually hunting species (Bolze, 1968).

As in such bird species the bill tip is overrepresented in the corresponding brain structures, it has been suggested—analogously to retinal organization—to be a tactile fovea. However, as there is no description of a bill-tip periphery compared to a central sensory area, this term is a bit misleading, and the strong central representation may be more related to the general importance of this mechanosensory system (but see the nasal "star" of the star-nosed mole for comparison).

In behavioral experiments Piersma et al. (1998) have shown that sandpipers (*Calidris canutus*, Charadriidae) feeding in coastal intertidal areas are able to detect immobile hard-shelled bivalves by inserting their beak repeatedly half a centimeter into the water-saturated sand. This object detection is suggested to be based on hydrodynamic reception of pressure gradients reflected by the bivalves buried in the sediment (Gerritsen and Meiboom, 1986; Piersma et al., 1998). However, the birds cannot discriminate between bivalves and hard inanimate objects such as stones. Consequently, sandpipers are rarely found searching for food in areas where the sand contains many stones.

MONOTREMATA

The terrestrial echidnas *(Tachyglossus aculeatus and Zaglossus bruijni)* and the aquatic platypus *(Ornithorhynchus anatinus)* share many characteristics with reptiles (e.g., shell-covered eggs incubated and hatched outside the mother's body) (see the section by Pihlström in chapter 1 in this volume for a brief overview of mammalian diversity). Although with respect to sensory systems the conspicuous bill of the platypus is primarily associated with electroreception, it is also covered by numerous mechanosensory organs, so-called push rods (Andres and von Düring, 1990). The density of push rods is highest around the labial margins of the bill and decreases caudally and toward the middle portion of the bill (Manger and Pettigrew, 1996). Push rods are compacted columns of epidermal cells with their tips at the skin surface where they appear as small white domes (Pettigrew et al., 1998). Although unique to the skin of monotremes, they resemble the touch papillae of reptiles, the papillae of the bill-tip organ of birds (see above), and the Eimer's organs in the skin of the snout of the star-nosed mole (Catania, 1999; see also the section below on Rodentia and other small semiaquatic mammals). The little dome at the skin surface is the tip of an epidermal rod that is—in contrast to the push rods of terrestrial echidnas (Iggo et al., 1996)—not tethered to the surrounding tissue and thus is free to rotate about its base. At the base of each push rod there is a cluster of encapsulated nerve endings. In the push rods of echidnas some of these mechanoreceptors have been shown to respond to high-frequency vibration, like eutherian Pacinian corpuscles. Others are slowly adapting (SA mechanoreceptors), showing response characteristics like Merkel cell neurite complexes and Ruffini endings (Iggo et al., 1996). Due to this pattern of innervation and its movability relative to adjacent tissues, this complex mechanosensory structure could provide a foraging platypus with mechanosensory information about an object having physical contact with the bill skin. However, similar to a hair sensor this structure also allows the transduction of medium disturbances such as low-amplitude water motions produced by potential prey (e.g., the crustacean *Cherax quadricarinatus*). In this regard Pettigrew et al. (1998) propose a functional correlation between the mechanosensory detection of water movements via push rods and electroreception. As a natural moving stimulus source such as the tail flick of a shrimp inevitably produces both electrical signals and water movements (e.g., pressure waves or vortices), this electromechanosensory system would be suited to measure the distance of the prey from the time interval between the arrival of the electrical impulse and the later arrival of the water disturbance. However, though some measurements of the latencies between electrical and mechanical signals generated by the same

moving prey item suggest that these should be perceptible for a platypus, corresponding psychophysical or electrophysiological studies have not been conducted yet.

EUTHERIAN MAMMALS, GENERAL

Specialized tactile organs in placental aquatic mammals are sinus hairs, also called vibrissae (colloquially "whiskers"). In semiaquatic mammals, like many rodents, cats, or viverrids vibrissae are generally well developed, at least with respect to the number and strength of their vibrissal hair shafts.

Vibrissal hair shafts transfer mechanical information into the vibrissal follicle sinus complex (F-SC) (after the nomenclature of Rice et al. [1986]), where the hair grows from a dermal papilla at the base of the follicle (Fig. 18.5). Inside the capsule, the hair is anchored to the dermal papilla and has a second fulcrum at the mouth of the follicle (Rice et al., 1986). A deflection of the outer hair shaft is proposed to be transformed into a contrarotating bowing action of the hair shaft inside the follicle (Lichtenstein et al., 1990), resulting in a compression of mechanoreceptors on the leading edge of the bending shaft, whereas on all other sides around the shaft stretch is the effective stimulus.

Generally, F-SCs are intricately structured and densely innervated mechanosensory organs, though the functional significance of most F-SC structures is not known yet. Although there is some information on the structure of vibrissal follicles in different orders of aquatic or semiaquatic mammals (Goldschmidt-Lange, 1976; Ling, 1977), sufficient descriptions of their morphology are available only for some pinniped species (*Zalophus californianus* [Stephens et al., 1973], *Phoca hispida* [Hyvärinen and Katajisto, 1984; Hyvärinen, 1989], *Erignathus barbatus* [Marshall et al., 2006]), the Florida manatee (*Trichechus manatus latirostris* [Reep et al., 2001, 2002]), and the Australian water rat (*Hydromys chrysogaster* [Dehnhardt et al., 1999]). Characteristic for all

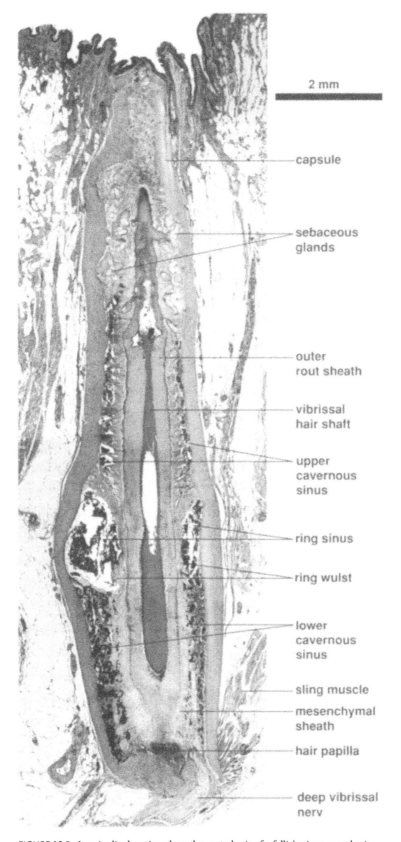

2 mm

capsule

sebaceous
glands

outer
rout sheath

vibrissal
hair shaft

upper
cavernous
sinus

ring sinus

ring wulst

lower
cavernous
sinus

sling muscle

mesenchymal
sheath

hair papilla

deep vibrissal
nerv

FIGURE 18.5. Longitudinal section along the central axis of a follicle sinus complex in the facial skin of a harbor seal (*Phoca vitulina*, Phocidae).

sinus hairs are the blood sinuses that are absent in normal body hair. Although in some species almost the entire F-SC is filled by a trabeculated cavernous sinus (Rhesus monkey [van Horn, 1970], brush-tailed possum [Hollis and Lyne, 1974], Tammar wallaby [Marotte et al., 1992]), most species studied so far possess a bipartite blood sinus composed of a ring sinus close to the apical end of the follicle and a cavernous sinus located below it. Pinnipeds seem to be the only group of mammals possessing an additional cavernous sinus situated above the ring sinus.

CETACEA

Even the hairless baleen whales have about 100 immobile and very thin (0.3 mm in diameter) vibrissae along their lower and/or upper jaw, whereas most odontocete species loose these hairs prenatally or a short time after birth (Ling, 1977). However, several follicle crypts remain on each side of the dorsal rostrum (e.g., *Phocoena phocoena*, 2; *Delphinus delphis*, 10). While in species such as *Tursiops* or *Phocoena* these follicle crypts appear like pinpricks, the diameter of the follicle mouth in, for example, *Sotalia guianensis* and *Kogia breviceps* is considerable (Fig. 18.6A). Because vibrissal follicle crypts in *Sotalia* appear as hot spots in infrared thermographies, they have been considered to be functional sensory units (Mauck et al., 2000). Recent psychophysical experiments revealed that these follicle crypts no longer serve a mechanosensory function. Instead, similar to the platypus, *Sotalia guianensis* can use its highly innervated follicle crypts to detect weak electric fields (Wilkens and Hofmann, chapter 20 in this volume).

Among odontocetes, freshwater dolphins seem to be an exception. At least in the Amazon River dolphin *(Inia geoffrensis)* adults have tactile hairs, but these are irregularly distributed on both the upper and the lower jaw, and it still has to be shown whether these bristles are true sinus hairs (see below for the relevance of sinuses).

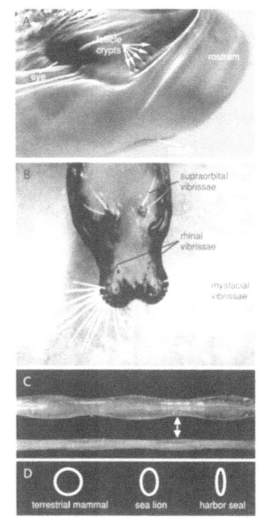

FIGURE 18.6. (A) Conspicuously well-developed vibrissal follicle crypts in the odontocete *Kogia breviceps*. (B) Facial vibrissae of the harbor seal *Phoca vitulina*. (C) Characteristic corrugated surface pattern *(top)* and extreme flattening *(bottom)* of vibrissal hair shafts of harbor seals. (D) Cross sections of vibrissae. The *white double arrow* in C marks a narrowed area of the wide side of a hair shaft and the corresponding area of the thin side of a vibrissal hair shaft, which is slightly thickened.

PINNIPEDIA

The facial vibrissae of pinnipeds (Fig. 18.6B) can be divided into the very mobile mystacial vibrissae on the muzzle, immobile supraorbital vibrissae above each eye, and two immobile rhinal vibrissae vertically situated on the back of the muzzle (only in phocids). In terrestrial mammals a typical vibrissal hair shaft is round in cross section, while those of eared seals

(Otariidae) and walruses (Odobenidae), as well as those of some phocid species such as the bearded seal *(Erignathus barbatus)* and the Monk seals *(Monachus* spp.), are oval (Fig. 18.6D). As in terrestrial mammals, vibrissal hair shafts of all eared seals, walruses, the bearded seal, and Monk seals are smooth in outline. In contrast, those of all other phocid species are extremely flattened and have waved surfaces (Fig. 18.6C and D) (Watkins and Wartzok, 1985; Hyvärinen, 1989; Dehnhardt and Kaminski, 1995). Although we have no information about the significance of these characteristics of vibrissal hair shafts in different species, differences in hair structure can be expected to be associated with a quantitatively and/or qualitatively different response to tactile stimulation. Especially for the structural specialization found for vibrissae of most phocid seals, it is conceivable that it determines their vibratory properties and that these are tuned to the physical demands of water. The significance of the biomechanical properties of vibrissal hair shafts for the process of mechanical transduction has largely been disregarded so far. As has been shown for the filiform hairs of arthropods (Devarakonda et al., 1996), studies on hair mechanics and its effect on stimulus transduction are urgently required for a better understanding of vibrissal sensation.

F-SCs of pinnipeds show an outstanding degree of innervation. With 1000 to 1600 myelinated axons counted in the deep vibrissal nerve (a branch of the infraorbital branch of the trigeminal nerve) innervating each F-SC, innervation density of vibrissal follicles of the ringed seal *(Phoca hispida* [Hyvärinen and Katajisto, 1984; Hyvärinen, 1989]) and the bearded seal *(Erignathus barbatus* [Marshall et al., 2006]) exceeds that calculated for well-endowed terrestrial species such as the laboratory rat and cat by a factor of 10. This means that in the bearded seal, for example, the entire mystacial vibrissal array (240 single hairs) is innervated by about 320,000 myelinated nerve fibres. Although the number of Merkel cell–neurite complexes (up to 20,000 per F-SC) indicates that this is by far the dominating sensory element, there are also encapsulated endorgans (lanceolate endings, 1000 to 4000 per F-SC; lamellated endings, 100 to 400 per F-SC) and numerous small free nerve endings in the ring sinus and the lower cavernous sinus area (Dehnhardt et al., 2003).

As many pinniped species forage on the sea bottom (Lindt, 1956; Fay, 1982; Härkönen, 1987), haptic vibrissal information could be essential for the detection and identification of benthic prey. First evidence that the vibrissae of harbor seals and gray seals represent a haptic system came from single-unit recordings at the infraorbital branch of the trigeminal nerve (Dykes, 1975), suggesting that the animals could use their vibrissae for the haptic recognition of the surface texture, shape, and size of objects. In accordance with these neurophysiological results, psychophysical experiments have shown that species of all three pinniped families (walruses [Kastelein and van Gaalen, 1988], California sea lions [Dehnhardt, 1990, 1994; Dehnhardt and Dücker, 1996], harbor seals [Dehnhardt and Kaminski, 1995; Dehnhardt et al., 1997; Dehnhardt et al., 1998a]) are indeed capable of identifying these object properties by active touch with their mystacial vibrissae. Efficiency and haptic resolving power of the vibrissal systems of the species tested are in all ways comparable to that of the prehensile hands of some primate species (Carlson et al., 1989; Hille et al., 2001).

As long as seals and sea lions do not use their mystacial vibrissae for object exploration, the hairs usually lie more or less (differing by species) close against the skin surface of the snout (but see work by Miller [1975] for the communicative role of vibrissae in some pinniped species). When an animal starts to touch an object, it protracts its vibrissae to the most forward position, but, unlike the complex whisking behavior of many terrestrial rodents (Carvell and Simons, 1990), seals and sea lions do not move their protracted vibrissae to palpate the object. As whisking is also not known in other semiaquatic mammals such as the Australian water rat, it is conceivable that these

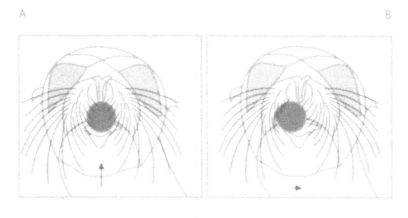

FIGURE 18.7. Haptic strategy of a California sea lion (Otariidae) for measuring the diameter of a Perspex cylinder. (A) Centering of the head with respect to the cylinder; (B) lateral head movements *(arrow)* cause a deflection of the vibrissae at the cylinder (for explanation, see text).

movements of the pliable hairs are impeded underwater due to the density of the medium, and thus whisking is futile in aquatic species (Dehnhardt et al., 1999). An exception might be the extraordinarily stiff vibrissae of walruses (Kastelein and van Gaalen, 1988). However, as relative motion between the tactile organ and the explored object is essential for the active touch process, seals and sea lions perform multiple short head movements while the vibrissae are in contact with the object. For haptic size discrimination in seals and sea lions, Dehnhardt (1994) and Dehnhardt and Kaminski (1995) suggested a model describing the respective roles of the two sensory subsystems, mechanoreception and kinaesthesis, both contributing to active touch achievements (Fig. 18.7). While searching for an object, the object is usually localized by contact with the long, most caudal vibrissae. After localization, seals and sea lions immediately center their head in relation to the object, so that it is covered by the shorter frontal vibrissae of both sides of the snout. Unlike humans sensing the angles of the finger joints when

measuring size differences with the thumb and index finger (kinaesthetic discrimination) (John et al., 1989), seals and sea lions carry out identical horizontal movements with the entire head. Due to the centered position of the head and identical head movements at objects differing in size, a larger object causes a correspondingly greater deflection of vibrissae than does a smaller object. Differences in deflection angle result in quantitatively different mechanical stimulations of follicle receptors, so that haptic size discrimination in these pinnipeds relies on mechanoreception, whereas its accuracy is codetermined by the kinaesthetic control of tactile head movements.

As described above, pinnipeds possess an upper cavernous sinus that takes about 60% of the total length of the follicle (Fig. 18.5). Therefore, the ring sinus area, where most mechanoreceptors are located, is inserted much deeper into the capsule than is usually found in terrestrial mammals. Based on studies on terrestrial mammals, the sinus system has primarily been considered to play a role in the process of mechanical transduction, for example, by amplifying mechanical stimulation inside the F-SC (Melaragno and Motagna, 1953; Woolsey et al., 1981; Rice et al., 1986; Lichtenstein et al., 1990). However, with regard to the special situation found for F-SCs in pinnipeds, Dehnhardt et al. (1998a) suggested a thermoregulatory function for at least the upper cavernous sinus (see below). While in humans a substantial decrease in skin temperature leads to severe deterioration of tactile sensitivity (Green, 1977; Green et al., 1979; Bolanowski et al., 1988; Gescheider et al., 1997), haptic difference thresholds for vibrissal texture discrimination in harbor seals remained the same at water temperatures of about 1°C as compared to 22°C (Fig. 18.8A). Thermographic examinations revealed that the skin areas of the head where the mystacial and supraorbital vibrissae are located show a substantially higher degree of thermal emission than do adjacent skin areas (e.g., about 2°C versus about 17°C, see Fig. 18.8B). This selective heating of vibrissal pads is suggested to be primarily a function of the upper cavernous blood sinus (Fig. 18.5), where no receptors can be found. Thus it may serve as a thermic insulator for the receptor area below it. An adequate heat supply is important not only for the physiology of mechanoreceptors but even more so for the mechanical properties of the surrounding tissue. Cooling of the vibrissal pads would increase tissue stiffness, which, in turn, would have a negative effect on the mobility of vibrissae and thus on the transduction of mechanical stimulation via the hair shaft to the receptors (Green et al., 1979). With regard to these effects of tissue cooling, the unusual fatty acid composition of the adipose tissue around the mystacial and supraorbital vibrissal follicles of phocids may represent an additional adaptation to low ambient temperatures. The excess low-melting-point monoenoic fatty acids found in the adipose tissue near the vibrissae is thought to maintain adequate fluidity and thus the mobility of the vibrissae under cold conditions (Käkelä and Hyvärinen, 1993, 1996).

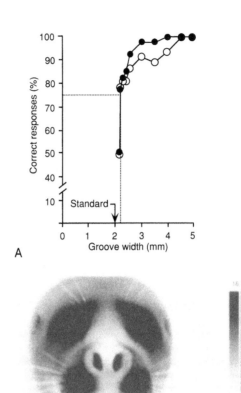

FIGURE 18.8. (A) Underwater haptic discrimination of grooved surfaces in a harbor seal at water temperatures of 1.2°C *(black circles)* and 22°C *(white circles)*. The seal was required to choose the surface of 2 mm groove width *(vertical arrow)* in each stimulus combination by pushing it out of its vertical position. The *horizontal dashed line* at 75% correct choices marks the defined difference threshold; the *vertical dashed line* indicates the size of the interpolated comparison stimulus at threshold. (B) Infrared thermogram showing the typical distribution of temperatures measured on the surface of a seal's face immediately after the animal had left water of 1°C.

After some earlier studies showed that the vibrissae of harbor seals respond to vibrations mediated by a rod directly contacting the hair (Dykes, 1975; Renouf, 1979; Mills and Renouf, 1986), Dehnhardt et al. (1998b) demonstrated their function as a hydrodynamic receptor system using dipole water movements generated by a vibrating sphere, a technique commonly used to test lateral line function in fish (Bleckmann, 1994). The constant-volume oscillating sphere was positioned 5 to 50 cm in front of the vibrissae of a harbor seal and generated water movements in the range 10 to 100 Hz (Fig. 18.9A). In

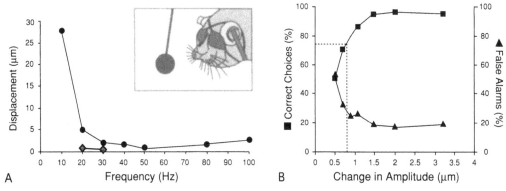

FIGURE 18.9. (A) Tuning curve of detection thresholds of a harbor seal (Phocidae) *(black circles)* and a California sea lion (Otariidae) *(gray diamonds)* for the detection of dipole water movements. At the beginning of a trial the respective test animal was outfitted with eye caps and head phones and stationed in front of the vibrating sphere. (B) Psychometric function of performance of a harbor seal *(squares)* discriminating the amplitude of dipole water movements of 40 Hz. Experiments were conducted using a go/no-go response paradigm (Dehnhardt et al., 1998b). In a single trial the seal was presented with a starting stimulus amplitude of 3 μm, which was either changed in amplitude after a few seconds or remained constant for the same amount of time (control trials). The seal indicated the detection of a change in amplitude by immediately leaving its position at the test apparatus (go response). During control trials, the seal was required to keep its position at its station for 10 seconds (no-go response). As a measure of the subject's response bias, the false-alarm rate *(triangles)* is calculated from trials in which the seal showed a go-response to a control stimulus (no change of the amplitude during a trial). The low false-alarm rates at stimulus differences that produce a high percentage of correct decisions demonstrate the reliability of the test animal's discrimination (see Schusterman and Johnson, 1975, for signal detection theory).

terms of particle displacement, absolute thresholds of the harbor seal (0.8 μm at 50 Hz) were of the same order of magnitude as those determined for the piscivorous sea snake *Lapemis curtus* (Westhoff et al., 2005), while thresholds of a California sea lion *(Zalophus californianus)* for hydrodynamic dipole stimuli of 20 and 30 Hz were even lower (Fig. 18.9A) (G. Dehnhardt, unpublished). However, unlike the sea snake tested by Westhoff et al. (2005), the harbor seal responded to the acceleration component of low-frequency hydrodynamic stimuli (up to 50 Hz). In exclusion tests, both seal and sea lion never responded when wearing a muzzle of wire mesh, which let the water movement pass but impeded whisker movements. The spectral sensitivity of the harbor seal's vibrissal system is well tuned to the frequency range of fish-generated water movements, and the shape of the tuning curve compares well to those determined for other aquatic animals equipped with a hydrodynamic sense (Fig. 18.9A) (compare with Bleckmann, 1994). Further experiments with hydrodynamic dipole stimuli revealed that a harbor

seal not only can detect but also can discriminate such water movements (Fig. 18.9B) (G. Dehnhardt, unpublished). Presented with a stimulus amplitude of 3 μm at 40 Hz, the animal could discriminate a change in amplitude of 0.8 μm (Weber fraction 0.26). While these results with hydrodynamic dipole stimuli provided physiological evidence for a hydrodynamic receptor system in pinnipeds, sensory ecology experiments were needed to show whether the vibrissal system could in fact be used for the detection and, even more important, tracking of hydrodynamic trails. As mentioned in Chapter 17, the wakes of fishes persist for several minutes (Hanke et al., 2000; Hanke and Bleckmann, 2004), thus representing trackable hydrodynamic trails of considerable length. In order to generate well-controlled hydrodynamic trails, Dehnhardt et al. (2001) used a miniature submarine and showed that a blindfolded Harbor seal can use its mystacial vibrissae to detect and track trails as long as 40 meters (Fig. 18.10). This was the very first evidence that hydrodynamic information can be used for long-distance object location.

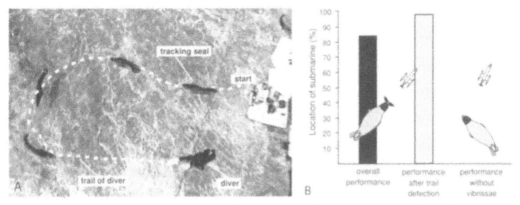

FIGURE18.10. (A) Harbor seal starting to track the trail of a human skin diver after the diver arrived at his final position. The seal protracts its mystacial vibrissae to its most forward position, and this characterizes the search for the hydrodynamic trail. Frame-by-frame analysis of video recordings revealed that the seal's swimming path exactly matched that of the skin diver. (B) Performance of a harbor seal on tracking hydrodynamic trails generated by a miniature submarine, as indicated by the successful location of the submarine's final position.

Hypothetically, hydrodynamic trail following can be considered to be never-ending, as long as a fish is swimming and natural hydrodynamic events such as currents do not disturb a trail. The arrangement of mystacial vibrissae allows a seal to perform simultaneous multiple-point velocity measurements in the wake of a swimming fish from which the three-dimensional vorticity can be derived (Fig. 18.11). In general, hydrodynamic trail following provides an explanation for how pinnipeds may successfully hunt in dark and murky waters and represents a new mechanism of spatial orientation in the aquatic environment.

SIRENIA

Manatees and dugongs seem to be true vibrissal specialists. These obligate herbivore marine mammals possess about 3000 postcranial sinus hairs, which are very thin and flexible. Although distributed all over the body, there is a dorsoventral decrease in hair density (Reep et al., 2002). For the Florida manatee it has been demonstrated that their numerous (about 2000) facial vibrissae show a distinct morphological differentiation (Marshall et al., 1998b; Reep et al., 1998; Reep et al., 2001). The entire outer muzzle is densely covered with quite flexible vibrissae, whereas extraordinarily rigid and actively moveable vibrissae are located in six pads on the upper lip, the oral cavity, and the lower jaw.

Together with the highly mobile muscular lips, the different types of facial vibrissae of manatees and dugongs form a haptic system unique among marine mammals. Sirenians use their vibrissal-muscular complex, consisting of the rigid perioral bristles and the muscular lips, in a prehensile manner and thus process a complex interaction of mechanosensory and kinesthetic information during haptic exploration and object manipulation. (Marshall et al., 1998a, 1998b). These prehensile properties of the vibrissal-muscular complex are primarily used to grasp and bring food into the oral cavity when feeding on submerged or floating vegetation (Marshall et al., 2000). Similar to the employment of mystacial vibrissae in pinnipeds, Antillean manatees *(Trichechus manatus manatus)* (Bachteler and Dehnhardt, 1999) and Florida manatees *(T. m. latirostris)* (Bauer et al., 2005) use the thin bristlelike hairs of the oral disc for the discrimination of textured surfaces. This is consistent with the results by Marshall et al. (1998b) suggesting that the bristlelike hairs of the oral disc are involved in the tactile exploration of objects. It is conceivable that the two different types of facial hairs of sirenians serve different functions in active touch processes. However, Bachteler and Dehnhardt (1999) suggest that in many cases the two hair types might be

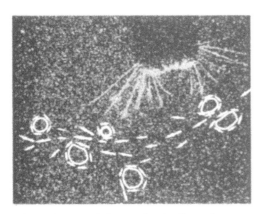

FIGURE 18.11. A technique used to visualize and measure the complex flow fields in the wake of moving objects is particle image velocimetry (PIV) (see Adrian, 1991), based on neutrally buoyant particles seeded into the water. A thin layer of the seeded water is illuminated with a laser and is filmed from a direction perpendicular to this layer. Here, the single frame from video recordings during PIV illustrates water movements in a hydrodynamic trail generated by a finlike paddle. A harbor seal approaches the trail and detects the vortices with its mystacial vibrissae that are also illuminated by the laser.

employed synergistically in a two-stage haptic process, with the bristlelike hairs providing initial sensory information before an object is grasped and manipulated for further evaluation by means of the perioral bristles.

The vibrissae found on the entire postcranial body of sirenians are suggested to represent a hydrodynamic receptor system analogous to the lateral line in fish, conveying significant hydrodynamic information from approaching animals, water currents, or large stationary objects (Reep et al., 2001). However, although this is most likely, experimental results supporting this hypothesis are still missing.

RODENTIA AND OTHER SMALL SEMIAQUATIC MAMMALS

The cooling effect of water on the tactile sensitivity already discussed for pinnipeds might be especially severe for small semiaquatic mammals such as the Australian water rat (*Hydromys chrysogaster*) that is suggested to be thermally instable (Fanning and Dawson, 1980; Dawson and Fanning, 1981). One effective way to respond to their thermally hostile

environment, obviously realized by all aquatic mammals studied so far, is an increase in F-SC size. While for example the difference in F-SC length, as one parameter of size, between the Australian water rat and the terrestrial rat is considerable (6.3 versus 2.4 mm), the size difference between F-SCs of both species becomes impressive with respect to F-SC volume (31 mm^3 versus 4.4 mm^3 [Dehnhardt et al., 1999]). Since an F-SC's volume strongly correlates with the volume of its blood-filled sinus system, this supports the hypothesis that the vibrissal sinus represents an effective system of heat supply to the follicle.

Despite major differences in body size, the numbers of axons of the deep vibrissal nerves are remarkably similar among the terrestrial rodent species studied so far (Rice et al., 1986). Therefore, it is of special interest that at the level where the deep vibrissal nerve is passing through the capsule, the number of nerve fibers in deep vibrissal nerve cross sections of the semiaquatic Australian water rat (540 axons in large caudal F-SCs) exceeds that of the terrestrial rat by a factor of at least 2.5 (Dehnhardt et al., 1999). This high F-SC innervation already indicates that the vibrissal system is of special importance for this semiaquatic mammal.

Similar to crocodilians (Fig. 18.3) some small semiaquatic mammals that primarily forage in aquatic habitats may use their vibrissae to detect surface water waves. Following a sit and wait strategy the Australian water rat, the African water rat *(Colomys goslingi)*, the otter civet *(Cynogale benettii)*, and the fishing cat *(Felis viverrina)* can be observed to crouch stock-still at the bank of a river or lake with the tips of their conspicuously well developed vibrissae immersed in water (Dieterlen and Stazner, 1981; Seidensticker and Lumpkin, 1991; Heydon and Ghaffar, 1997). For the Australian water rat, psychophysical experiments (S. Meier, H. Bleckmann, W. Hanke, and G. Dehnhardt, unpublished) have shown that these piscivorous rodents respond to minute water motions when the tips of their numerous vibrissae have contact to the water surface (Fig. 18.12). This supports the hypothesis

FIGURE 18.12. An Australian water rat detecting surface water waves with the tips of its vibrissae immersed into the water.

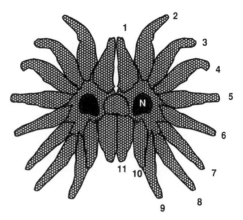

FIGURE 18.13. Frontal view of the bilaterally symmetrical organized "star" of the star-nosed mole, showing the 11 paired rays (numbered from 1 to 11) densely covered with more than 25,000 dome-shaped Eimer's organs. Abbreviation: N, nostril. Adapted from Catania and Kaas (1997).

that *Hydromys* and probably other mammals hunting on aquatic prey from land use their vibrissae to detect prey-borne water motions. The fact that the behavior corresponding to this sensory function of vibrissae has been observed in diverse mammalian orders suggests that the water surface is an important and reliable niche for hydrodynamic information.

A unique and highly specialized mechanosensory system primarily used for the haptic detection and identification of prey items is the nasal "star" of the star-nosed mole (*Condylura cristata*, Talpidae, Insectivora). Condylura is a small (body length about 12 cm) subterranean mammal inhabiting wet lowland areas of North America, where it digs tunnels while foraging (Nowak, 1991). However, different from other mole species, some of these tunnels may directly lead into the water of streams and lakes. Hence, the star-nosed mole is a good swimmer and diver, and a considerable amount of its food consists of aquatic invertebrates found by active touch on the bottoms of its aquatic habitats. The nose of *Condylura* is surrounded by 22 fleshy appendages arranged like a star around the nostrils (Fig. 18.13). Each of these tentaclelike appendages is densely covered with mechanosensory units called Eimer's organs, which resemble the reptilian touch papillae (Fig. 18.1), avian papillae of the bill-tip organ, and the push rods

of the monotreme platypus (Catania [1996] has described the ultrastructure of Eimer's organs of the star-nosed mole). About 25,000 dome-shaped Eimer's organs supplied with more than 100,000 nerve fibers cover the star that measures only 1 cm in diameter.

When the star-nosed mole is searching for prey, the rays of the star are in continuous motion, a prerequisite for efficient haptic information processing. A mole touches the star (tactile sensitive surface about 1 m^2) to different areas of the environment about 13 times per second and thus may investigate 46 m^2 per hour (Catania and Remple, 2004). After initial contact with a food item a mole needs an average of 227 ms to identify and consume it, which makes the star-nosed mole the fastest-eating mammal known (Catania and Remple, 2005). During this handling time a mole touches an object for only 25 ms before it decides whether the object is edible, or it should continue searching. As neurons in the mole's somatosensory cortex need 12 ms to respond to tactile stimulation of the star (Sachdev and Catania, 2002) and the corresponding afferent motor command needs about 5 ms to reach the sensory system, there remain only 8 ms for the brain of a mole to process the haptic information, what is suggested to be the speed limit for haptic information processing.

If any of the 22 rays of the star contacts a potential prey item, the mole immediately moves its head in a way that the object contacts the lowermost central pair of rays (ray 11 in Fig. 18.13). This detailed exploration of an object with the most central sensors in the array is consistent with the model suggested for haptic size measurements by pinnipeds using their mystacial vibrissae (Fig. 18.7) and might be a general principle in bilaterally symmetrical organized mechanosensory systems. As the Eimer's organs on the ventral central pair of tactile rays are particularly well supplied with nerve fibres, and, compared to peripheral rays, they show a stronger representation (covering more space) in the cerebral cortex, Catania and Kaas (1997) and Catania (1999) suggest it to be the somatosensory fovea of the system, a functional analogy between vision and touch already mentioned for the bill-tip organ of aquatic birds (see above).

In comparison to primarily terrestrial mole species, which show no extension of a nasal tactile surface and have only 1000 to 2000 Eimer's organs on the tip of the snout, the mechanosensory star of *Condylura* with its increased sensory surface, high receptor density, and fast information processing capacity is of particular importance for this small diving mole when searching for prey underwater. Due to the relatively low oxygen stores and usually high metabolic rates of small divers, their diving cycles and thus foraging time must be short (Kooyman, 1989), which in consequence may have favored the evolution of the highly efficient mechanosensory star and appropriate behavioral strategies to optimize foraging success.

EVOLUTION OF MECHANORECEPTION

This chapter shows that cutaneous mechanoreception is far from being a merely receptive sense; instead it enables animals as diverse as reptiles, birds, and mammals to actively seek environmental information. Particularly with regard to the perception of water movements, there is good evidence that in adaptation to

their aquatic life style all classes of secondarily aquatic tetrapods developed a hydrodynamic sense, either for the reception of water surface waves or for hydrodynamic midwater events. The fact that, for instance, the dome pressure receptors of crocodilians are also present in fossils from the Jurassic period suggests that these semiaquatic predators might have already used hydrodynamic information at the water surface many millions of years ago. As does the widespread occurrence of hydrodynamic receptor systems in primarily aquatic invertebrates and fish, this immediate rediscovery of the hydrodynamic information channel when returning to an aquatic lifestyle underlines its accessibility and importance (Soares, 2002). The similar morphology (with corresponding functional modifications) of reptilian touch papillae and dome pressure receptors, avian papillae of the bill-tip organ, push rods of the monotreme platypus, and Eimer's organs of the star-nosed mole and many other species of the family Talpidae suggests a convergent evolution of these complex mechanosensory structures when recapturing an anciently lost sensory niche. However, we are only at the beginning of understanding the significance and function of hydrodynamic sensory systems in secondarily aquatic tetrapods.

With regard to the special adaptations mechanosensory systems probably have undergone in response to the special needs of particular ecological niches, the pinniped vibrissal system might provide an interesting, though still not well understood, example. Manifold variations can be found regarding the morphology of the single vibrissal hair shaft (length, diameter, surface structure, pliability), the three-dimensional arrangement of vibrissae on the muzzle of different pinniped species, as well as the total number of mystacial and supraorbital vibrissae. These variations can be expected to be associated with quantitatively and/or qualitatively different responses to tactile stimulation. However, up to now we could only speculate how these differences may have evolved as an adaptation to the respective ecological niche. While a walrus might need its stiff vibrissae mainly to search for prey items at the sea bottom and might have to detect even small differences in the substrate indicating the presence of prey, many other species might use their longer and more pliable vibrissae for detection of hydrodynamic stimuli in the open water column. Further studies relating various aspects of the natural behavior of pinnipeds to the special characteristics of their vibrissal systems might lead to a better understanding of this kind of mechanosensory information in the context of ecology and evolution.

LITERATURE CITED

Adrian, R. J. 1991. Particle imaging techniques for experimental fluid mechanics. Annual Review of Fluid Mechanics 23:261–304.

Andres, K. H., and M. von Düring. 1990. Comparative and functional aspects of the histological organization of cutaneous receptors in vertebrates; pp. 1–17 in W. Zenker and W. Neuhuber (eds.), The Primary Afferent Neuron: A Survey of Recent Morpho-functional Aspects. Plenum Press, New York.

Bachteler, D., and G. Dehnhardt. 1999. Active touch performance in the Antillean manatee: evidence for a functional differentiation of facial tactile hairs. Zoology 102:61–69.

Bauer G. B., J. C. Gaspard, D. E. Colbert, J. B. Leach, S. A. Stamper, D. Sarko, J. D. Hammelman, A. Schmieg, D. Mann, and R. Reep 2005. Tactile discrimination by Florida manatees, *Trichechus manatus latirostris*. Paper presented at the 16th Biennial Conference on the Biology of Marine Mammals, San Diego, California, Dec. 12–16.

Bleckmann, H. 1994. Reception of Hydrodynamic Stimuli in Aquatic and Semiaquatic Animals. Fischer Verlag, Stuttgart, Jena, New York.

Bolanowski, S. J., G. A. Gescheider, R. T. Verrillo, and C. M. Checkosky. 1988. Four channels mediate the mechanical aspects of touch. Journal of the Acoustical Society of America 84:1680–1694.

Bolze, G. 1968. Anordnung und Bau der Herbstschen Körperchen in Limicolenschnäbeln im Zusammenhang mit der Nahrungsfindung. Zoologischer Anzeiger 181:313–355.

Carlson, S., H. Tamila, I. Linnankoski, A. Pertovaara, and A. Kehr. 1989. Comparison of tactile discrimination ability of visually deprived and normal monkeys. Acta Physiologia Scandinavia 135:405–410.

Carvell, G. E., and D. J. Simons. 1990. Biometric analysis of vibrissal tactile discrimination in the rat. Journal of Neuroscience 10:2638–2648.

Catania, K. C. 1996. Ultrastructure of the Eimer's organ of the star-nosed mole. Journal of Comparative Neurology 365:343–354.

Catania, K. C. 1999. A nose that looks like a hand and acts like an eye: the unusual mechanosensory system of the star nosed mole. Journal of Comparative Physiology A 185:367–372.

Catania, K. C., and J. H. Kaas. 1997. Somatosensory fovea in the star-nosed mole: behavioral use of the star in relation to innervation patterns and cortical representation. Journal of Comparative Neurology 387:215–233.

Catania, K. C., and F. E. Remple. 2004. Tactile foveation in the star-nosed mole. Brain, Behavior, and Evolution 63:1–12.

Catania, K. C., and F. E. Remple. 2005. Asymptotic prey profitability drives star-nosed moles to the foraging speed limit. Nature 433:519–522.

Dawson, T. J., and F. D. Fanning. 1981. Thermal and energetic problems of semiaquatic mammals: a study of the Australian water rat, including comparisons with the Platypus. Physiological Zoology 54:285–296.

Dehnhardt, G. 1990. Preliminary results from psychophysical studies on the tactile sensitivity in marine mammals; pp. 435–446 in J. A. Thomas and R. A. Kastelein (eds.), Sensory Abilities of Cetaceans. Plenum Press, New York.

Dehnhardt, G. 1994. Tactile size discrimination by a California sea lion *(Zalophus californianus)* using its mystacial vibrissae. Journal of Comparative Physiology A 175:791–800.

Dehnhardt, G., and G. Dücker. 1996. Tactual discrimination of size and shape by a California sea lion *(Zalophus californianus)*. Animal Learning and Behavior 24:366–374.

Dehnhardt, G., and A. Kaminski. 1995. Sensitivity of the mystacial vibrissae of harbour seals *(Phoca vitulina)* for size differences of actively touched objects. Journal of Experimental Biology 198:2317–2323.

Dehnhardt, G., M. Sinder, and N. Sachser. 1997. Tactual discrimination of size by means of mystacial vibrissae in harbour seals: in air versus underwater. Zeitschrift für Säugetierkunde 62:40–43.

Dehnhardt, G., B. Mauck, and H. Hyvärinen. 1998a. Ambient temperature does not affect the tactile sensitivity of mystacial vibrissae in harbour seals. Journal of Experimental Biology 201: 3023–3029.

Dehnhardt, G., B. Mauck, and H. Bleckmann. 1998b. Seal whiskers detect water movements. Nature 394:235–236.

Dehnhardt, G., H. Hyvärinen, and A. Palviainen. 1999. Structure and innervation of the vibrissal follicle-sinus complex in the Australian water rat, *Hydromys chrysogaster*. Journal of Comparative Neurology 411:550–562.

Dehnhardt, G., B. Mauck, W. Hanke, and H. Bleckmann. 2001. Hydrodynamic trail-following in Harbor seals *(Phoca vitulina)*. Science 193: 102–104.

Dehnhardt, G., B. Mauck, and H. Hyvärinen. 2003. The functional significance of the vibrissal system of marine mammals; pp. 127–135 in K. I. Baumann, I. Moll, and Z. Halata (eds.), The Merkel Cell: Structure, Development, Function, and Cancerogenesis. Springer-Verlag, Berlin.

Devarakonda, R., F. G. Barth, and J. A. C. Humphrey. 1996. Dynamics of arthropod filiform hairs. 4. Hair motion in air and water. Philosophical Transactions of the Royal Society of London B 351:933–946.

Dieterlen, F., and B. Stazner. 1981. The African rodent *Colomys goslingi* Thomas and Wroughton, 1907 (Rodentia: Muridae): a predator in limnetic ecosystems. Zeitschrift für Säugetierkunde 46:369–383.

Dykes, R. W. 1975. Afferent fibers from mystacial vibrissae of cats and seals. Journal of Neurophysiology 38:650–662.

Fanning, F. D., and T. J. Dawson. 1980. Body temperature variability in the Australian water rat, *Hydromys chrysogaster*, in air and in water. Australian Journal of Zoology 28:229–238.

Fay, F. H. 1982. Ecology and biology of the Pacific walrus, *Odobenus rosmarus divergens* Illiger. U.S. Fish and Wildlife Service, Washington, DC.

Gentle, M. J., and J. Breward. 1986. The bill tip organ of the chicken *(Gallus gallus* var. *domesticus)*. Journal of Anatomy 145:79–85.

Gerritsen, A. F. C., and A. Meiboom. 1986. The role of touch in prey density estimation by *Caldris alba*. Zoologischer Anzeiger 181:313–355.

Gescheider, G. A., J. M. Thorpe, J. Goodarz, and S. J. Bolanowski. 1997. The effects of skin temperature on the detection and discrimination of tactile stimulation. Somatosensory and Motor Research 14:181–188.

Goldschmidt-Lange, U. 1976. The morphological differences of the structure of the facial vibrissae of various mammals [in German]. Zoologischer Anzeiger 196:417–427.

Green, B. G. 1977. The effect of skin temperature on vibrotactile sensitivity. Perception and Psychophysics 21:243–248.

Green, B. G., S. J. Lederman, and J. C. Stevens. 1979. The effect of skin temperature on the perception of roughness. Sensory Processes 3:327–333.

Hanke, W., and H. Bleckmann. 2004. The hydrodynamic trails of *Lepomis gibbosus* (Centrarchidae), *Colomesus psittacus* (Tetraodontidae) and *Thysochromis ansorgii* (Cichlidae) investigated with scanning particle image velocimetry. Journal of Experimental Biology 207:1585–1596.

Hanke, W., C. Brücker, and H. Bleckmann. 2000. The ageing of the low-frequency water disturbances caused by swimming goldfish and its possible relevance to prey detection. Journal of Experimental Biology 203:1193–1200.

Härkönen, T. J. 1987. Seasonal and regional variations in the feeding habits of the harbour seal, *Phoca vitulina*, in the Skagerrak and Kattegat. Journal of Zoology (London) 213:535–543.

Heydon, M. J., and N. Ghaffar. 1997. Records of Otter civets *(Cygnogale bennettii)* from northern Borneo. Small Carnivore Conservation 16:27.

Hille, P., C. Becker-Carus, G. Dücker, and G. Dehnhardt. 2001. Haptic discrimination of size and texture in squirrel monkeys *(Saimiri sciureus)*. Somatosensory and Motor Research 18:50–61.

Hollis, D. E., and A. G. Lyne. 1974. Innervation of vibrissa follicles in the marsupial *Trichosurus vulpecula*. Australian Journal of Zoology 22:263–276.

Hyvärinen, H. 1989. Diving in darkness: whiskers as sense organs of the ringed seal *(Phoca hispida saimensis)*. Journal of Zoology 218:663–678.

Hyvärinen, H., and H. Katajisto. 1984. Functional structure of the vibrissae of the ringed seal *(Phoca hispida* Schr.). Acta Zoologica Fennica 171:27–30.

Iggo, A., J. E. Gregory, and U. Proske. 1996. Studies of mechanoreceptors in skin of the snout of the echidna *Tachyglossus aculeatus*. Somatosensory and Motor Research 13:129–138.

John, K. T., A. W. Goodwin, and I. Darian-Smith. 1989. Tactile discrimination of thickness. Experimental Brain Research 78:62–68.

Käkelä, R., and H. Hyvärinen. 1993. Fatty acid composition of fats around the mystacial and superciliary vibrissae differs from that of blubber in the Saimaa ringed seal *(Phoca hispida saimensis)*. Comparative Biochemistry and Physiology 105:547–552.

Käkelä, R., and H. Hyvärinen. 1996. Site-specific fatty acid composition in adipose tissues of several northern aquatic and terrestrial mammals. Comparative Biochemistry and Physiology 115:501–514.

Kastelein, R. A., and M. A. van Gaalen. 1988. The sensitivity of the vibrissae of a pacific walrus (*Odobenus rosmarus divergens*). Aquatic Mammals 14:123–133.

Kooyman, G. L. 1989. Diverse Divers: Physiology and Behavior. Springer-Verlag, Berlin.

Krueger, L. E. 1982. Tactual perception in historical perspective: David Katz's world of touch; pp. 1–53 in W. Schiff and E. Foulke (eds.), Tactual Perception: A Sourcebook. Cambridge University Press, New York.

Lederman, S. J., and R. L. Klatzky. 1987. Hand movements: a window into haptic object recognition. Cognitive Psychology 19:342–368.

Lichtenstein, S. H., G. E. Carvell, and D. J. Simons. 1990. Responses of rat trigeminal ganglion neurons to movements of vibrissae in different directions. Somatosensory and Motor Research 7:47–65.

Lindt, C. C. 1956. Underwater behaviour of the southern sea lion, *Otaria jubata*. Journal of Mammalogy 37:287–288.

Ling, J. K. 1977. Vibrissae of marine mammals; pp. 387–415 in R. J. Harrison (ed.), Functional Anatomy of Marine Mammals. Academic Press, London.

Manger, P. R., and J. D. Pettigrew. 1996. Ultrastructure, number, distribution and innervation of electroreceptors and mechanoreceptors in the bill skin of the platypus, *Ornithorhynchus anatinus*. Brain, Behavior, and Evolution 48:27–54.

Marotte, L. R., F. L. Rice, and P. M. E. Waite. 1992. The morphology and innervation of facial vibrissae in the tammar wallaby, *Macropus eugenii*. Journal of Anatomy 180:401–417.

Marshall, C. D., L. A. Clark, and R. L. Reep. 1998a. The muscular hydrostat of the Florida manatee (*Trichechus manatus latirostris*): a functional morphological model of perioral bristle use. Marine Mammal Science 14:290–303.

Marshall, C. D., G. D. Huth, V. M. Edmonds, D. L. Halin, and R. L. Reep. 1998b. Prehensile use of perioral bristles during feeding and associated behaviors of the Florida manatee (*Trichechus manatus latirostris*). Marine Mammal Science 14:274–289.

Marshall, C. D., P. S. Kubilis, G. D. Huth, V. M. Edmonds, D. L. Halin, and R. L. Reep. 2000. Food-handling ability and feeding-cycle length of manatees feeding on several species of aquatic plants. Journal of Mammalogy 81:649–658.

Marshall, C. D., H. Amin, K. Kovacs, and C. Lydersen. 2006. Microstructure and innervation of the vibrissal follicle-sinus complex in the bearded seal, *Erignathus barbatus* (Pinnipedia: Phocidae). Anatomical Record 288A:13–25.

Mauck, B., U. Eysel, and G. Dehnhardt. 2000. Selective heating of vibrissal follicles in seals (*Phoca vitulina*) and dolphins (*Sotalia fluviatilis guianensis*). Journal of Experimental Biology 203:2125–2131.

Melaragno, H. P., and W. Motagna. 1953. The tactile hair follicles in the mouse. Anatomical Record 115:129–149.

Miller, E. H. 1975. A comparative study of facial expressions of two species of pinnipeds. Behaviour 53:268–284.

Mills, F. H. J., and D. Renouf. 1986. Determination of the vibration sensitivity of harbour seals (*Phoca vitulina*) vibrissae. Journal of Experimental Marine Biology and Ecology 100:3–9.

Nebel, S., D. L. Jackson, and R. W. Elner. 2005. Functional association of bill morphology and foraging behaviour in calidrid sandpipers. Animal Biology 55:235–243.

Necker, R. 2000. The somatosensory system; pp. 57–69 in C. G. Whittow (ed.), Sturkies Avian Physiology. Academic Press, New York.

Nowak, R. M. 1991. Walker's mammals of the world, 5th Ed. The Johns Hopkins University Press, Baltimore.

Pettigrew, J. D., P. R. Manger, and S. L. B. Fine. 1998. The sensory world of the platypus. Philosophical Transactions of the Royal Society of London B 353:1199–1210.

Piersma, T., R. Van Aelst, K. Kurk, H. Berkhoudt, and L. R. M. Maas. 1998. A new pressure sensory mechanism for prey detection in birds: the use of principles of seabed dynamics? Proceedings of the Royal Society of London B 265:1377–1383.

Povel, D., and J. v. d. Kooij. 1997. Scale sensillae of the file snake (Serpentes: Acrochordidae) and some other aquatic and burrowing snakes. Netherlands Journal of Zoology 47:443–456.

Reep, R. L., C. D. Marshall, M. L. Stoll, and D. M. Whitaker. 1998. Distribution and innervation of facial bristles and hairs in the Florida manatee (*Trichechus manatus latirostris*). Marine Mammal Science 14:257–273.

Reep, R. L., M. L. Stoll, C. D. Marshall, B. L. Homer, and D. A. Samuelson. 2001. Microanatomy of facial vibrissae in the Florida manatee: the basis for specialized sensory function and oripulation. Brain, Behavior, and Evolution 58:1–14.

Reep, R. L., C. D. Marshall, and M. L. Stoll. 2002. Tactile hairs on the postcranial body in Florida manatees: a mammalian lateral line? Brain, Behavior, and Evolution 59:141–154.

Renouf, D. 1979. Preliminary measurements of the sensitivity of the vibrissae of Harbour seals (*Phoca vitulina*) to low frequency vibrations. Journal of Zoology 188:443–450.

Renouf, D. 1991. Sensory reception and processing in Phocidae and Otariidae; pp. 345–394 in D. Renouf (ed.), The Behaviour of Pinnipeds. Chapman and Hall, London.

Rice, F. L., A. Mance and B. L. Munger. 1986. A comparative light microscopic analysis of the sensory innervation of the mystacial pad. I. Innervation of vibrissal follicle-sinus complexes. Journal of Comparative Neurology 252:154–174.

Sachdev, R. N., and K. C. Catania. 2002. Receptive fields and response properties of neurons in the star-nosed mole's somatosensory fovea. Journal of Neurophysiology 87:2602–2611.

Schusterman, R. J., and B. W. Johnson. 1975. Signal probability and response bias in California sea lions. Psychological Record 25:39–45.

Seidensticker, J., and S. Lumpkin. 1991. Great Cats: Majestic Creatures of the Wild. Rodale Press, Emmaus, PA.

Soares, D. 2002. Neurology: an ancient sensory organ in crocodilians. Nature 417:241–242.

Stephens, R. J., I. J. Beebe, and T. C. Poulter. 1973. Innervation of the vibrissae of the California sea lion, *Zalophus californianus*. Anatomical Record 176:421–442.

von Düring, M. 1975. The ultrastructure of cutaneous receptors in the skin of *Caiman crocodilus*. Abhandlungen der Rheinisch-Westfälischen Akademie der Wissenschaften 53:123–134.

von Düring, M., and M. R. Miller. 1979. Sensory nerve endings of the skin and deeper structures of reptiles; pp. 407–441 in C. Gans (ed.), Biology of the Reptilia, Vol. 9, Neurobiology A. Academic Press, London.

van Horn, R. N. 1970. Vibrissae structure in the rhesus monkey. Folia Primatologica 13:241–285.

Watkins, W. A., and D. Wartzok. 1985. Sensory biophysics of marine mammals. Marine Mammal Science 1:219–260.

Westhoff, G., B. G. Fry, and H. Bleckmann. 2005. Sea snakes *(Lapemis curtus)* are sensitive to low-amplitude water motions. Zoology 108:195–200.

Woolsey, T. A., D. Durham, R. M. Harris, D. J. Simons, and K. L. Valentino. 1981. Somatosensory development. Development and Perception 1:259–292.

Magnetoreception and Electroreception

19

Magnetoreception

Michael H. Hofmann and Lon A. Wilkens

Unlike the other senses, magnetoreception has been investigated in very few animal species. Although this sense is found from mud bacteria to mammals (e.g., Kirschvink et al., 2001) and even in plants (Galland and Pazur, 2005), it is not clear whether this sensory system is widely used or whether it is present only in some specialized groups. The limited information on the magnetic sense is largely due to the fact that magnetic fields penetrate the body and receptors can be located practically anywhere inside the body. Furthermore, the animal's response to magnetic fields is usually not immediately apparent and affects in many cases only its general migratory route. Technical difficulties are probably responsible for the limited data on magnetosensory animals, and this makes it impossible to speculate about the evolution or distribution of magnetoreception in animals.

This chapter gives a short overview of the physics of the Earth's magnetic fields, followed by a description of animals known to sense these fields. The last part of the chapter focuses on sea turtles and cetaceans, the only secondarily aquatic tetrapods for which information on a magnetic sense is available.

THE PHYSICS OF THE EARTH'S MAGNETIC FIELD

The magnetic field of the Earth provides important information for many animals. Convection currents in the Earth's core are the main source of the magnetic fields. Fields leave the magnetic poles and travel around the Earth to the opposite pole. Thus, field lines on the Earth's surface can be used to determine compass headings relative to the magnetic poles. In addition, magnetic field vectors show an inclination that depends on the latitude (Fig. 19.1). At the South Pole, field lines are vertical, but their angle changes gradually to

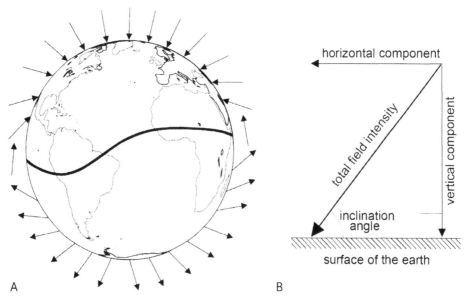

FIGURE 19.1. (A) Diagram of the earth's magnetic field. The *thick line* shows the position of the magnetic equator. The *arrows* indicate the local magnetic field vector. Its direction is horizontal at the magnetic equator and vertical at the poles. The inclination angle depends on the latitude. (B) Vector components and inclination angle of a magnetic field vector. The field has a total intensity and inclination angle. The total intensity can be determined by measuring either the horizontal or vertical component of the field vector if the inclination angle is known. Alternatively, if the intensity is known, the inclination angle can be determined from either the vertical or horizontal component. After Lohmann et al. (1999).

reach a horizontal orientation at the magnetic equator. Then, their orientation changes again, and they enter the surface at the North Pole vertically. This inclination can be used to determine the latitude. The resulting field is the main magnetic field, which varies little in amplitude on the Earth's surface. Local magnetic fields caused by rock formations in the Earth's crust can distort the main field. These distortions are called magnetic anomalies. They can modify both the local amplitude and the orientation of the main field. Although such anomalies disturb long-range navigation, they can also be used for short-range orientation.

MAGNETOSENSORY ANIMALS

The presence of a magnetic sense has been confirmed in many animals, and there is some evidence that this sense occurs even more widely. Mud bacteria were among the first animals known to respond to magnetic fields (Blakemore, 1975; Kalmijn and Blakemore, 1977). They contain iron rich deposits that enable them to orient and swim toward the North Pole (Bazylinski and Frankel, 2004). Due to the inclination of the magnetic field, they also swim downward unless they are over the equator. Magnetic fields can thus be used to detect up and down, which is a plausible function in bacteria.

A number of invertebrates are known or suspected to be magnetosensory. Bees (Lindauer and Martin, 1972; Walker and Bitterman, 1989) and mealworm beetles (Vácha and Soukopová, 2004) are known to respond to magnetic fields, and ferromagnetic material has been found in other social insects such as ants (Acosta-Avalos et al., 1999) and termites (Maher, 1998). Two aquatic invertebrates are also known to respond to magnetic fields. Lobsters migrating in a straight path deviate from the original course after manipulating the horizontal component of the magnetic field

(Lohmann et al., 1995). The marine mollusk *Tritonia diomedea* makes on- and offshore migrations that are guided by magnetoreceptors (Lohmann et al., 1991).

Migratory vertebrates are also known to use magnetic cues. Some elasmobranchs make extensive migrations in the open ocean and have been shown to navigate over featureless sandy bottoms. Their extremely sensitive electrosensory system has been known for some time, but only recently has behavioral evidence led to the hypothesis that they can sense magnetic fields by measuring the electric fields induced by moving through the Earth's magnetic field (Kalmijn, 1978, 1982, 1984). Among bony fishes, salmon (Quinn, 1980; Quinn et al. 1981; Quinn and Brannon, 1982) and trout (Mann et al., 1988; Walker et al., 1997; Diebel et al., 2000) are well investigated with respect to a magnetic sense, and some evidence has been put forward for eels (Nishi et al., 2004; van Ginneken et al., 2005) and tuna (Walker et al., 1984). Many amphibians migrate between breeding and hibernation sites and it has been shown that magnetic cues are used for navigation (Phillips, 1986; Phillips et al., 2001; Freake and Phillips, 2005).

BIRDS

The best-investigated vertebrates with respect to magnetoreception are birds. Numerous behavioral studies have shown that many migratory birds use the compass sense for navigation (Mouritsen and Ritz, 2005; Wiltschko and Wiltschko, 2006). In addition, nonmigratory birds such as the pigeon can use magnetic information for homing. Birds probably have two independent receptor mechanisms. One involves magnetite-based receptors that have been identified in the beak (Fleissner et al., 2003). The other is a chemical reaction in the retina that can be influenced by magnetic fields (Wiltschko et al., 2004). Specialized photopigments can absorb photons and produce radical pairs. These radical pairs can be converted into triplets, depending on their alignment within

the magnetic field. The orientation of the receptors varies in a systematic way in different parts of the hemispherical retina, and this activity pattern could give information on the orientation of the magnetic field.

MAMMALS

An interesting example in a mammal is the blind mole rat (Kimchi and Terkel, 2001; Němec et al., 2001). This animal spends its entire life in earthen tunnels that have no exits to the surface. Without vision, and with very limited auditory capabilities, they are able to navigate accurately in the extensive tunnel system, guided by their magnetic sense.

MAGNETOSENSE IN AQUATIC TETRAPODS

As research proceeds, it is likely that more species exploiting magnetic information will be identified. Navigation is probably the major function of this sense, and this is especially important when the use of other sensory systems is limited. This is the case in many of the magnetosensory animals that migrate in the dark or in featureless open oceans. Many secondarily aquatic tetrapods face this problem. They live in the ocean and navigate over long distances; visual cues are limited in open waters, and magnetic cues could aid in orientation.

TURTLES

There is convincing evidence for magnetoreception in cheloniid sea turtles. Green turtles *(Chelonia mydas)* and loggerheads *(Caretta caretta)* make extensive migrations between their feeding grounds and nesting areas (Lohmann et al., 1999). Loggerheads, for example, hatching on Japan's beaches travel through the Pacific Ocean to their feeding areas in Baja California and return to Japan to reproduce (Bowen et al., 1995). During their journey, many sea turtles take a straight path over distances that can exceed hundreds of kilometers. This remarkable ability has led to the assumption

Southern Gyre

FIGURE 19.2. Migratory route within the Atlantic gyre of loggerhead turtles (*Caretta caretta*, Cheloniidae). The three orientation diagrams show the mean orientation angle of individual turtles *(dots)* tested in magnetic fields with the same properties as present at three different locations within the gyre. At each location, turtles show a different orientation preference, and this preference is optimal to enable the turtles to stay within the gyre. Reprinted with permission from Lohmann et al., Science 294:364–366. Copyright, 2001, AAAS.

that they use a compass sense for navigation. Early studies showed that sea turtles can obtain directional information from the vertical component of the Earth's magnetic field (Lohmann and Lohmann, 1993, 1994a, 1994b, 1996a, 1996b; Lohmann, 1991; Light et al., 1993). Due to the inclination of magnetic fields, field vectors can be divided into a horizontal and a vertical component. By manipulating each component separately, it is possible to find out which component is used for orientation. Birds (Wiltschko and Wiltschko, 2006) and sea turtles (Lohmann, 1991; Light et al., 1993) use the vertical, and salmon (Quinn et al., 1981) and mole rats (Marhold et al., 1991) the horizontal

component. Sea turtles born on Florida's beaches can thus find the eastbound direction that takes them into the Gulf Stream and brings them across the Atlantic.

Somewhere along its course, the Gulf Stream divides, and the turtles have to swim south to avoid being dragged into the North Atlantic stream that would transport them into the cold waters around England or Scandinavia. While passing West Africa, the Gulf Stream turns west to complete the cycle that originated just east of the Gulf of Mexico. To follow that path, sea turtles have to change headings and swim westbound before they are dragged too far south out of the main

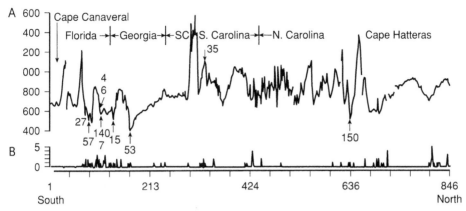

FIGURE 19.3. (A) Relative magnetic field strength (in nanoteslas) of segments along the coast from Cape Canaveral (Florida), through Cape Hatteras (North Carolina). (B) Histogram with the number of stranding events of cetaceans. *Numbers with arrows* in A indicate the number of individuals that composed a single stranded group in one event. After Kirschvink et al. (1986), Company of Biologists, Ltd.

streams that can return them to their nesting sites (Fig. 19.2). Thus, at each location in the Atlantic, sea turtles have to maintain different heading directions to avoid being swept out of the main stream. It has been shown that sea turtles can use the inclination angle and intensity of magnetic fields to determine their location within the Atlantic gyre and to adjust their headings accordingly (Lohmann et al., 1999, 2001). Thus, sea turtles not only rely on a compass sense but can also obtain positional information from features of the local magnetic fields.

CETACEANS

Another secondarily aquatic animal group that is thought to have a magnetic sense is cetaceans. Cetaceans may use the magnetic fields as a compass mechanism or they may use local anomalies as landmarks. Initial evidence for this came from studies that correlated stranding patterns with local geomagnetic anomalies (Klinowska, 1985a, 1985b, 1988, 1990; Cornwell-Huston, 1986; Kirschvink et al., 1986; Kirschvink, 1990). By measuring local magnetic field strengths and comparing them with known stranding sites, it became clear that stranding occurs primarily at sites that show local magnetic minima (Fig. 19.3). In a later study, Walker et al. (1992) suggested that

fin whales use geomagnetic anomalies as cues for their migrations in the North Atlantic. Such anomalies are present at the Atlantic floor (Gross, 1996) and could guide the whales on their north-south migrations. Dolphins (Delphinidae) also undertake long journeys, but there are no distinct migratory routes as for mysticetes. Although stranding patterns may correlate with magnetic anomalies, foraging routes of free-ranging dolphins are apparently not influenced by geomagnetic cues (Hui, 1994).

Possible sensory mechanisms are magnetite-based receptors where magnetic particles embedded in a mechanoreceptor change their position according to the ambient magnetic field. The presence of magnetic material in an animal is often evidence for magnetoreception. Magnetite has been detected in the brain of the common dolphin *(Delphinus delphis)* (Zoeger et al., 1981) and humpback whales *(Megaptera novaeangliae)* (Fuller et al., 1985). The material was found in the dura mater of the brain and consisted of small anisotropic particles. Scanning electron microscope pictures of the dolphin suggested the presence of nerve fibers around the particle, which is evidence for a sensory function. Although the material is magnetically soft, it should experience a torque in the geomagnetic field or possibly deform. The presence of magnetic material in an animal is not proof, however, of magnetoreception, as many organs contain variable amounts of magnetite, including the human brain (Kirschvink et al., 1992; Dunn et al., 1995), heart, spleen, and liver. Mechanisms not relying on the presence of magnetic material may also be involved in magnetic field detection, as proposed by Gerrits and Kastelein (1990). Their model consists of a population of spontaneously active neurons whose axons are organized in a circular loop. If moved through a magnetic field, the induced electric fields could be large enough to change the spike rate of the neurons.

CONCLUSIONS

The magnetic sense is probably the least understood sensory system. It is used in many animals for long-range navigation, particularly in situations where other senses are of limited use. This applies especially to the aquatic environment. Hence, it is not a surprise that aquatic tetrapods such as sea turtles and cetaceans use magnetic field information for navigation during their extensive migrations. Further investigation may help us to understand the special problems that aquatic tetrapods face to navigate and survive in an environment that they once left.

ACKNOWLEDGMENTS

We thank the editors, Hans Thewissen and Sirpa Nummela, for inviting us to contribute to this volume.

LITERATURE CITED

Acosta-Avalos, D., E. Wajnberg, P. S. Oliveira, I. Leal, M. Farina, and D. M. S. Esquivel. 1999. Isolation of magnetic nanoparticles from *Pachycondyla marginata* ants. Journal of Experimental Biology 202:2687–2692.

Bazylinski, D. A., and R. B. Frankel. 2004. Magnetosome formation in procaryotes. Nature Reviews in Microbiology 2:217–230.

Blakemore, R. P. 1975. Magnetotactic bacteria. Science 190:377–379.

Bowen, B. W., F. A. Abreu-Grobois, G. H. Balazs, N. Kamezaki, C. J. Limpus, and R. J. Ferl. 1995.

Trans-Pacific migrations of the loggerhead turtle *(Caretta caretta)* demonstrated with mitochondrial DNA markers. Proceedings of the National Academy of Sciences USA 92:3731–3734.

Cornwell-Huston, C. J. 1986. An analysis of cetacean mass strandings and the coastal geomagnetic topography along the U.S. eastern seaboards and eastern Gulf of Mexico. M.S. thesis, Boston University, Boston.

Diebel, C. E., R. Proksch, C. R. Green, P. Neilson, and M. M. Walker. 2000. Magnetite defines a vertebrate magnetoreceptor. Nature 406:299–302.

Dunn, J. R., M. Fuller, J. Zoeger, J. Dobson, F. Heller, J. Hammann, E. Caine, and B. M. Moskowitz. 1995. Magnetic material in the human hippocampus. Brain Research Bulletin 36:149–153.

Fleissner, G., E. Holtkamp-Rotzler, M. Hanzlik, M. Winklhofer, G. Fleissner, N. Petersen, and W. Wiltschko. 2003. Ultrastructural analysis of a putative magnetoreceptor in the beak of homing pigeons. Journal of Comparative Neurology 458:350–360.

Freake, M. J., and J. B. Phillips. 2005. Light-dependent shift in bullfrog tadpole magnetic compass orientation: evidence for a common magnetoreception mechanism in anuran and urodele amphibians. Ethology 111:241–254.

Fuller, M., W. S. Goree, and W. L. Goodman. 1985. An introduction to the use of SQUID magnetometers in biomagnetism; pp. 103–151, in J. L. Kirschvink, D. S. Jones, and B. J. MacFadden (eds.), Magnetite Biomineralization and Magnetoreception in Organisms: A New Biomagnetism. Plenum Press, New York.

Galland, P., and A. Pazur. 2005. Magnetoreception in plants. Journal of Plant Research 118:371–389.

Gerrits, N. M., and R. A. Kastelein. 1990. A potential neural substrate for geomagnetic sensitivity in cetaceans; pp. 31–38 in J. A. Thomas and R. A. Kastelein (eds.), Sensory Abilities of Cetaceans: Laboratory and Field Evidence. Plenum Press, New York.

Gross, M. G. 1996. Oceanography: A View of the Earth, 7th Ed. Prentice-Hall, Upper Saddle River, NJ.

Hui, C. A. 1994. Lack of association between magnetic patterns and the distribution of free-ranging dolphins. Journal of Mammalogy 75:399–405.

Kalmijn, A. J. 1978. Electric and magnetic sensory world of sharks, skates, and rays; pp. 507–528 in E. S. Hodgson and R. F. Mathewson (eds.), Sensory Biology of Sharks, Skates, and Rays. U.S. Government Printing Office, Washington DC.

Kalmijn, A. J. 1982. Electric and magnetic field detection in elasmobranch fishes. Science 218:916–918.

Kalmijn, A. J. 1984. Theory of electromagnetic orientation: a further analysis; pp. 525–560 in L. Bolis, R. D. Keynes, and S. H. P. Maddrell (eds.), Comparative Physiology of Sensory Systems. Cambridge University Press, Cambridge.

Kalmijn, A. J., and R. P. Blakemore. 1977. Geomagnetic orientation in marine mud bacteria. Proceedings of the International Union of Physiological Science 13:364.

Kimchi, T., and J. Terkel. 2001. Magnetic compass orientation in the blind mole rat *Spalax ehrenbergi*. Journal of Experimental Biology 204:751–758.

Kirschvink, J. L. 1990. Geomagnetic sensitivity in cetaceans: an update with live stranding records in the United States; pp. 639–649 in J. A. Thomas and R. A. Kastelein (eds.), Sensory Abilities of Cetaceans: Laboratory and Field Evidence. Plenum Press, New York.

Kirschvink, J. L., A. E. Dizon, and J. E. Westphal. 1986. Evidence from strandings for geomagnetic sensitivity in cetaceans. Journal of Experimental Biology 120:1–24.

Kirschvink, J. L., A. Kobayashi-Kirschvink, and B. J. Woodford. 1992. Magnetite biomineralization in the human brain. Proceedings of the National Academy of Science USA 89:7683–7687.

Kirschvink, J. L., M. M. Walker, and C. E. Diebel. 2001. Magnetite-based magnetoreception. Current Opinion in Neurobiology 11:462–467.

Klinowska, M. 1985a. Cetacean live stranding sites relate to geomagnetic topography. Aquatic Mammals 11:27–32.

Klinowska, M. 1985b. Cetacean live stranding dates relate to geomagnetic disturbances. Aquatic Mammals 11:109–119.

Klinowska, M. 1988. Cetacean navigation and the geomagnetic field. Journal of Navigation 4:52–71.

Klinowska, M. 1990. Geomagnetic orientation in cetaceans: behavioural evidence; pp. 651–663 in J. A. Thomas and R. A. Kastelein (eds.), Sensory Abilities of Cetaceans: Laboratory and Field Evidence. Plenum Press, New York.

Light, P., M. Salmon, and K. L. Lohmann. 1993. Geomagnetic orientation of loggerhead sea turtles: evidence for an inclination compass. Journal of Experimental Biology 182:1–10.

Lindauer, M., and H. Martin. 1972. Magnetic effect on dancing bees; pp. 559–567 in S. R. Galler, K. Schmidt-König, G. J. Jacobs, and R. E. Belleville (eds.), Animal Orientation and Navigation. National Aeronautic and Space Administration, Washington, DC.

Lohmann K. J. 1991. Magnetic orientation by hatchling loggerhead sea turtles *(Caretta caretta)*. Journal of Experimental Biology 155:37–49.

Lohmann K. J., and C. M. F. Lohmann. 1993. A light-independent magnetic compass in the leatherback sea turtle. Biological Bulletin 185:149–151.

Lohmann K. J., and C. M. F. Lohmann. 1994a. Acquisition of magnetic directional preference in hatchling loggerhead sea turtles. Journal of Experimental Biology 190:1–8.

Lohmann K. J., and C. M. F. Lohmann. 1994b. Detection of magnetic inclination angle by sea turtles: a possible mechanism for determining latitude. Journal of Experimental Biology 194:23–32.

Lohmann K. J., and C. M. F. Lohmann. 1996a. Orientation and open-sea navigation in sea turtles. Journal of Experimental Biology 199:73–81.

Lohmann K. J., and C. M. F. Lohmann. 1996b. Detection of magnetic field intensity by sea turtles. Nature 380:59–61.

Lohmann, K. J., A. O. D. Willows, and R. B. Pinter. 1991. An identifiable molluscan neuron responds to changes in earth-strength magnetic fields. Journal of Experimental Biology 161:1–24.

Lohmann, K. J., N. D. Pentcheff, G. A. Nevitt, G. D. Stetten, R. K. Zimmer-Faust, H. E. Jarrard, and L. C. Boles. 1995. Magnetic orientation of spiny lobsters in the ocean: experiments with undersea coil systems. Journal of Experimental Biology 198:2041–2048.

Lohmann, K. J., J. T. Hester, and C. M. F. Lohmann. 1999. Long-distance navigation in sea turtles. Ethology Ecology and Evolution 11:1–23.

Lohmann, K. J., S. D. Cain, S. A. Dodge, and C. M. F. Lohmann. 2001. Regional magnetic fields as navigational markers for sea turtles. Science 294:364–366.

Maher, B. A. 1998. Magnetite biomineralization in termites. Proceedings of the Royal Society of London B 265:733–737.

Mann S, N. H. C. Sparks, M. M. Walker, and J. L. Kirschvink. 1988. Ultrastructure, morphology and organization of biogenic magnetite from Sockeye salmon, *Oncorhynchus nerka*: implications for magnetoreception. Journal of Experimental Biology 140:35–49.

Marhold, S., H. Burda, and W. Wiltschko. 1991. Magnetkompaßorientierung und Richtungspräferenz bei subterranen Graumullen *Cryptomys hottentotus* (Rodentia). Verhandlungen der deutschen zoologischen Gesellschaft 84:354.

Mouritsen, H., and T. Ritz. 2005. Magnetoreception and its use in bird navigation. Current Opinion in Neurobiology 15:406–414.

Němec, P., J. Altmann, S. Marhold, H. Burda, and H. H. A. Oelschläger. 2001. Neuroanatomy of magnetoreception: the superior colliculus involved in magnetic orientation in a mammal. Science 294:366–368.

Nishi, T., G. Kawamura, and K. Matsumoto. 2004. Magnetic sense in the Japanese eel, *Anguilla japonica*, as determined by conditioning and electrocardiography. Journal of Experimental Biology 207:2965–2970.

Phillips, J. B. 1986. Two magnetoreception pathways in a migratory salamander. Science 233:765–767.

Phillips, J. B., M. E. Deutschlander, M. J. Freake, and S. C. Borland. 2001. The role of extraocular photoreceptors in newt magnetic compass orientation: parallels between light-dependent magnetoreception and polarized light detection in vertebrates. Journal of Experimental Biology 204:2543–2552.

Quinn, T. P. 1980. Evidence for celestial and magnetic compass orientation in lake migrating sockeye salmon fry. Journal of Comparative Physiology 137:243–248.

Quinn, T. P., and E. L. Brannon. 1982. The use of celestial and magnetic cues by orienting sockeye salmon smolts. Journal of Comparative Physiology 147:547–552.

Quinn, T. P., R. T. Merrill, and E. L. Brannon. 1981. Magnetic field detection in sockeye salmon. Journal of Experimental Zoology 217:137–142.

Vácha, M., and H. Soukopová. 2004. Magnetic orientation in the mealworm beetle *Tenebrio* and the effect of light. Journal of Experimental Biology 207:1241–1248

van Ginneken, V., B. Muusze, J. Klein Breteler, D. Jansma, and G. van den Thillart. 2005. Microelectronic detection of activity level and magnetic orientation of yellow European eel, *Anguilla anguilla* L., in a pond. Environmental Biology of Fishes 72:313–320.

Walker, M. M., and M. E. Bitterman. 1989. Honeybees can be trained to respond to very small changes in the geomagnetic field intensity. Journal of Experimental Biology 145:489–494.

Walker, M. M., J. L. Kirschvink, S. R. Chang, and A. E. Dizon. 1984. A candidate magnetic sense organ in the yellowfin tuna, *Thunnus albacares*. Science 224:751–753.

Walker, M. M., J. L. Kirschvink, G. Ahmed, and A. E. Dizon. 1992. Evidence that fin whales respond to the geomagnetic field during migration. Journal of Experimental Biology 171:67–78.

Walker, M. M., C. E. Diebel, C. V. Haugh, P. M. Pankhurst, J. C. Montgomery, and C. R. Green. 1997. Structure and function of the vertebrate magnetic sense. Nature 390:371–376.

Wiltschko, R., and W. Wiltschko. 2006. Magnetoreception. Bioessays 28:157–168.

Wiltschko, W., A. Möller, M. Gesson, C. Noll, and R. Wiltschko. 2004. Light-dependent magnetoreception in birds: analysis of the behaviour under red light after pre-exposure to red light. Journal of Experimental Biology 207:1193–1202.

Zoeger, J., J. R. Dunn, and M. Fuller. 1981. Magnetic material in the head of the common Pacific dolphin. Science 213:892–894.

20

Electroreception

Lon A. Wilkens and Michael H. Hofmann

The electrosense is a remarkable but fundamental sensory modality that appeared early in vertebrate evolution as a mechanism with both near-field and far-field applications in the conductive aquatic environment. In fish, where the electrosense is widely represented, prey and predators are detected at close range in response to the direct current (DC) and low-frequency alternating current electric fields they generate, in contrast to the long-distance navigational cues detected as a consequence of the Earth's magnetic field and electrochemical potentials. These same signals are available to aquatic tetrapods, although electrosensory taxa are limited to the more aquatic urodele amphibians, caecilians, the primitive egg-laying mammals, and, possibly, to a marine mammal, the dolphin *Sotalia guianensis*.

THE PHYSICS OF ELECTRORECEPTION

Electric signals are present in the aquatic environment from both animate and inanimate sources. The physical sources include geomagnetic phenomena (magnetic storms, earthquakes, lightning) that produce telluric or earth currents of limited or unknown biological consequence, and electric fields that result from motion relative to the Earth's magnetic field, and which potentially underlie orientational and, also potentially, navigational capabilities in certain animals. Tidal currents and streams flowing through the Earth's magnetic field are considered a passive electric field source for animals either adrift in the flowing water or fixed to the substrate, whereas an animal swimming through the Earth's magnetic field is considered to be active. Water currents will induce velocity-dependent voltage gradients perpendicular to the direction of the flow and magnetic field that in turn generate electric currents. These currents have the potential to signal flow velocity and direction for the animal (Kalmijn, 1974). For actively swimming organisms, magnetically induced currents may also provide for compass orientation (Kalmijn, 1974). Electric fields also originate from the electrochemical properties of the physical environment, including the boundaries between waters and/or substrates that differ in salinity, pH, temperature, and chemical composition and give rise to electrochemical gradients or streaming currents. These geomagnetic and electrochemical sources present uniform electric fields relative to the size of the receiving animal.

Animals themselves are ubiquitous sources of nonuniform, dipolar or multipolar electric fields in the aquatic environment. These range from the relatively weak "skin potentials" of all organisms to the high-voltage discharges of electric fish. Electrochemical potentials arise passively across the integumental boundaries that separate the dissimilar internal and external fluids of all animals, in addition to the active electrogenic potentials from ion transport, the energetic potentials prominent for epithelial tissues of the gills and oral cavities. These steady DC potentials range from 10 to 1000 μV cm^{-1}, as measured close to the skin surface of vertebrate as well as invertebrate organisms, where the larger potentials (more than 500 μV) are associated with fish and crustaceans (Kalmijn, 1972; Peters and Bretschneider, 1972; Taylor et al., 1992). The DC potentials are further modulated by the impedance changes associated with respiratory and locomotory movements, and by nerve and muscle potentials (e.g., heartbeat or tail flicks). The relevance of these electric fields is unequivocal for prey detection, as evident in the electrically guided feeding strikes of sharks toward flatfish (Kalmijn, 1971), and planktonic prey captures in paddlefish (Wilkens et al., 2003). Similarly, predator avoidance and mate attraction are well documented in stingrays in response to animate sources (Tricas et al., 1995; Sisneros et al., 1998).

THE ELECTROSENSE IN AQUATIC TETRAPODS

The primitive ampullary electrosense is characteristic of the early cyclostome and cartilaginous fish and has been retained only in the primitive taxa of ray-finned (actinopterygian) bony fish, for example, paddlefish, sturgeon, and bichirs (Wilkens and Hofmann, 2005). The ampullary system features a ciliary-based sensory epithelium housed in skin canals of various length. Nevertheless, homologous ampullary receptors remain within the evolutionary lineage of lobe-finned (sarcopterygian) fish ancestral to the tetrapods. Thus, caecilians and representative electrosensory amphibians, along with coelacanths and lungfish, have ampullary receptors equivalent to those better known in sharks and rays, the ampullae of Lorenzini. Electrosensory organs are absent in the remaining, more terrestrial tetrapod anurans, frogs, and toads, as well as in nearly all higher vertebrates. A notable exception is the egg-laying platypus, along with the closely related monotreme echidnas. The platypus and echidnas have reinvented the electrosense by adapting mucus-secreting glands in the skin to form electroreceptors.

AMPHIBIANS

The existence of the ampullary electrosense in amphibians was reported initially in salamanders (Fritzsch, 1981) (see the section by Hetherington and Thewissen in chapter 1 for a discussion of amphibian diversity). Unlike fish, amphibian electroreceptors are found only on the head, for example, in the axolotl *Ambystoma mexicanum* (Northcutt, 1992), as is also the case in the larval caecilian, *Ichthyophis kohtaoensis* (Himstedt and Fritzsch, 1990). In those salamanders whose metamorphosis from larvae leads to a terrestrial existence, the lateral line nerve and corresponding organs, including the electroreceptors, are lost. Where they remain, the amphibian electrosense is passive and is believed to help in locating prey, although the blind cave salamander, *Euproctus asper*, shows both orientation toward and avoidance of electrical stimuli (Schlegel, 1997). The electrosense has been lost in the anurans, frogs and toads.

PLATYPUS

The ducklike bill of the platypus, *Ornithorhynchus anatinus* (Monotremata), was relatively recently shown to be sensitive to weak electrical fields (Scheich et al., 1986). A morphological description of receptors in the bill had previously hinted at the existence of an electrosensory function (Andres and von During, 1984). It was also known that the platypus forages in streams at night with its eyes, ears, and nose closed and that it had a voracious appetite for aquatic crustaceans, worms, and insects, which it excavates from the stream bed. This, plus the fact that the platypus consumes half its weight in food daily, led to early speculation (Burrell, 1927) for an unknown sixth sense to explain foraging success. The prominent bill, with its extensive array of receptors, both electro- and mechosensitive, is a striking example of convergent evolution in parallel with the flat, elongated rostrum of the paddlefish, both of which function as an electrosensory antenna (Pettigrew and Wilkens, 2003).

The skin of the platypus bill contains both mucous and serous glands. Each receives innervation, but only the mucus gland has been confirmed as electrosensory. The mucus gland is a coiled subdermal secretory tube that connects with a papillary invagination of the skin (Andres and von During, 1984). This juncture receives a cuff of sensory innervation quite unlike the structure of electrosensory organs in fish, where primary sensory afferents are innervated by epithelial receptors. In the platypus a dozen or more sensory cells penetrate in radial fashion into the keratinocyte layers lining the papillary duct (Fig. 20.1), each terminating in a bulbous portion with a terminal filament that projects to within 20 μm of the duct lumen (Manger et al., 1995). A series of orthogonal arborizations arise at the base of the terminal filament and

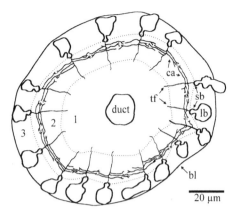

FIGURE 20.1. Diagram of transverse section through papillary segment of mucus gland duct of a platypus (Monotremata). Sensory fibers penetrate the duct wall (basal lamina, bl) with large bulbous (lb) and small bulbous (sb) segments in an outer layer of germinative keratinocytes (3). Terminal filaments (tf) project into the loosely packed periluminal layer (1). Collateral sensory filaments form a circumferential arbor (ca) in the densely packed intermediate layer (2) of keratinocytes. Figure modified after Manger et al. (1995).

form a neural plexus that encircles the duct. It is likely that the sensory cells interact via these arborizations in some fashion, as yet undetermined, to enhance sensitivity. Since the terminal filaments fall short of the duct lumen it is hypothesized that the filaments form a low-resistance pathway for current flow through the dense, high-resistance keratin layers (Manger et al., 1995). The electroreceptor afferents in the platypus enter the brain via the trigeminal nerve, which attests to their independent origin, although their cathode-excitatory response polarity is the same as the ampullary receptors of fish. Most afferent fibers are spontaneously active, firing at a frequency between 20 and 50 Hz and modulated up or down by the electric stimulus (Gregory et al., 1989).

Electroreception with respect to prey detection in the platypus has been carefully investigated in the laboratory. Familiar prey—worms, insects, and crustaceans—produce low-frequency potentials associated with rapid escape movements, with a maximum for an atyid

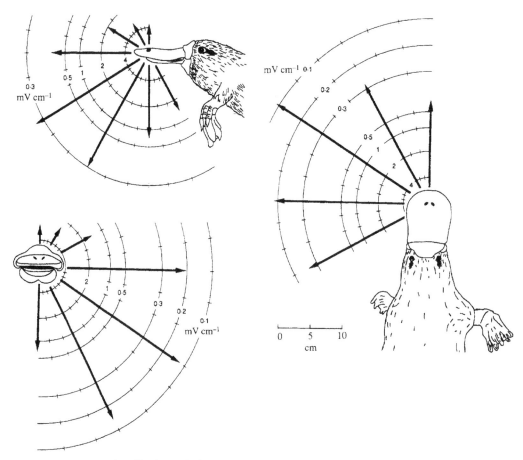

FIGURE 20.2. Directionality of head saccade elicited by constant-amplitude electrical stimuli from a hand-held dipole (shrimp simulator). The simulator was advanced toward a resting platypus, and its vertical and horizontal position relative to the bill was recorded at the point where phase-locked head saccades were observed. Threshold distance is indicated by length of *arrow* (in centimeters) and corresponding field strength by *circles* (millivolts per centimeter) From Manger and Pettigrew (1995).

shrimp of 1900 μV cm^{-1}; most had smaller potentials from 100 to 800 μV cm^{-1} (Taylor et al., 1992). Behavioral tests have shown these potentials to be within the sensitivity threshold of the platypus (Manger and Pettigrew, 1995). These experiments have demonstrated that sensitivity is highly directional, with a maximum of 50 μV cm^{-1} for signals at the preferred azimuth and elevation with respect to the bill, that is, ventrolaterally in front of the animal (Fig. 20.2). A square wave stimulus in the tank of a swimming platypus elicits a head saccade in the direction of the source, causing the animal to reorient, swimming toward the unseen electrodes. Indeed, swimming is accompanied by lateral

head saccades as the platypus scans its environment while patrolling for prey. Whereas a platypus will investigate and eventually habituate to a weak DC signal, the electrically evoked saccade cannot be suppressed, even after thousands of presentations. Thus, the platypus is clearly electrosensory, and all evidence points to the use of this sense in feeding.

The platypus bill as a sensory organ contains approximately 40,000 mucus gland electroreceptors, arranged in parasagittal stripes along the upper and lower bill (Fig. 20.3). In addition, the bill contains upward of 60,000 push-rod mechanoreceptors interspersed among the electroreceptors, whose presence suggests a

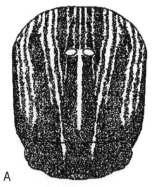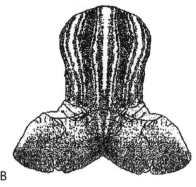

A   B

FIGURE 20.3. Diagram of the upper (A) and lower (B) surface of the bill of the platypus. Cutaneous mucus glands (black dots) form parasagittal stripes on the bill. Reprinted from Manger and Pettigrew, Brain, Behav., Evol. 1996; 48:27–54, S. Karger AG, Basel.

role in distance measurement for prey complementing the directionality of the electrosense. Mechanoreceptive and electroreceptive afferent projections onto the neocortex in the brain are interdigitated anatomically so as to present the possibility that they cooperate in prey detection. For example, crustacean tail flicks produce hydrodynamic stimuli that are within the sensitivity range of the mechanoreceptors, in addition to electrical pulses (Pettigrew, 1999). The proximity of afferents terminating in the cortex that signal relatively instantaneous electrical signals alongside their delayed mechanical counterparts is suggestive of a sophisticated neural mechanism for distance as well as directional measurement for prey animals.

DOLPHINS

It was recently discovered that at least one dolphin species may have electroreceptors that are used to detect prey fishes hiding in the sand. Dolphins are known to be able to locate fish that are hiding deep in the sandy sediment (Rossbach and Herring, 1997). Recent behavioral experiments showed that marine dolphins, *Sotalia guianensis* (Cunha et al., 2005), could be trained to respond to small electric fields that are delivered through metal electrodes (G. Dehnhardt, pers. comm.). Threshold values were estimated at 4.6 μV cm^{-1}. This is much higher than thresholds measured in elasmobranchs in seawater, but it may be sufficient

to locate larger fish at close range. Covering the hair follicles in the dolphins' rostrum with plastic sheaths prevented them from detecting electric fields. Adult dolphins have a series of hair follicles in the rostrum, but the hair or vibrissa is missing (Fig. 20.4). Early studies suggested that the follicles are vestigial (Yablokov and Klezeval, 1969; Ling, 1977), but newer investigations show that they are apparently well supplied by blood vessels (Mauck et al., 2000) and that they are heavily innervated by nerve fibers (G. Dehnhardt, pers. comm.). Normally, hair follicles are innervated by mechanoreceptors that sense deflections of the hair, but this function is useless without the hair. The hair follicles are therefore suggested to function as electroreceptors. There are no studies in other dolphins, but it seems unlikely that only one species would be electrosensory.

COMPARATIVE EVOLUTION OF THE TETRAPOD ELECTROSENSE

The plesiomorphic, ampullary-based electrosense has survived by way of the lobe-finned members of the bony fish and is represented in the more aquatic amphibians, salamanders, and newts. The early loss of the electrosense in ray-finned fish, specifically, its disappearance in neopterygians, is hypothesized to have been the result of genetic drift or negative selection when linked to other traits (Northcutt, 1986).

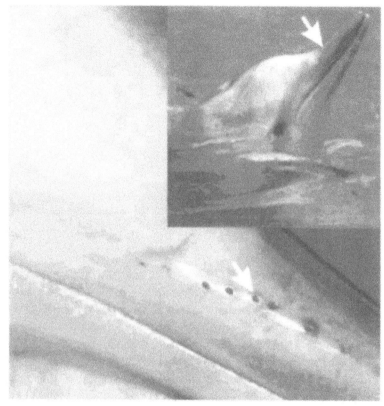

FIGURE 20.4. Head and detail of snout of the dolphin *Sotalia* showing a series of well developed follicles that lack hairs (*arrows*) and are suspected of being electroreceptors. From Mauck et al. (2000). Reproduced with permission of the Company of Biologists.

As a consequence, the large majority of fish, the teleosts, lack the electrosense. Nevertheless, several teleost taxa have reacquired the electrosense, taking advantage of the conductive properties of the aquatic environment. The teleost exceptions include catfish, where the electrosense involves only passive detection by nonhomologous ampullary organs, and the two sister groups of weakly electric fish that feature electric organs and whose electric sense is active. In the neopterygian weakly electric fish, the electroreceptors feature a receptor epithelium with apical microvilli in tuberous organs distinct from the original ciliary receptors of ampullary organs. In amphibians, electrosense was retained in some groups, and lost in others.

The electrosense was further selected against in tetrapods as amniotes became fully terrestrial. Among the secondarily aquatic higher vertebrates, only the egg-laying monotremes and at least one species of dolphin have been shown to be electroreceptive, again as a result of reinventing nonhomologous electrosensory receptors. The semiaquatic platypus, which feeds exclusively in the water, is the sole extant representative of a platypus lineage that extends back to the Cretaceous, more than 100 million years ago (Pettigrew, 1999). The two remaining extant monotremes, the spiny echidnas, also possess glandular electroreceptors and respond to electric fields (Proske et al., 1998). The electrosense presumably arose much earlier, as the echidnas are recent offshoots of the platypus lineage, having split from the platypus line relatively recently (20 to 30 million years ago). The electrosense is greatly reduced in these terrestrial animals. The long-billed echidna

(*Zaglossus brujni*) of wet tropical forests has approximately 2000 mucus receptors, whereas the short-billed echidna *(Tachyglossus aculeatus)* has only 400 electroreceptors and presumably has limited opportunity to utilize the electrosensory modality in its arid environment.

Dolphins are different from other marine mammals because they lack vibrissae. Vibrissae are very important mechanosensors in many mammals and are well developed in pinnipeds. The lack of this sensory system in dolphins may have led to the evolution of electroreception.

Several other secondarily aquatic tetrapods have seemed likely candidates for electrosensitivity. The star-nosed mole, which feeds on aquatic annelids, insect larvae, and crustaceans, was "tentatively" concluded to detect prey electrically (Gould et al., 1993). The 22 fleshy fingers of its nose, heavily endowed with Eimer's organs, seemed likely candidates but are now recognized as being exquisite mechanosensory organs (Catania, 1995). The desman, *Galemys pyrenaicus*, its European relative, similarly lacks the electrosense (Schlegel and Richard, 1992). It is tempting to speculate that river otters, like the platypus, may have developed the electrosense since they feed in streams and lakes, searching the bottom with their eyes and nose closed. However, there are no reports yet that otters, or the other secondarily aquatic pinnipeds and sirenians, all in possession of highly sensitive mechanosensory vibrissae, are electroreceptive.

LITERATURE CITED

Andres, K. H., and M. von During. 1984. The platypus bill: a structural and functional model of a pattern-like arrangement of cutaneous sensory receptors; pp. 81–89 in A. Iggo (ed.), Sensory Receptor Mechanisms. World Scientific Publishing, Singapore.

Burrell, H. 1927. The Platypus. Angus and Robertson, Sydney.

Catania, K. D. 1995. Structure and innervation of the sensory organs on the snout of the star-nosed mole. Journal of Comparative Neurology 351:536–548.

Cunha, H. A., V. M. F. da Silva, J. Lailson-Brito Jr., M. C. O. Santos, P. A. C. Flores, A. R. Martin, A. F. Azevedo, A. B. L. Fragoso, R. C. Zanelatto, and A. M. Sole-Cava. 2005. Riverine and marine ecotypes of *Sotalia* dolphins are different species. Marine Biology 148:449–457.

Fritzsch, B. 1981. The pattern of lateral-line afferents in urodeles. Cell and Tissue Research 218: 581–594.

Gould, E., W. McShea, and T. Grand. 1993. Function of the star in the star-nosed mole, *Condylura cristata*. Journal of Mammology 74:108–116.

Gregory, J. E., A. Iggo, A. K. McIntyre, and U. Proske. 1989. Receptors in the bill of the platypus. Journal of Physiology 400:349–366.

Himsted, W., and B. Fritzsch. 1990. Behavioural evidence for electroreception in larvae of the caecilian *Ichthyophis kohtaoensis* (Amphibia, Gymnophiona). Zoologische Jahrbücher für Physiologie 94:486–492.

Kalmijn, A. J. 1971. The electric sense of sharks and rays. Journal of Experimental Biology 55:371–383.

Kalmijn, A. J. 1972. Bioelectric fields in sea water and the function of the ampullae of Lorenzini in elasmobranch fishes. Scripps Institution of Oceanography Reference Series 72–83:1–21.

Kalmijn, A. J. 1974. The detection of electric fields from inanimate and animate sources other than electric organs; pp. 147–200 in A. Fessard (ed.), Handbook of Sensory Physiology, Vol. III-3. Springer-Verlag, New York.

Ling, J. K. 1977. Vibrissae of marine mammals; pp. 387–415 in R. J. Harrison (ed.), Functional Anatomy of Marine Mammals. Academic Press, London.

Manger, P. R., and J. D. Pettigrew. 1995. Electroreception and the feeding behaviour of platypus (*Ornithorhynchus anatinus*: Monotremata: Mammalia). Philosophical Transactions of the Royal Society of London B 347:359–381.

Manger, P. R., J. D. Pettigrew, J. R. Keast, and A. Bauer. 1995. Nerve terminals of mucous gland electroreceptors in the platypus (*Ornithorhynchus anatinus*). Proceedings of the Royal Society of London B 260:13–19.

Mauck, B., U. Eysel, and G. Dehnhardt. 2000. Selective heating of vibrissal follicles in seals (*Phoca vitulina*) and dolphins (*Sotalia fluviatilis guianensis*). Journal of Experimental Biology 203:2125–2131.

Northcutt, R. G. 1986. Electroreception in nonteleosts bony fishes; pp. 257–285 in T. H. Bullock and W. Heiligenberg (eds.), Electroreception. John Wiley and Sons, New York.

Northcutt, R. G. 1992. Distribution and innervation of lateral line organs in the axolotl. Journal of Comparative Neurology 325:95–123.

Peters, R. C., and F. Bretschneider. 1972. Electric phenomena in the habitat of the catfish *Ictalurus nebulosus* LeS. Journal of Comparative Physiology 81:345–362.

Pettigrew, J. D. 1999. Electroreception in monotremes. Journal of Experimental Biology 202:1447–1454.

Pettigrew, J. D., and L. A. Wilkens. 2003. Paddlefish and platypus: parallel evolution of passive electroreception in a rostral bill organ; pp. 420–434 in S. P. Collin and N. J. Marshall (eds.), Sensory Processing in Aquatic Environments. Springer-Verlag, New York.

Proske, U., J. E. Gregory, and A. Iggo. 1998. Sensory receptors in monotremes. Philosophical Transactions of the Royal Society of London B 353: 1199–1210.

Rossbach, K. A., and D. L. Herring, 1997. Underwater observations of benthic-feeding bottlenose dolphins *(Tursiops truncatus)* near Grand Bahama Island, Bahamas. Marine Mammal Science 13:498–504.

Scheich, H., G. Langner, C. Tidemann, R. B. Coles, and A. Guppy. 1986. Electroreception and electrolocation in platypus. Nature (London) 319: 401–402.

Schlegel, P. A. 1997. Behavioral sensitivity of the European blind cave salmander, *Proteus anguinus*, and a Pyrenean newt, *Euproctus asper*, to electrical fields in water. Brain, Behavior, and Evolution 49:121–131.

Schlegel, P. A., and P. B. Richard. 1992. Behavioral evidence against possible subaquatic electrosensitivity in the Pyrenean desman *Galemys pyrenaicus* (Talpidae, Mammalia). Mammalia 56:527–532.

Sisneros, J. A., T. C. Tricas, and C. A. Leur. 1998. Response properties and biological function of the skate electrosensory system during ontogeny. Journal of Comparative Physiology A 183:87–99.

Taylor, N. G., P. R. Manger, J. D. Pettigrew, and L. S. Hall. 1992. Electromyogenic potentials of a variety of platypus prey items: an amplitude and frequency analysis; pp. 216–224 in M. L. Augee (ed.), Platypus and Echidnas. The Royal Society of New South Wales, Sydney.

Tricas, T. C., S. W. Michael, and J. A. Sisneros. 1995. Electrosensory optimization to conspecifics signals for mating. Neuroscience Letters 202:29–131.

Wilkens, L. A., and M. H. Hofmann. 2005. Behavior of animals with passive low-frequency electrosensory systems; pp. 229–263 in T. H. Bullock, C. D. Hopkins, A. N. Popper, and R. R. Fay (eds.), Electroreception. Springer Handbook of Auditory Research. Springer-Verlag, New York.

Wilkens, L. A., M. H. Hofmann, and W. Wojtenek. 2003. The electric sense of the paddlefish: a passive system for the detection and capture of zooplankton prey. Journal of Physiology (Paris) 96:363–377.

Yablokov, A. V., and G. A. Klezeval. 1969. Whiskers of whales and seals and their distribution, structure and significance; pp. 48–81 in S. E. Kleinenberg (ed.), Morphological Characteristics of Aquatic Mammals. Izdatel'stvo Nauka, Moscow.

21

Toward an Integrative Approach

J. G. M. Thewissen and Sirpa Nummela

The senses are an array of intricate organs that are functionally reasonably well understood and of great importance to the animal. During the transition from land to water, the stimuli for most sense organs changed greatly, and the organs had to adapt to the watery environment. Morphological evolution is best studied when selection pressures are high and taxa undergo major changes, and the transition from land to water therefore offers a unique opportunity to evolutionary biologists.

A variety of unrelated terrestrial tetrapods have entered the water independently (see chapter 1), and changes in their sense organs can thus be compared in unrelated groups and taxonomic effects evaluated. Integrating the study of a sensory system across clades is a common and powerful tool in the study of evolution, and several chapters in this book take this approach explicitly (e.g., Georgi and Sipla in chapter 15).

Taken together, and by definition, the sense organs provide an organism with all the information from the outside world. As anatomical and physiological entities, they differ greatly: they are located in different parts of the animal, have different anatomical and embryological relations, and differ functionally at the cell and tissue level, as well as in neural hardware. However, because all sense organs are dedicated to gathering information and because they are energetically expensive, they complement each other and coevolve: expansion of the capabilities of one sense is commonly accompanied by the reduction of the capabilities of another. For evolutionary purposes then, the sense organs are best studied in an integrated fashion, across all senses.

Although the sense organs are commonly studied across clades, it is rare that they are

TABLE 21.1
Inferred Body Length and Eyeball Height of Eocene Cetaceans

	BODY LENGTH (MM)	EYEBALL HEIGHT (MM)	MUSEUM NUMBER (SKULL)	MUSEUM NUMBER (BODY LENGTH)
Pakicetus	1900	30	HGSP 96234	Madar, 2007
Ambulocetus	3040	45	HGSP 18507	HGSP 18507
Kutchicetus	2690	30	IITR SB 2647	IITR SB 2907
Remingtonocetus	4500	20	IITR SB 2770	Gingerich, 1998 Uhen, 2004
Indocetus/ Rodhocetus	2860	35	LUVP 11034	Gingerich et al., 2001

studied integratively across the senses. In this chapter, we advocate this dual integrative approach to the study of sensory evolution in general, and in particular for clades that change their sensory environment in similar ways, such as those moving from land to water. As editors of this volume, we see great opportunity for such integrative research. The greatest impediment to this approach is that the sense organs are complex, and that individual researchers often are forced to focus on a single sense organ. In this book, we brought together basic information and reviews for all sense organs of aquatic tetrapods, in the hope that it will afford readers easy access to general knowledge about all the senses and empower them to study sensory evolution integratively across senses as well as integratively across taxa.

As an example of such an approach, we present below some preliminary results of our own research on sensory evolution of the group of mammals we know best, the cetaceans. Cetaceans moved from land to water in the Eocene, and their fossil record is well known (Thewissen and Williams, 2002). Evolution of several sense organs throughout the Eocene has been well described, for example, hearing (Nummela et al., 2004) and balance (Spoor et al., 2002). Below, we add new information on two additional sensory organs, vision and mechanoreception, interpreted in a comparative

context. Finally, we integrate our results and discuss sensory evolution in early cetaceans comprehensively.

VISION IN CETACEANS, PINNIPEDS, AND ICHTHYOSAURS

The Mesozoic ichthyosaurs were fast, dolphin-like hunters of the open ocean. They had large eyes, and vision was probably their most important sense organ (Motani et al., 1999). In this volume, Kröger and Katzir (chapter 9) suggested that they used these to track bioluminescent prey of the deep ocean. Some modern pinnipeds also have large eyes, but experimental work reviewed by Dehnhardt and Mauck (chapter 18 in this volume) indicates that investigated pinnipeds use multiple sensory modalities in finding prey. Modern toothed whales (odontocetes) rely mostly on echolocation, and vision is less important, even though they do have good eyesight (Wartzok and Ketten, 1999; Mass and Supin, 2001) and have a variety of adaptations for aquatic vision (Walls, 1942; Kröger and Katzir, chapter 9 in this volume).

Ichthyosaurs, pinnipeds, and cetaceans offer an opportunity to study vision comparatively and make evolutionary inferences using data from the fossil record. In Eocene cetaceans, orbit size and position varied greatly (Nummela et al., 2006), and none of these early whales

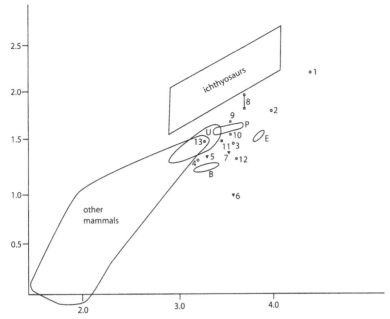

FIGURE 21.1. Eye size in ichthyosaurs and mammals. Range for ichthyosaurs is from Motani et al. (1999). Most mammal data are from Ritland (1982), datum for *Balaena mysticetus* is from Zhu (1998), and datum for *Dorudon* is based on data from Uhen (2004) and Kellogg (1936). Abbreviations: For multipoint ranges, B, bears; E, elephants; P, pinnipeds; U, ungulates; Single data points, 1, *Balaenoptera musculus* (blue whale); 2, *Balaena mysticetus* (bowhead whale); 3, Delphinidae (dolphin species from Ritland, 1982); 4, *Phocoena* (porpoise species from Ritland, 1982); 5, *Tapirus* (tapir); 6, *Trichechus* (manatee); 7, rhinoceros species from Ritland, 1982; 8, *Dorudon* (estimates based on two specimens measured from illustrations); 9, *Ambulocetus natans*; 10, *Indocetus ramani*; 11, *Kutchicetus minimus*; 12, *Remingtonocetus*, and 13, *Pakicetus*.

echolocated. Table 21.1 provides summary information for eye and body sizes of several Eocene whales.

Ritland (1982) presented detailed information on the eye size of modern mammals, based on measurements of eyes in cadaver specimens. Motani et al. (1999) provided data on ichthyosaurs, based on measurements of the sclerotic ring of the eye, which is a good indicator of eyeball size (Kröger and Katzir, chapter 9 in this volume). Figure 21.1 compiles data on eye size. Our data on fossil cetaceans is based on measuring orbit sizes, which is probably a reasonable estimator of eyeball size in animals with well-developed bony orbits, such as archaeocetes. It is not a good estimator in modern cetaceans, where the bones that surround the orbit are severely modified. For instance, the cetacean with the smallest eyes is *Platanista*, the Ganges river dolphin, which has eyes 5 mm in diameter (Purves and Pilleri, 1973). The orbit of *Platanista*, on the other hand is 13 mm (Fitzgerald, 2006). As a result, the data on modern cetaceans in Figure 21.1 is based on cadaver specimens, not skulls.

Figure 21.1 elucidates size evolution of the cetacean eye in several regards. The earliest whales, pakicetids, had eyes that were similar in size to the eyes of their terrestrial relatives, the ungulates. Pakicetids were mostly terrestrial and probably lived by wading and feeding in shallow water (Thewissen et al., 2001). Whales with eyes of this size persisted in the Eocene: *Ambulocetus*, *Kutchicetus*, and *Indocetus* had eyes that did not differ greatly from pakicetids' in size. *Ambulocetus* (Thewissen et al., 1994, 1996), *Kutchicetus* (Bajpai and Thewissen,

2000), and *Indocetus* (Bajpai and Thewissen, 1998) lived in nearshore marine environments. Figure 21.1 suggests that eyesize was not modified by evolution in these whales as they evolved to be more aquatic.

Remingtonocetus had eyes that were significantly smaller than those of its Indian contemporaries *Indocetus* and *Kutchicetus* (Bajpai and Thewissen, 1998). Habitats at the localities where these animals are found varied from clear to muddy water (Bajpai et al., 2006), and it is possible that vision was not an efficient sense in the waters in which *Remingtonocetus* lived. Although, the data is poor (based on measurements on illustrations of skulls), Late Eocene *Dorudon* had eyes that are larger than what is expected for mammals of its size. Its eyes approach the range of eye sizes of ichthyosaurs, and the data suggest that these whales underwent selection for eye size. It is likely that *Dorudon* was a visual predator.

Taken together, these data imply that Eocene cetaceans had a variety of visual specializations and varied more significantly in eye size than most other groups of mammals. Vision was clearly an important sense in some, but not all, early cetaceans, an observation that is underscored by the recent discovery of an early baleen whale (Fitzgerald, 2006) with enormous eyes. Further study of these patterns may elucidate their evolutionary pattern in more detail.

MECHANOSENSE IN CETACEANS, SHOREBIRDS, AND CROCODYLIANS

Dehnhardt and Mauck (chapter 18 in this volume) reviewed the specialized mechanoreceptors that occur on the rostrum of charadriiform birds, crocodylians, and the mechanoreceptive function of the vibrissae of a variety of mammals. Although the microanatomy and function of these sensory organs differ greatly among the aquatic tetrapods that have them, all take advantage of the noncompressive and dense nature of water, which allows movements at some distance from the observer to be registered by its skin. While the detailed histology of these organs cannot be evaluated in fossils, there are some bony correlates: either the sensory receptors are housed in distinct depressions in the bone (Fig. 18.1 in this volume), or the nerves and vessels associated with these structures leave their impressions on the surface of the bone.

The morphology of the rostrum of pakicetid cetaceans is unlike that of other early whales. Parts of the snout show extreme pitting and grooving (Fig. 21.2A and B) of the kind that is suggestive of the scars that vessels and nerves leave as they traverse the surface of bone. Pakicetids are similar to manatees in some respects. Manatees have large and well-defined fields of facial vibrissae (Reep et al., 2001), and the areas where these occur are densely pitted, similar to the bill-tip organ in aquatic birds (Dehnhardt and Mauck, chapter 18 in this volume) (Fig. 21.2C). Moreover, adjacent areas of the snout show many vascular depressions (Fig. 21.2D).

Although these observations need to be quantified and interpreted in the context of a broader comparative sample, the pitting in the earliest cetaceans is suggestive of enhanced mechanosensory abilities. It is possible then that early cetaceans went through an evolutionary stage where mechanoreception with the tip of the snout was important, more important than in their ancestors and their descendants.

THE SENSORY LANDSCAPE OF EARLY CETACEANS

With our understanding of the evolution of hearing (Nummela et al., 2004; Nummela, chapter 13 in this volume), balance (Spoor et al., 2002; Spoor and Thewissen, chapter 16 in this volume), and chemical senses (Pihlstrom, chapter 7 in this volume), and the discussion of vision and mechanosense above, evolution of the sensory landscape in cetaceans is coming into focus.

Figure 21.3 lists, along the *x* axis, the families of Eocene cetaceans and the two modern suborders by their approximate position on

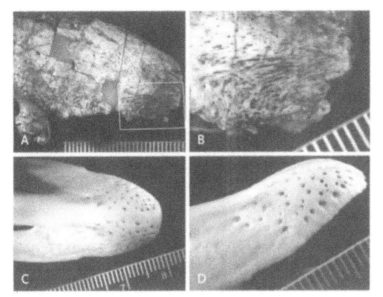

FIGURE 21.2. Bone texture of the snout in two aquatic tetrapods. (A, B) Lateral view of the premaxilla of a pakicetid cetacean (specimen number H-GSP 18467, tooth visible is third upper incisor) (Thewissen and Hussain, 1998) in which the tip of the premaxilla is pitted and grooved, and grooves are filled with red sediment, which emphasizes their depth. B is an enlargement of the area in the rectangle in A. (C, D) Dorsal and lateral view of the beak of the duck *Samotheria*, showing pitting (in C) and grooving (near inferior edge in D) related to the bill-tip organ. Scale is in millimeters.

the cetacean cladograms, which coincides with their approximate time of first appearance. This figure is a crude representation of sensory evolution, the x axis showing time and the y axis a rather arbitrary gauge of the relative importance of a sensory modality compared to the same sensory modality in the cetaceans just below and above it on the cladograms. However, taken together the axes do show where the senses are changing, and this figure suggests patterns of change in sensory integration. The figure can thus be used to formulate hypotheses regarding sensory evolution that can then be tested with quantitative data.

CHEMICAL SENSES AND BALANCE

During the Eocene (from pakicetids to dorudontids in Fig. 21.3), aquatic adaptations increased (Thewissen and Williams, 2002), and some sense organs—olfaction, vomeronasal sense, and balance—underwent a general decline in importance. For olfaction and vomeronasal sense, this pattern is expected and is repeated in other aquatic groups (e.g., Schwenk, chapter 5 in this volume), whereas for balance, this pattern is unique and just one of several possible ways in which the organ of

balance can adapt to the aquatic environment (Spoor and Thewissen, chapter 16 in this volume).

HEARING

Hearing presents a more complicated pattern, partly because there are at least three different sound transmission mechanisms through the cetacean ear in the Eocene (Nummela et al., 2004; Nummela, chapter 13 in this volume). The first of these, tympanic hearing, is the sound transmission mechanism used by land mammals. It functions poorly in water and shows a general decline in early cetaceans and disappears when cetaceans become obligately aquatic. Bone conduction is a sound-conducting mechanism that can used, under special conditions, in land mammals too. Eocene cetaceans improved on this mechanism and made it their main sound transmission mechanism underwater early on. However, bone-conducted sound also has severe limitations, first and foremost the lack of directional sound perception. Improved bone conduction was accomplished by making some changes in the middle ear. These changes were exapted for a novel mode of hearing, "odontocete hearing," in dorudontids, to be perfected in odontocetes.

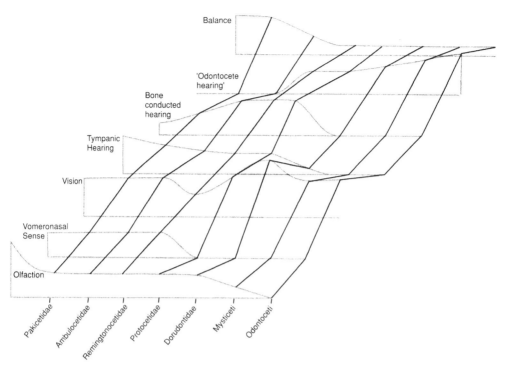

FIGURE 21.3. The evolving sensory landscape of cetaceans. Shown are five families of Eocene cetaceans, with the two modern suborders (see chapter 1). Eocene forms are arranged according to their order on the cetacean cladogram (which matches approximately the first appearance in the fossil record). Different sense organs and/or sensory mechanisms are shown as gray polygons, changing over time. The y axis is arbitrary, and no comparisons can be made between different polygons, although slopes do indicate shifts in the importance of a sense organ.

Summarizing hearing evolution, we suggest that after an experimental phase during which an existing hearing mode (bone conduction) was adapted for improved underwater hearing, an exaptation to a novel sound-transmission mode occurred, and this may have given cetaceans enormous potential to evolve new prey-detection mechanisms (echolocation).

VISION

The complicated evolutionary picture presented by vision was explained above. Here too, there is evidence for an experimental phase, where some groups increase vision (dorudontids), and others reduce it (remingtonocetids). In addition to the eye-size considerations above, it is also clear that eye position varies (Nummela et al., 2006). Further study is necessary to follow this pattern in the ancestors of the modern suborders.

MECHANOSENSE

Sensory reception in the skin is not plotted in Figure 21.3. The reason for this is that it has not been studied in detail in most cetaceans and cannot be evaluated in most fossil forms. However, the data presented above suggests that at least in pakicetid cetaceans, mechanoreception at the tip of the rostrum was significant, and that such adaptations of the rostrum were not present in most other Eocene whales. Among modern forms, mechanoreception data in cetaceans was summarized by Dehnhardt and Mauck (chapter 18 in this volume) and, as explained by Wilkens and Hofmann (chapter 20 in this volume), mechanosensory receptors (vibrissae) were exapted in some modern cetaceans as electroreceptors.

SENSORY EVOLUTION INTEGRATION

Taken together, it is obvious that the sensory landscape of cetaceans changed completely as they passed the land-water threshold. In less than 10 million years, some sense organs that were important all but disappeared, whereas others completely changed their transmission mechanism. Some of these changes are mimicked in the evolution of other aquatic tetrapods, and some are unique to cetaceans.

Pakicetids, the earliest cetaceans, had sense organs similar to their land ancestors in most regards and probably spent much of their time on land. However, changes in their ear for improved bone conduction, and adaptations of the rostrum lodging mechanoreceptive structures at the tip of the snout suggest that these were important sense organs. In addition, the position of the eyes on top of the head suggests that the eyes may have been used to look up, out of the water, while the animal was submerged. Cranial remains are poorly preserved for ambulocetids, but available data suggest that their sensory landscape was similar to that of pakicetids, even though they were clearly more aquatic and lived near the shore in the ocean (Thewissen et al., 1994; Madar et al., 2002; Madar, 2007).

Remingtonocetids are characterized by changes in the ear that would have resulted in improved underwater hearing, and a reduction of the size of the eyes and semicircular canals. In addition to a general trend toward more aquatic behaviors, it appears that the ears were important sense organs for remingtonocetids, although sound transmission underwater was clearly less sophisticated than in modern whales.

Protocetids are mostly different from remingtonocetids in the presence of larger eyes. Based on middle ear data, their hearing is similar to that of remingtonocetids, although computed tomography study of the inner ears, similar to that by Ketten's on modern whales, may show important differences. Protocetids were probably visual predators, unlike remingtonocetids. Protocetids are a diverse group, and it is also possible that the sensory landscape of different protocetid species varied.

The sensory landscape of dorudontids is similar to that of modern cetaceans, except in the relatively large eyes, a feature also present in early mysticetes (Fitzgerald, 2006). Their sound-receiving mechanism was perfected and similar to that of many modern odontocetes, although dorudontids did not echolocate. As discussed by Nummela (chapter 13 in this volume), mysticete sound transmission remains poorly understood, and clarifying it will benefit understanding of dorudontid sound transmission.

CONCLUSION

Studying sense organ evolution comparatively across unrelated groups, and integratively across all senses is likely to greatly benefit our understanding of the evolution of the terrestrial to aquatic transition in all groups of vertebrates that have made this transition. Although based, in part, on preliminary data, a definite pattern of changes in the sensory landscape of cetaceans emerges. This pattern can be used to test specific hypotheses of sensory evolution in the future.

ACKNOWLEDGMENTS

We thank the contributors to this book, whose writings inspired us to think about sensory landscape evolution and whose courage to boldly interpret their data will form a spark that electrifies the field of sensory-evolution research. Part of the research discussed in this chapter was funded by NSF grants to the senior author. We thank S. Bajpai for access to fossels in his care.

LITERATURE CITED

Bajpai, S., and J. G. M. Thewissen. 1998. Middle Eocene cetaceans from the Harudi and Subathu formations of India; pp. 213–234 in J. G. M. Thewissen (ed.), The Emergence of Whales: Evolutionary Patterns in the Origin of Cetacea. Plenum Press, New York.

Bajpai, S., and J. G. M. Thewissen. 2000. A new, diminutive whale from Kachchh (Gujarat, India) and its implications for locomotor evolution of cetaceans. Current Science (New Delhi) 79: 1478–1482.

Bajpai, S., J. G. M. Thewissen, V. v. Kapur, B. N. Tiwari, and A. Sahni. 2006. Eocene and Oligocene sirenians (Mammalia) from Kachchh, India. Journal of Vertebrate Paleontology 26:400–410.

Fitzgerald, E. M. G. 2006. A bizarre new toothed mysticete (Cetacea) from Australia and the early evolution of baleen whales. Proceedings of the Royal Society B 273:2955–2963.

Gingerich, P. D. 1998. Paleobiological perspective on Mesonychia, Archaeoceti, and the origin of whales; pp. 423–450 in J. G. M. Thewissen (ed.), The Emergence of Whales: Evolutionary Patterns in the Origin of Cetacea. Plenum Press, New York.

Gingerich, P. D. M. ul Haq, I. S. Zalmout, I. H. Khan, and M. S. Malkani. 2001. Origin of whales from early artiodactyls: hands and feet of Eocene Protocetidae from Pakistan. Science 293:2239–2242.

Kellogg, R. 1936. A review of the Archaeoceti. Carnegie Institute of Washington, DC, 1–366.

Madar, S. I. 2007. Postcranial osteology of the Pakicetidae (Cetacea). Journal of Paleontology, 81:176–200.

Madar, S. I., J. G. M. Thewissen, and S. T. Hussain. 2002. Additional holotype remains of *Ambulocetus natans* (Cetacea, Ambulocetidae) and their implications for locomotion in early whales. Journal of Vertebrate Paleontology 22:405–422.

Mass, A. M., and A. Y. Supin. 2001. Vision; pp. 1280–1293 in W. F. Perrin, B. Würsig, and J. G. M. Thewissen (eds.), Encyclopedia of Marine Mammals. Academic Press, San Diego.

Motani, R., B. M. Rothschild, and W. Wahl Jr. 1999. Large eyeballs in diving ichthyosaurs. Nature 402:747–748.

Nummela, S., J. G. M. Thewissen, S. Bajpai, S. T. Hussain, and K. Kumar. 2004. Eocene evolution of whale hearing. Nature 430:776–778.

Nummela, S., S. T. Hussain, and J. G. M. Thewissen. 2006. Cranial anatomy in the Pakicetidae (Cetacea, Mammalia). Journal of Vertebrate Paleontology 26:746–759.

Purves, P. E., and G. Pilleri. 1973. Observations on the ear, nose, and eye of *Platanisti indi*. Investigations on Cetacea 5:13–57.

Reep, R. L., M. L. Stoll, C. D. Marshall, B. L. Horner, and D. A. Samuelson. 2001. Microanatomy of facial vibrissae in the Florida manatee: the basics for specialized sensory function and oripulation. Brain, Behavior, and Evolution 58:1–14.

Ritland, S. 1982. The allometry of the vertebrate eye. Ph.D. Dissertation, University of Chicago, Chicago.

Spoor, F., S. Bajpai, S. T. Hussain, K. Kumar, and J. G. M. Thewissen. 2002. Vestibular evidence for the evolution of aquatic behaviour in early cetaceans. Nature 417:163–166.

Thewissen, J. G. M., and S. I. Hussein. 1998. Systematic review of the Pakicetidae, early and middle Eocene Cetacea (Mammalia) from Pakistan and India. Bulletin of the Carnegie Museum of Natural History 34:220–238.

Thewissen, J. G. M., and E. M. Williams. 2002. The early evolution of Cetacea (whales, dolphins, and porpoises). Annual Review of Ecology and Systematics 33:73–90.

Thewissen, J. G. M., S. T. Hussain, and M. Arif. 1994. Fossil evidence for the origin of aquatic locomotion in archaeocete whales. Science 263:210–212.

Thewissen, J. G. M., S. I. Madar, and S. T. Hussain. 1996. *Ambulocetus natans*, an Eocene cetacean (Mammalia) from Pakistan. Courier Forschungs-Institut Senckenberg, 190:1–86.

Thewissen, J. G. M., E. M. Williams, L. J. Roe, and S. T. Hussain. 2001. Skeletons of terrestrial cetaceans and the relationship of whales to artiodactyls. Nature 413:277–281.

Uhen, M. D. 2004. Form, function, and anatomy of *Dorudon atrox* (Mammalia, Cetacea): an archaeocete from the middle to late Eocene of Egypt. Papers on Paleontology, University of Michigan 34:1–222.

Walls, G. L. 1942. The vertebrate eye and its adaptive radiation. Cranbrook Institute of Science Bulletin 19:1–785.

Wartzok, D., and D. R. Ketten. 1999. Marine mammal sensory systems; pp. 117–175 in J. E. Reynolds III and S. A. Rommel (eds.), Biology of Marine Mammals. Smithsonian Institution Press, Washington, DC.

Zhu, Q. 1998. Studies on the eyes of the Bowhead whale *(Balaena mysticetus)*, ringed seal *(Phoca hispida)*, and caribou *(Rangifer tardigradus)*. Ph.D. Dissertation, Chinese Academy of Sciences, Bejing.

INDEX

Plesiosaurs
    gustation, 73–74
    vomeronasal system, 73–74
Plethodon cinereus, 58
Pleurodeles, 56
Polarity, chemical stimuli, 38
Pseudorca crassidens, 159
Puffinus pacificus, 154, 158
Puffinus puffinus, 164

Rana, 153
Rana catesbeiana, 162, 192–193, 202
Rana pipiens, 158
Rana temporaria, 162
Rattus norvegicus, 264
Reduced major axis (RMA), semicircular canal
        morphology analysis in mammals, 259, 262,
        264–265
Refraction, light, 114, 117–118
Remingtonocetus, 275, 334–336
Reptiles
    ancestry, 3, 8, 14, 16, 20–21
    auditory system comparative anatomy and function
        crocodilians, 201–203
        ichthyosaurs, 203–204
        lizards, 200
        mosasaurs, 200
        sauropterygians, 204
        snakes, 200–201
        turtles, 195–200
    chemosensory system comparative anatomy and
            function
        fossil reptiles
            mosasaurs, 73
            phytosaurs, 73
            plesiosaurs, 73–74
        functional and evolutionary patterns, 74–76
        nasal cavity comparative anatomy, 67–70
        olfaction and vomeronasal system, 70–71
        oral cavity comparative anatomy, 71–73
        overview, 66–67
    extant secondarily aquatic reptiles
        archosaurians, 20
        crocodylians, 20–21
        ichthyopterygia, 20
        mesosaurs, 14
        phtyosaurs, 21
        sauropterygia, 16
        squamates, 21
        table, 10–14
        thallatosauria, 21
        turtles, 14, 16
    eye anatomy and function
        crocodilians, 127

ichthyosaurs, 127–128
squamates, 125–127
turtles, 124–125
magnetoreception in turtles, 319–321
phylogeny, 5–8
semicircular canals
    anatomy
        crocodilians, 239–241
        squamates, 236–237
        turtles, 237–238
    aquatic amniote anatomy and function
        crocodilians, 245–247
        squamates, 244
        turtles, 244–245
    evolutionary patterns across aquatic amniotes,
        248–254
    function, 242–244
    vestibular system sensory endorgans,
        235–236
Retina
    anatomy and function, 123–124, 149–151
    evolution and species differences, 167–168
    ganglion cells
        density, 158–159
        temporal and spatial summation, 147, 166
        types and topographies, 159–161, 166–167
        visual acuity by species, 158–159, 161–162, 166
    photoreceptors
        cone types, 151–152
        pigments, 151, 153–155, 162–164
        rod noise and body temperature, 156, 165–166
        sensitivity and temporal summation, 152, 156, 165
Rhea, 242
Rhinophrynus dorsalis, 53
Rhomaleosaurus, 204
RMA, *see* Reduced major axis
Rodentia
    chemical sensing, 101
    mechanoreception, 308–310
    vestibular system, 270–271, 281
    vomeronasal system, 101
Rod
    noise and body temperature, 156, 165–166
    pigments, 151, 153–155, 162–163
    sensitivity and temporal summation, 152, 156

Saccule, vestibular system, 236
Salamandra, 56
Salamandra salamandra, 56
Sauropterygia, phylogeny, 16
Scattering, light, 116–117
Sciurus vulgaris, 270
Sclera, comparative anatomy and function, 124
Sclerotic ring, comparative anatomy and function, 123

Lightning Source UK Ltd.
Milton Keynes UK
UKHW031829240223
417617UK00001B/11